New Frontiers in Petroleum Science

New Frontiers in Petroleum Science

Editor: Natalie Mitchell

STATES
ACADEMIC PRESS
www.statesacademicpress.com

States Academic Press,
109 South 5th Street,
Brooklyn, NY 11249, USA

Visit us on the World Wide Web at:
www.statesacademicpress.com

ISBN: 978-1-63989-377-5 (Hardback)

Cataloging-in-Publication Data

New frontiers in petroleum science / edited by Natalie Mitchell.
p. cm.
Includes bibliographical references and index.
ISBN 978-1-63989-377-5
1. Petroleum. 2. Petroleum engineering. 3. Petroleum--Geology. I. Mitchell, Natalie.
TN860 .N49 2022
553.28--dc23

Table of Contents

Preface

Petroleum is a mixture of hydrocarbons which occurs naturally beneath the Earth's surface. It is yellowish-black in color and is commonly known as crude oil. Petroleum is formed when immensely large quantities of dead organisms, primarily algae and zooplankton get buried underneath sedimentary rocks and is subjected to very high heat and pressure. Petroleum science is a multidisciplinary field which employs the principles of geology, drilling technology, geostatistics, chemistry, railway engineering, logistics and transport, and economics. Petroleum and its products are widely used for various purposes such as heating, power generation, powering vehicles, etc. This book unfolds the innovative aspects of petroleum science which will be crucial for the progress of this field in the future. It elucidates new techniques and their applications in a multidisciplinary manner. This book aims to equip students and experts with the advanced topics and upcoming concepts in this area.

Various studies have approached the subject by analyzing it with a single perspective, but the present book provides diverse methodologies and techniques to address this field. This book contains theories and applications needed for understanding the subject from different perspectives. The aim is to keep the readers informed about the progresses in the field; therefore, the contributions were carefully examined to compile novel researches by specialists from across the globe.

Indeed, the job of the editor is the most crucial and challenging in compiling all chapters into a single book. In the end, I would extend my sincere thanks to the chapter authors for their profound work. I am also thankful for the support provided by my family and colleagues during the compilation of this book.

Editor

Preface

1

Performance Model for Vertical Wells with Multi-stage Horizontal Hydraulic Fractures in Water Flooded Multilayer Reservoirs

Gao Dapeng[1,3], Liu Yuewu[1]*, Ye Jigen[2], Hu Yunpeng[2], Hou Shaoji[1], Li Qi[1], Han Guofeng[1]

[1]Institute of Mechanics, Chinese Academy of Sciences
[2]PetroChina Research Institute of Petroleum Exploration & Development, Beijing, China
[3]School of Earth and Space Science, Peking University, Beijing, China

ABSTRACT

For the characteristics of horizontal fractures in shallow low-permeability oil layers after hydraulic fracturing in multilayer reservoirs, horizontal fractures are taken equivalent to an elliptical cylinder with the reservoir thickness using the equivalent permeability model; then, upon the elliptic seepage theory, the seepage field which has led by a vertical well with horizontal fractures is divided into two parts: (1) radial flow from the external formation to the equivalent area of horizontal fracture and (2) elliptic flow in the equivalent area of horizontal fracture. The loss of pressure caused by threshold pressure gradient, material balance in the reservoir, and multi-well pressure superposition principle are synthesized to calculate the performance. Finally, separate-layer multi-stage horizontal fractured well performance is deduced by summing the performance of high-permeability oil layers and fractured thin low-permeability oil layers. Low-permeability thin oil layers in Xing Shu-Gang oilfield are taken as practical cases, and the well space limits and economic reservoir thickness limits are calculated by the performance model; the relationship among recovery and productivity intensity, and the ratio of thin low-permeability oil layers thickness to the total thickness are also discussed.

Keywords: Multilayer Reservoir, Low-permeability Thin Oil Layer, Horizontal Fracture, Five-spot Waterflood Pattern, Performance

INTRODUCTION

Low-permeability reservoirs have become main potentials of multi-layer heterogeneous reservoirs, which have reached extreme high water cut. Hydraulic fracture is used frequently to stimulate low-permeability reservoirs, including two types of fractures: (1) horizontal fractures perpendicular to wellbore and (2) vertical fractures parallel to wellbore. Horizontal fractures are easier to be created in shallow formations, whose vertical stress may be less than horizontal stress [1-3]. Multi-stage fracturing (MSF) has been applied widely to multilayer reservoirs, and horizontal fractures may be generated in shallow oil layers along the vertical wellbore. It is an all-in-one technique for perforating, fracturing, isolating a

***Corresponding author**
Gao Dapeng
Email: gaodapeng2009@163.com

zone, and moving up the wellbore to repeat this process at next desired productive interval [4]. Many authors have studied the performance of vertical wells with vertical fractures [5-7]. However, the performance model of vertical wells with horizontal fractures has been investigated occasionally, and most studies focus on numerical solutions. W. Sung (1987) compared the performance of producers with vertical and horizontal fractures by numerical reservoir simulation [8], and Peter Valko (1998) studied the transient behavior of finite conductivity horizontal fractures [9]; Lasen Leif (2011) studied the pressure in multilayer reservoirs influenced by horizontal fractures [10]. However, few studies have been devoted to an analytical solution for the productivity of vertical wells with horizontal fractures considering oil-water two-phase flow in regular waterflooding patterns. In this paper, an accurate productivity model for vertical wells with horizontal fractures is proposed in conjunction with threshold pressure, material balance, and pressure superposition, for the sake of making better engineering decisions about the development of the reservoirs with complex hydraulic fracturing.

Flow Field Analysis of a Vertical Well with Horizontal Fractures

Model Assumptions

Figure 1a shows the separate-layer multi-stage horizontal fractures which have been studied in this paper, and the following assumptions are made to develop the model.

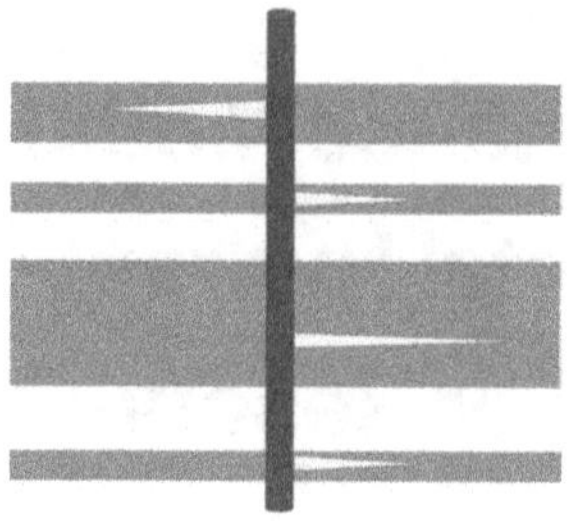

a. Separate-layer multi-stage hydraulic fracturing well

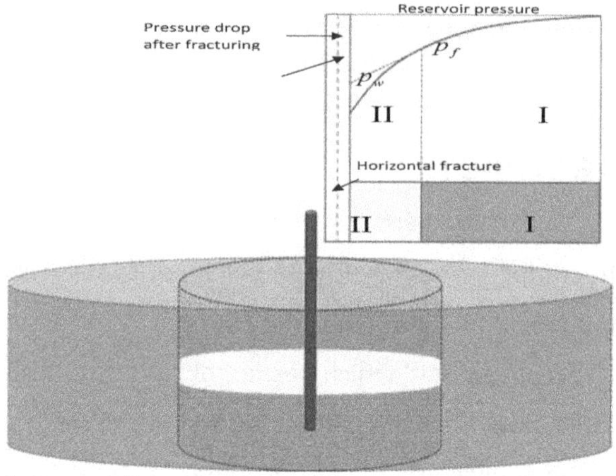

b. Permeability equivalent model for one horizontal fracture in one low-permeability oil layer

Figure 1: A Separate-layer multi-stage hydraulic fracturing well and a permeability equivalent model.

1. A horizontal fracture is regarded as an elliptical cylinder with a rectangular cross-section, and its height is assumed to be equal with the average fracture width; the fracture is also symmetric around the well.
2. Interlayers grow steadily between every two oil layers, so the vertical flow from upper or lower oil layers to fractures should be ignored.
3. The flow in the reservoir is assumed to be two-phase flow with incompressible fluid of constant viscosity, and capillary force and gravity are negligible.
4. For the sake of simplicity, flow field caused by a horizontal fracture could be equivalent to the one caused by one much larger elliptical cylinder with a height equal to the oil layer thickness, and the larger elliptical cylinder permeability could be calculated by the equivalent permeability model [11]; its shape and pressure distribution are shown in Figure 1b.

5. The flow in the formation with a horizontal fractured well can be divided into two parts: (1) radial flow in the outer zone and (2) elliptical flow in the inner zone. The outer zone is the external formation away from horizontal fracture, and the inner zone, the permeability of which is calculated by a weighted average method, is the equivalent zone for horizontal fractures.

MATHEMATICAL PROCEDURES

Radial Flow in Outer Zone

As Figure 2a shows, the isopiestic lines are a set of confocal elliptical closed lines in the outer zone. These isopiestic lines are more similar to the elliptical boundary when closer to the horizontal fractures, and they change from ellipses to circles gradually when they are away from the fracture.

Liu Yuetian (2000) transformed the radial flow in elliptical surface to parallel flow in a new surface by conformal transformation [12]. According to the transformation, the pressure in the outer zone can be expressed as [12]:

$$p = p_f + C_{SI}\frac{\mu_o B_o Q_{Po}}{2\pi K \bar{K}_{ro} h}\ln\frac{a+b}{a_f + b_f} \qquad (1)$$

where, p is the pressure in seepage field; P_f is the pressure of the fracture boundary; C_{SI} represents the conversion coefficient of two different units, and μ_o is the viscosity of oil underground; B_o is the oil volume factor, and Q_{po} is the oil production; K is reservoir permeability, and K_{ro} stands for the average oil relative permeability; h is reservoir thickness, and a and b are the half long axis and the minor axis of isopiestic lines respectively; a_f and b_f stand for the half long axis and the minor axis of the horizontal elliptical fracture respectively.

Elliptical Flow in Inner Zone

The seepage field from a source well to two parallel linear boundaries is studied by assuming two infinite linear constant pressure boundaries, which are 2d apart and their pressures are both p_e. The source well is located at the origin of coordinates, and these isopiestic lines in seepage field are similar to a series of coaxial elliptical. Assuming that the pressures on the long axis endpoint and the minor axis endpoint of both outer boundaries of horizontal fracture and wellbore are equal (which means $p_{af} = p_{bf} = p_f$ and $p_{aw} = p_{bw} = p_w$), according to pressure superposition principles, the pressure of the horizontal fracture equivalent area can be calculated by [12]:

$$p = p_f + C_{SI}\frac{\mu_o B_o Q_{Po}}{2\pi K_f^* \bar{K}_{ro} w_f}\ln\left(\tan\frac{\pi b_f}{4d^*}\right) - C_{SI}\frac{\mu_o B_o Q_{Po}}{2\pi K_f^* \bar{K}_{ro} w_f}\ln\left(\frac{ch\frac{\pi x}{2d^*} - \cos\frac{\pi y}{2d^*}}{ch\frac{\pi x}{2d^*} + \cos\frac{\pi y}{2d^*}}\right) \qquad (2)$$

where, K_f^* is the equivalent permeability in the inner zone, and it could be calculated by $K_f^* = \frac{K_f w_f + K_l}{h}$; d^* is the parameter determined by the long axis and minor axis of fracture and wellbore: $\tan\frac{\pi b_f}{4d^*} = \tanh\frac{\pi a_f}{4d^*}$, $\tan\frac{\pi b_w}{4d^*} = \tanh\frac{\pi a_w}{4d^*}$ x, y are x-coordinate and y-coordinate of an arbitrary point in flow field, and w_f denotes the average fracture width.

Pressure Loss Caused by Threshold Pressure Gradient in a Five-spot Waterflooding Pattern

The existence of threshold pressure gradient (TPG) has provoked much controversy in previous investigations [13-14]. Actually, low-permeability oil layers are developed more difficultly compared to high-permeability formations in waterflooded multilayer reservoirs, and threshold pressure gradient undoubtedly provides a quit reasonable interpretation for this difficulty. According to the pressure superposition principle, reservoir pressure with TPG factors in a symmetric five-spot waterflooding pattern can be expressed by [15]:

$$p = C_{SI}\frac{\mu_o B_o}{2\pi K\bar{K}_{ro}h}Q_{Po}\ln r_P + \lambda r_P - \sum_{m=1}^{n}\left(C_{SI}\frac{\mu_w B_w}{2\pi KK_{w\max}h}Q_{Iw}\ln r_{\mathrm{Im}} + \lambda r_{\mathrm{Im}}\right) + C \quad (3)$$

where, r_P is the distance from the oil well to field point (x,y), and r_{Im} denotes the distance from water well to field point; (x,y); n stands for the amount of water wells, and Q_{Iw} is water injection volume; μ_w represents water viscosity underground, and B_w and C is water volume factor and a constant respectively.

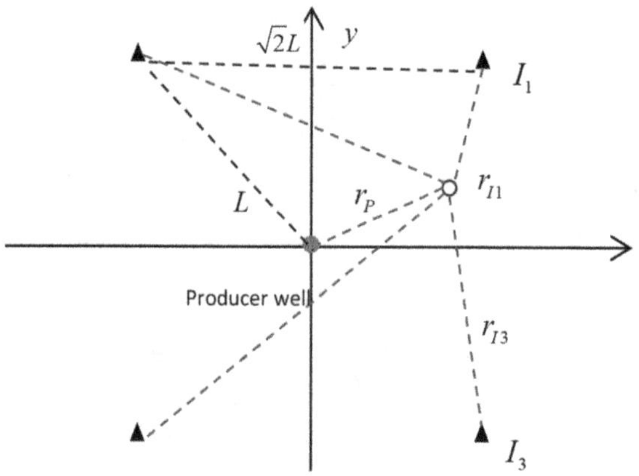

Figure 2: Square five-spot waterflooding pattern.

According to Equation 3, vertical producer well productivity without fractures can be calculated by the below equation:

$$Q_{po} = \frac{p_I - p_w - \beta(L - r_w)\lambda}{\left[C_{SI}\frac{\mu_o B_o}{2\pi K\bar{K}_{ro}h}\ln\left(\frac{L-r_w}{r_w}\right) + C_{SI}\alpha\left(1+\frac{\bar{K}_{rw}\mu_o B_o}{\bar{K}_{ro}\mu_w B_w}\right)\frac{\mu_w B_w}{2\pi KK_{w\max}h}\times \ln\frac{(L-r_w)^4}{\left(\sqrt{2}L - r_w\right)^2(2L-r_w)r_w}\right]} \quad (4)$$

where, r_w, L, α, and β represent wellbore radius, the distance between a producer well and an injector well, the ratio of water injector well number to producer well number, and a practical parameter respectively.

Performance of a Producer with One Horizontal Fracture in a Five-spot Pattern

For a water injector well, the main pressure drop is near its wellbore, so the influence of distal formation with horizontal fracture can be negligible. According to pressure superposition principles, the pressure drop caused by four water injector wells and one producer well (the five-spot waterflooding pattern) is superimposed respectively in the two zones, and then seepage field pressure can be calculated respectively; the pressure loss caused by TPG is not considered when the pressure in these two zones are discussed because the pressure loss is not influenced by different flow patterns; thus it is assumed to be equal with the one calculated in the regular five-spot pattern above.

1. Pressure in the outer zone

$$\begin{aligned} p &= p_e + \Delta p_I + \Delta p_P \\ &= p_w - C_{SI}\frac{\mu_o Q_{Po} B_o}{2\pi K_f^* \bar{K}_{ro}h}\ln\left(\tan\frac{\pi b_f}{4d^*}\right) \\ &+ C_{SI}\frac{\mu_o Q_{Po} B_o}{2\pi K_f^* \bar{K}_{ro}h}\ln\left(\frac{ch\frac{\pi x_w}{2d^*} - \cos\frac{\pi y_w}{2d^*}}{ch\frac{\pi x_w}{2d^*} + \cos\frac{\pi y_w}{2d^*}}\right) \\ &+ C_{SI}\frac{\mu_o Q_{Po} B_o}{2\pi K\bar{K}_{ro}h}\ln\frac{a+b}{a_f+b_f} \\ &+ \sum_{m=1}^{n}\left(p_I - C_{SI}\frac{\mu_w Q_{Iw} B_w}{2\pi K\bar{K}_{rw}h}\ln\frac{r_{\mathrm{Im}}}{r_w} - p_e\right) \end{aligned} \quad (5)$$

where, p_e is the initial reservoir pressure, and Δp_I denotes the pressure drop caused by water injector well; Δp_P is the pressure drop caused by producer well, and $\bar{K}_{rw}$ represents the average water relative permeability.

2. Pressure in the inner zone

$$p = p_w + C_{SI}\frac{\mu_o Q_{Po} B_o}{2\pi K_f^* \bar{K}_{ro} h}\cdot$$
$$\ln\left[\left(\frac{ch\frac{\pi x_w}{2d^*} - \cos\frac{\pi y_w}{2d^*}}{ch\frac{\pi x_w}{2d^*} + \cos\frac{\pi y_w}{2d^*}}\right) \div \left(\frac{ch\frac{\pi x}{2d^*} - \cos\frac{\pi y}{2d^*}}{ch\frac{\pi x}{2d^*} + \cos\frac{\pi y}{2d^*}}\right)\right]$$
$$+\sum_{m=1}^{n}\left(p_I - C_{SI}\frac{\mu_w Q_{Iw} B_w}{2\pi K \bar{K}_{rw} h}\ln\frac{r_{\mathrm{Im}}}{r_w} - p_e\right) \quad (6)$$

Hence injection-production pressure difference can be given as:

$$p_I - p_w = -C_{SI}\frac{\mu_o Q_{Po} B_o}{2\pi K_f^* \bar{K}_{ro} h}\ln\left(\tan\frac{\pi b_f}{4d^*}\right)$$
$$+C_{SI}\frac{\mu_o Q_{Po} B_o}{2\pi K_f^* \bar{K}_{ro} h}\ln\left(\frac{ch\frac{\pi x_w}{2d^*} - \cos\frac{\pi y_w}{2d^*}}{ch\frac{\pi x_w}{2d^*} + \cos\frac{\pi y_w}{2d^*}}\right)$$
$$+C_{SI}\frac{\mu_w Q_{Iw} B_w}{2\pi K K_{w\max} h}\ln\frac{\left(\sqrt{2}L - r_w\right)^2 (2L - r_w) r_w}{(L - r_w)^4}$$
$$+C_{SI}\frac{\mu_o Q_{Po} B_o}{2\pi K \bar{K}_{ro} h}\ln\frac{a_I + b_I}{a_f + b_f} \quad (7)$$

The oil production of a producer well with one horizontal fracture in the five-spot waterflooding pattern can be calculated by considering the pressure loss caused by TPG:

$$Q_{Po} = \frac{p_I - p_w - \beta\lambda\left(L - r_w - \frac{a_f + b_f}{2}\right)}{\left[\begin{array}{l} C_{SI}\frac{\mu_o B_o}{2\pi K \bar{K}_{ro} h}\ln\frac{a_I + b_I}{a_f + b_f} \\ -C_{SI}\frac{\mu_o B_o}{2\pi K_f^* \bar{K}_{ro} h}\ln\left(\tan\frac{\pi b_f}{4d^*}\right) \\ +C_{SI}\frac{\mu_o B_o}{2\pi K_f^* \bar{K}_{ro} h}\ln\left(\frac{ch\frac{\pi x_w}{2d^*} - \cos\frac{\pi y_w}{2d^*}}{ch\frac{\pi x_w}{2d^*} + \cos\frac{\pi y_w}{2d^*}}\right) \\ +C_{SI}\alpha\left(1 + \frac{\bar{K}_{rw}\mu_o B_o}{\bar{K}_{ro}\mu_w B_w}\right)\frac{\mu_w B_w}{2\pi K \bar{K}_{rw} h}\cdot \\ \ln\frac{(L - r_w)^4}{\left(\sqrt{2}L - r_w\right)^2 (2L - r_w) r_w} \end{array}\right]} \quad (8)$$

Similarly, water production is given by:

$$Q_{Pw} = \frac{p_I - p_w - \beta\lambda\left(L - r_w - \frac{a_f + b_f}{2}\right)}{\left[\begin{array}{l} C_{SI}\frac{\mu_o B_o}{2\pi K \bar{K}_{rw} h}\ln\frac{a_I + b_I}{a_f + b_f} \\ -C_{SI}\frac{\mu_o B_o}{2\pi K_f^* \bar{K}_{rw} h}\ln\left(\tan\frac{\pi b_f}{4d^*}\right) \\ +C_{SI}\frac{\mu_o B_o}{2\pi K_f^* \bar{K}_{rw} h}\ln\left(\frac{ch\frac{\pi x_w}{2d^*} - \cos\frac{\pi y_w}{2d^*}}{ch\frac{\pi x_w}{2d^*} + \cos\frac{\pi y_w}{2d^*}}\right) \\ +C_{SI}\alpha\left(1 + \frac{\bar{K}_{rw}\mu_o B_o}{\bar{K}_{ro}\mu_w B_w}\right)\frac{\mu_w B_w}{2\pi K \bar{K}_{rw} h}\cdot \\ \ln\frac{(L - r_w)^4}{\left(\sqrt{2}L - r_w\right)^2 (2L - r_w) r_w} \end{array}\right]} \quad (9)$$

Performance of the Five-spot Pattern after Separate-layerMulti-stageFracturing

The productivity of every low-permeability oil layer can be summed to calculate the whole productivity of vertical well with separate-layer multi-stage hydraulic fracturing using Equations 9-10. There is no need to take into account the influence of TPG for high-permeability oil layers in multilayer reservoirs. Assuming that all the low-permeability oil layers create horizontal fractures after separate-layer multi-stage hydraulic fracturing, the oil production of five-spot pattern after separate-layer multi-stage hydraulic fracturing can be calculated as defined below:

$$Q_{Po}^{Total} = \sum_{i=1}^{c} Q_{Poi} + \sum_{j=1}^{d} Q_{Poj} \quad (10)$$

where c is the sum of high-permeability oil layers, and d is the sum of low-permeability oil layers; i and j are number marks.

In a similar way, the water production of a producer well can be calculated as:

$$Q_{Pw}^{Total} = \sum_{i=1}^{c} Q_{Pwi} + \sum_{j=1}^{d} Q_{Pwj}, \quad (11)$$

Thus the water content reads:

$$f_w = \frac{Q_{Pw}^{all}}{Q_{Pw}^{all} + Q_{Po}^{all}} \times 100\% \quad (12)$$

On the basis of material balance, average oil saturation is calculated in different periods by assuming that liquid volume produced in a time step is equal to injection water volume, so oil saturation difference is equal to the ratio of oil production volume to reservoir pore volume; then the oil saturation in a new time step is calculated, and it is used to choose the oil and water relative permeability upon oil-water relative permeability curve. Oil production and water content are calculated for the next time step.

$$\Delta \overline{S}_w = -\Delta \overline{S}_o = \frac{Q_{Po} B_o \Delta t}{V_\phi} \quad (13)$$

Model Validation

According to the test data of producer X5-41-S737 in the thin low-permeability oil layer of Xing No.6 block which is located at Daqing oilfield, the results which are calculated by the performance model are validated. As shown in Figure 3, daily oil production has been calculated by our model and the test data of X5-41-S737 only differ slightly; decline trends are basic coincidence, so the performance model is reliable, and it could be used to design development projects for thin low-permeability oil layers in Daqing oilfield.

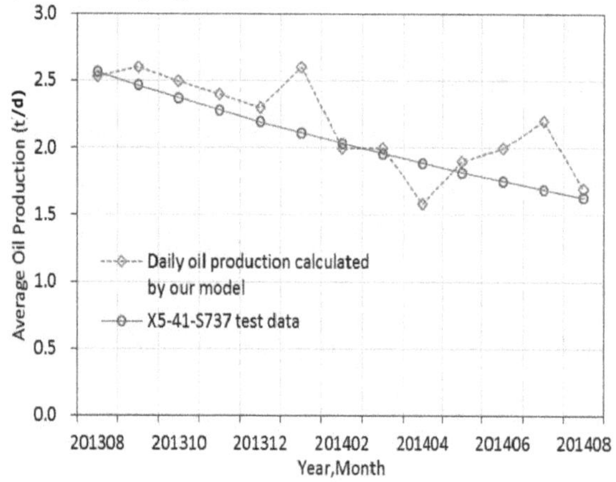

Figure 3: Daily oil production calculated by our model and the test data of X5-41-S737.

CASE STUDY

The Xing Shu-Gang oilfield is located in Daqing, China. It belongs to shallow sandstone multilayer reservoirs. Table 1 shows the basic parameters of the reservoirs, fractures, and fluids. Figure 4 displays the oil-water relative permeability curve for high-permeability oil layer and low-permeability oil layer respectively. The productivity of a producer well with multi-stage horizontal fractures is calculated, and the minimum well space, economic oil layer thickness, recovery, and productivity intensity are discussed.

Table 1: Physical parameters of reservoirs, fractures, and fluids.

Basic Parameters	High-permeability Oil Layers	Low-permeability Oil Layers
Porosity	0.25	0.22
Permeability (mD)	100～1000	5～10
Average half axis (m)	-	30
Fracture width (m)	-	0.003
Dimensionless fracture conductivity	-	0.3
Injection-production pressure difference (MPa)	20	20
Initial oil saturation	0.72	0.5
Oil volume factor	1.2	1.2
TPG (MPa/m)	0	0.025
Fluid viscosity(MPa.s)	7.8	7.8

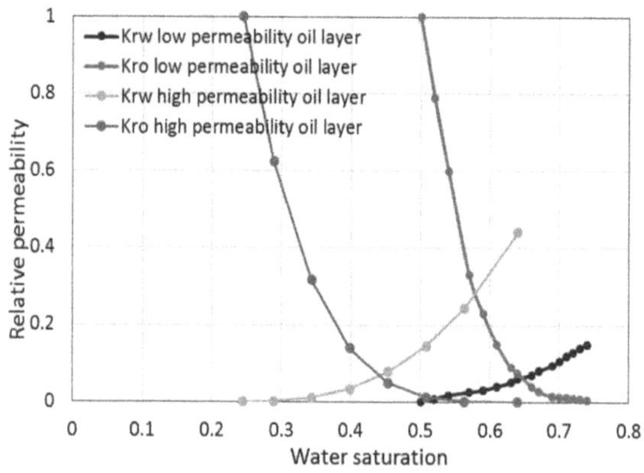

Figure 4: Relative permeability curves for both high-permeability and low-permeability oil layers.

Minimum Well Space and Economic Oil Layer Thickness

The minimum well space is the injection-production well space which producer well has the minimum oil production by waterflooding in the five-spot pattern. Economic oil layer thickness is the thickness of oil layer which could have economic production. Assuming that a vertical producer well only perforates some low-permeability oil layers, then the initial oil production under different initial water cut, well space, and oil layer thickness after hydraulic fracturing are calculated using Equation 10. Figures 5 and 6 show the results of initial oil production.

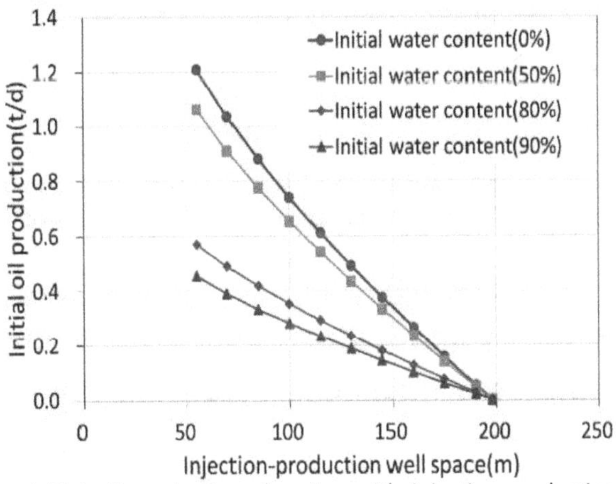

a. Initial oil production changing with injection-production well space in 6m oil layers without fracturing.

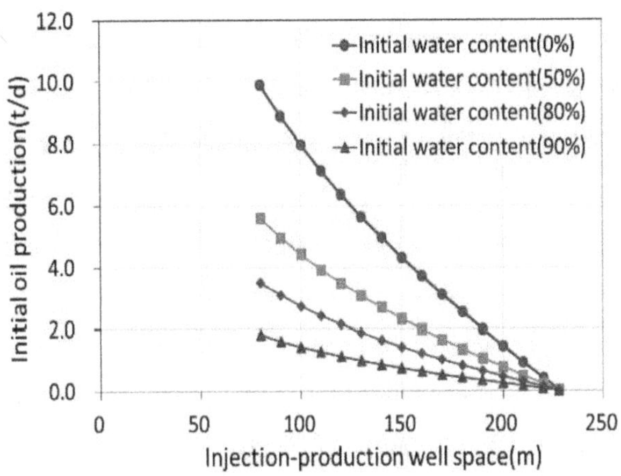

b. Initial oil production changing with oil layers thickness under 150m well space with fracturing.

Figure 5: The relationship between oil production and well space

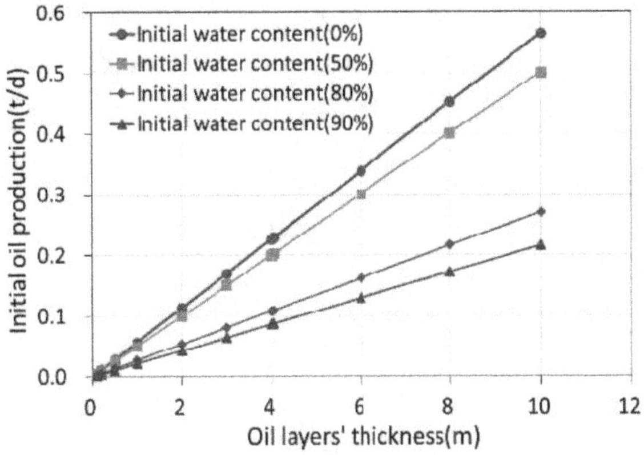

a. Initial oil production changing with oil layers thickness under 150m well space without fracturing.

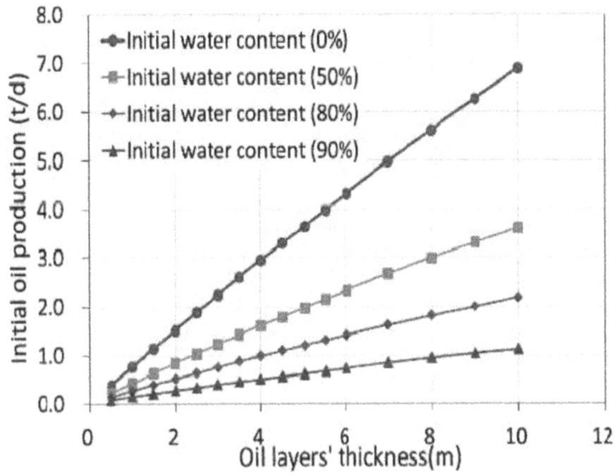

b. Initial oil production changing with oil layers thickness under 150m well space with fracturing.

Figure 6: The relationship between initial oil production and oil layers thickness.

As Figure 5 shows, the minimum well space is 198 m before hydraulic fracturing, and it extends to 228 m after hydraulic fracturing, so it is lengthened by 30 m because of horizontal fracturing. When the oil layer thickness is 6 m, the production-injection well space is 150 m, and the initial oil production will be less than 0.4 t/d under any initial water content; thus it has not reached the economic productivity of Xing Shu-Gang oil field (1.5 t/d each producer); however, oil production increases to more than 2 t/d after hydraulic fracturing with an initial water content less than 50%.

As shown in Figure 6, the initial oil production enhances with increasing oil layer thickness. Even if oil layer thickness increases to 10 m, the initial oil production will be as low as 0.57 t/d without fracturing under an initial water content of 0%; thus it has not reached the economic oil production. After hydraulic fracturing, the initial oil production increases to 6.9 t/d with a water content of 0%; however, when the oil-layer thickness decreases to 2 m, the initial oil production is generally less than 1.5 t/d, which causes the oil layer to lose economic recoverable values.

The Relationship between Recovery and Ratio of Low-permeability Oil Layers Thickness to the Total Thickness

Assuming that the vertical producer well perforates both high-permeability and low-permeability oil layers, the relationship between recovery and ratio of low-permeability oil layers thickness to total thickness is analyzed as follows. Figure 7 shows that with increasing the ratio, its recovery first drops, levels off, and then increases rapidly. This may be because of the fact that oil production is lower, and the water content increases more slowly from low-permeability oil layers than from high-permeability oil layers. When the ratio is low, high-permeability oil layers would play a more important role in providing oil production. In addition to, when the ratio is low, injection water flows mainly through high-permeability oil layers, and the producer well reaches extreme water content (98%) suddenly. However, the total recovery of the multilayer reservoir is low too because the recovery of these low-permeability oil layers maintains quite low. When the ratio is high, low-permeability oil layers would play a more important role in oil production. Although some injection water flows through high-permeability oil layers, which does not cause the producer well to reach the extreme water content (98%) suddenly, the total recovery enhances.

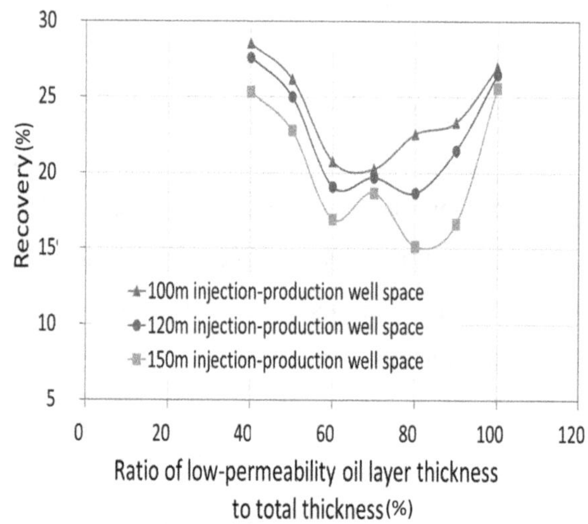

Figure 7: The relationship between recovery and ratio of low-permeability oil layer thickness to total thickness.

The Relationship between Productivity Intensity and Ratio of Low-permeability Oil Layers Thickness to Total Thickness

Assuming that the vertical producer well perforates both high-permeability and low-permeability oil layers, the relationship between productivity intensity and the ratio of low-permeability oil layers thickness to total thickness is analyzed as follows. Figure 8 shows that productivity intensity gradually decreases with increasing the ratio, but the productivity intensity gradually decreases with increasing well space. That is because the productivity intensity of low-permeability oil layer is weaker than the high-permeability oil layer; also, if the ratio becomes higher, the productivity intensity of the whole multilayer reservoir becomes weaker. With increasing well space, the supply of water injector well becomes weaker, and then the productivity intensity of the whole multilayer reservoir lessens.

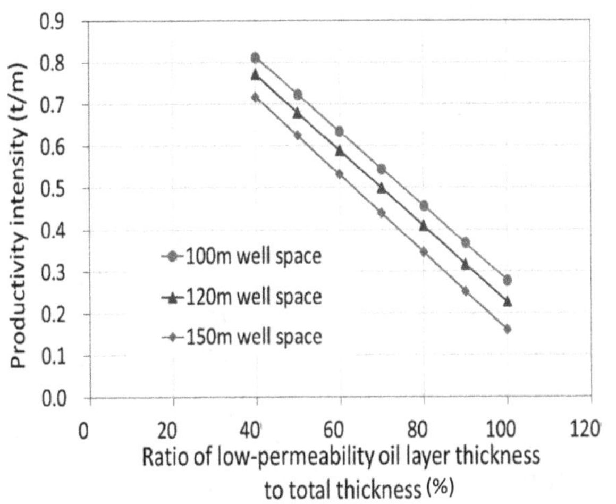

Figure 8: The relationship between productivity intensity and the ratio of low-permeability oil layer thickness to total thickness.

CONCLUSIONS

1. The performance model of vertical well with separate-layer multi-stage horizontal fractures is derived by considering the loss of pressure caused by TPG, material balance, and pressure superposition.
2. A high-precision analytical solution method for this performance model is presented by assuming the average decrement of water saturation in the reservoir, so the evaluation process of the performance has been simplified.
3. The minimum well space and economic oil layer thickness are calculated by the performance model for the multilayer reservoir in Xing Shu-Gang oilfield. When the total low-permeability oil layer thickness is less than 10 m, the initial oil production without fracturing is less than 1.5 t/d, so it will lose the recoverable value. When the well space is less than 150 m with initial water content less than 50%, the initial oil production after fracturing is more than 2 t/d, which has the recoverable value. When the total thickness of low-permeability oil layer decreases to 2 m, the initial oil production after fracturing is less than 1.5 t/d, and it will lose the recoverable value.
4. The relationship among recovery, productivity intensity, and the thickness ratio of low-permeability oil layers to high-permeability oil layers is studied. With increasing the thickness of low-permeability oil layer, first recovery first decreases, second levels off, and finally increases rapidly; however, the productivity intensity gradually drops.

ACKNOWLEDGEMENTS

The authors thank Mr. Shi Chengfang, Mr. Tian Changbing, and others who have been working at PetroChina Research Institute of Petroleum Exploration and Development (RIPED) for providing data, guidance and assistance.

REFERENCES

1. Ren L., Zhao J. Zh., Hu Y. Q. et al., "Affecting Factor Analysis of Productivity for Horizontal Fractured Wells," *Journal of Southwest Petroleum University (Science & Technology Edition)*, **2010**, *32*, 77-78.
2. Zhang G. S., Tang G. C., Wang X. Y. et al., "A New Productive Method of Horizontal Fracture Wells in Hydraulic Fracturing," *Journal of Oil and Gas Technology*, **2009**, *31*, 121-124.
3. Yang Y., Li X. F., Lu W. et al., "Application and Derivation of Vertical Well Deliverability Equation for Shallow Layer and Low Permeability Reservoirs with Horizontal Fracture," *Science Technology and Engineering*, **2013**, *13*, 7015-7020.
4. McDaniel B. W., "Review of Current Fracture Stimulation Techniques for Best Economics in Multilayer, Lower-permeability Reservoirs," SPE 98025, **2005**.
5. Prats M., Aime M., Hazebroek P., and Strickler W. R., "Effect of Vertical Fractures on Reservoir Behavior-compressible-fluid Case," SPE 98, **1962**.
6. Friehauf K. E., Suri A., and Sharma M. M., "A Simple and Accurate Model for Well Productivity for Hydraulically Fractured Wells," *SPE Production & Operations*, **2010**, *25*, 453-460.

7. Deng Y. E. and Liu C. Q., "Analysis of Pressure of Nonlinear Flow through Low-permeability Reservoir with Vertically Fractured Well Producing," *Petroleum Exploration and Development*, **2003**, *30*, 81-83.
8. W. S. and Ertekin T., "Performance Comparison of Vertical and Horizontal Hydraulic Fractures and Horizontal Boreholes in Low Permeability Reservoirs: A Numerical Study," SPE 16407, **1987**.
9. Valko, Peter, "Transient Behavior of Finite Conductivity Horizontal Fractures," SPE 38436, **1998**.
10. Leif L., "Horizontal Fractures in Single and Multilayer Reservoirs," SPE 147004, **2011**.
11. Qi Zh., "Oil Production Engineering Theory and Design," *China University of Petroleum Press*, **2000**, 251-293.
12. Liu Y. T. and Ge J. L., "Analytic Solution for Fluid Flow though A Circular Anisotropic Formation," *Journal of the University of Petroleum*, **2000**, *24*, 40-43.
13. Baoquan Z., Linsong Ch., and Chunlan L., "Low Velocity Nonlinear Flow in Ultralow Permeability Reservoir," *Journal of Petroleum Science and Engineering*, **2011**, *80*, 1-6.
14. Wang X. D., Hao M. Q., and Han Y. X., "Implication of the Threshold Gradient and its Application," *Acta. Petrolei. Sinica.*, **2013**, *34*, 188-191.
15. Xiangyan K., "Advanced Porous Flow," University of Science and Technology of China Press, **2010**, 110-560.

2

3D and 4D Seismic Data Integration in Static and Dynamic Reservoir Modeling

Mohammad Emami Niri*

Institute of Petroleum Engineering, University of Tehran, College of Engineering, Tehran, Iran

ABSTRACT

Reservoir modeling is the process of generating numerical representations of reservoir conditions and properties on the basis of geological, geophysical, and engineering data measured on the Earth's surface or in depth at a limited number of borehole locations. Therefore, reservoir modeling requires an incorporation of the data from a variety of sources, along with an integration of knowledge and skills from various disciplines. In particular, recent advances in 3D and time-lapse 4D seismic data acquisition, processing, and quantitative interpretation have led to an increasing use of seismic data in the reservoir modeling processes. This paper provides an overview of static and dynamic reservoir modeling and outlines the key roles of 3D and time-lapse 4D seismic data in reservoir characterization and model building/updating processes. The review focuses on the methods, workflows, and challenges in the incorporation of 3D/4D seismic data into the static and dynamic reservoir model building/updating processes.

Keywords: Reservoir Model, Static and Dynamic Properties, 3D Seismic, Time-Lapse 4D Seismic

INTRODUCTION

Subsurface reservoir model estimation is a fundamental practice in many geoscience disciplines, including geothermal studies, hydrology and ground water analyses, exploration and recovery of fossil fuel energy resources, and CO_2 geo-sequestration among others. In particular, numerical reservoir models play a central role throughout a hydrocarbon field's life cycle. During exploration, appraisal, development, and production stages of any field life cycle, reservoir models are widely used to broaden the available knowledge of the geological, geophysical, and engineering components of the reservoir.

In general, there are two categories of reservoir properties: first, time-independent static properties such as porosity, lithology, and shale content, and second, time-varying dynamic fluid flow properties such as fluid saturation and pore pressure. Static reservoir models simulate time-independent reservoir properties; typically, such models are constructed from static data that have been measured or interpreted on a single occasion (once in time); these models include well logs, core measurements, sedimentology and stratigraphy interpretations, and baseline (pre-production) 3D

***Corresponding author**

Mohammad Emami Niri

Email: emami.m@ut.ac.ir

seismic surveys. Usually, such models are used for initializing the dynamic reservoir modeling process, which involves the dynamic simulation of fluid flows within a reservoir. Typical dynamic datasets in hydrocarbon energy applications include historical production data (flow rates or volumes and pressure data) and time-lapse 4D seismic data. Borehole measurements are often the main information source for reservoir modeling; however, boreholes are sparsely distributed, and these measurements are not sufficiently informative to yield accurate and detailed representations of the whole 3D reservoir volume. There are many non-unique models that fit the sparse well data despite the huge challenge posed by extrapolating these data to an entire reservoir volume. Due to their excellent spatial coverage, 3D seismic data play a key role in defining the structure and geometry of the reservoir, and in setting constraints to variations in reservoir properties. To produce realistic models of reservoir lithofacies and corresponding petrophysical properties while avoiding non-physical results at the same time, 3D seismic information should be actively incorporated into the static reservoir property of modeling process. On the other hand, 4D seismic data are powerful constraints on dynamic reservoir models because of its valuable information relating to production-induced reservoir changes such as fluid movements and pressure and saturation changes.

This paper is organized in two main sections as follows. In section 1, the fundamentals of static reservoir modeling and 3D seismic data are provided and followed by a review of the methods and challenges in 3D seismic data incorporation into static reservoir models. In section 2, the basics of dynamic reservoir modeling and time-lapse 4D seismic data are described and followed by a review of the methods and challenges in 4D seismic data incorporation into dynamic reservoir models [1].

EXPERIMENTAL PROCEDURES

Static Reservoir Modeling

Generating a static reservoir model that is consistent with geological knowledge and pre-production seismic data is a fundamental step of reservoir characterization and performance forecasting [2]. Currently, 3D static reservoir models are commonly used in [3]:

- estimating reserves;
- targeting new producer or injector locations;
- performing uncertainty and risk analysis;
- geosteering (i.e. well-path steering during drilling);
- providing a basis for production forecasting and cost estimation when coupled with reservoir hydrodynamic simulators; and
- providing a basis for rock mechanics modeling and fracture analysis.

To construct a static reservoir model, the reservoir-specific input data are the key to achieve the real representation of the reservoir conditions. These input data include [4]:

- horizons and fault surfaces interpreted from 2D or 3D seismic data to define the size, shape, and geometrical framework of the reservoir container (including the top and base of the different intervals);
- geological information such as sedimentology maps and analogue outcrop data;
- forward sediment modeling datasets such as geometrical data, stacking patterns, and spatial information for porosity and permeability;

• complete well datasets, including wellhead coordinates, top markers, well paths, logs, and interpretations;
• core measurements, where applicable;
• seismic data, including seismic amplitude maps, seismic attribute volumes (for reservoir property modeling), and seismic velocity cubes (for time to depth conversion process);
• Other geophysical measurements (gravity, magnetic etc.), where applicable; and
• Engineering data, for example in the hydrocarbon energy applications, which includes historical production, drill stem test (DST), and repeat formation test (RFT).

A typical workflow for generating a static reservoir model involves the following key steps [5-6]: (1) define the geometric structure and framework of the reservoir, (2) construct cellular grids within the reservoir, and (3) create reservoir property models.

Three-dimensional static reservoir models may include reservoir properties in two categories: discrete reservoir properties (e.g. lithofacies, fluid volume, and hydraulic flow units) and continuous reservoir properties (e.g. net to gross, porosity, permeability, and initial fluid saturations). As shown in Figure 1, the construction of a static reservoir property model is typically performed in a two-stage sequential approach: modeling reservoir lithofacies to define the main lithologies and/or flow units, and modeling petrophysical property to simulate the petrophysical properties facies-by-facies or layer-by-layer. Both of these steps can be performed using either deterministic or stochastic techniques [7].

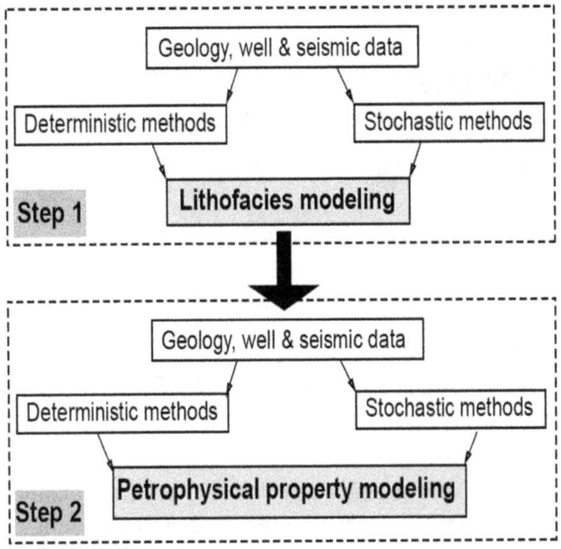

Figure 1: A two-stage sequential approach to reservoir lithofacies and petrophysical property modeling [1].

However, it should be noted that deterministic techniques cannot effectively describe the reservoir conditions and interdependency of reservoir properties. This is mainly due to: (i) the complexity of reservoir heterogeneity and spatial variations of the reservoir properties, and (ii) sparseness in the sampling of a reservoir at limited numbers of well locations, leading to incomplete information about the reservoir. In this context, stochastic methods provide a framework for generating a set of reservoir models (not merely one model, as is the case with deterministic methods) to account for the uncertainties and spatial variations in discrete and continuous reservoir properties [8-10].

3D Seismic Fundamentals

The main objective of seismic exploration is to provide an accurate representation of subsurface geological features. Seismic reflection data is typically acquired by transmitting energy into the specific portion of the Earth and recording the reflected energy from the subsurface geological boundaries (Figure 2).

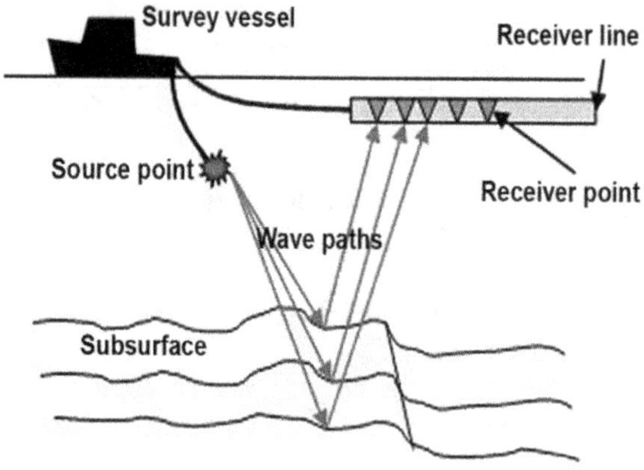

Figure 2: Seismic reflection data are typically acquired by transmitting energy into the Earth and recording the reflected energy from subsurface geological boundaries.

When processed, these data produce images that effectively evaluate the potential targets. Two-way travel times and amplitudes, or any other attributes of the reflected energy can also be used to derive useful information about subsurface geological features and their properties.

Seismic surveys conventionally involve two-dimensional (2D) and three-dimensional (3D) surveys. Two-dimensional seismic data is typically acquired along a single seismic recording line; therefore, it is displayed as a single vertical cross-section through the subsurface (e.g. Figure 3a).

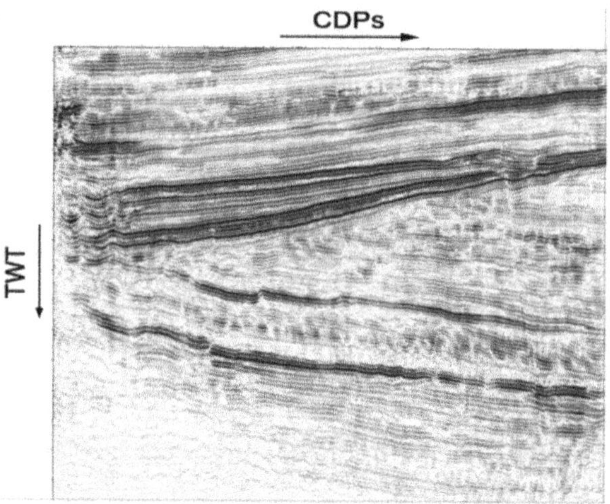

(a)

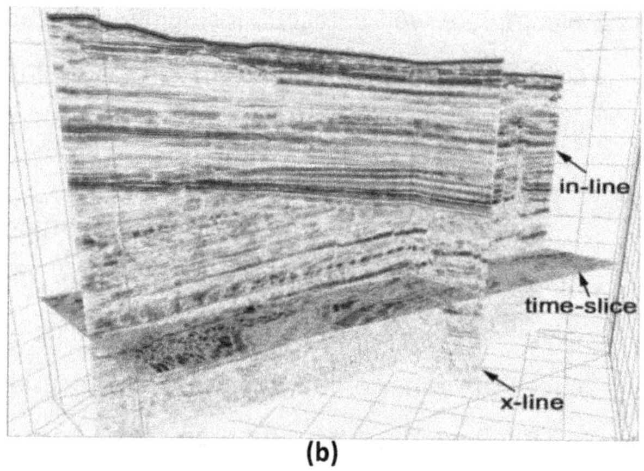

(b)

Figure 3: (a) A 2D post-stack seismic section, and (b) a 3D seismic visualization along an in-line, an x-line, and a time slice.

The nearest seismic recording line may typically be kilometers away. Two-dimensional seismic data are generally used for regional or large-area assessments, and provide information about the location and orientation of subsurface geological features [11]. The critical problems with 2D seismic surveys are (i) coverage problem (2D seismic data only cover the thin cross-sections of the subsurface) and (ii) sideswipe or off line reflection problem (2D seismic cannot distinguish features which are not on the plane of the 2D seismic section) [12].

Three-dimensional seismic surveys solve both the coverage and the sideswipe problems [13]. They are typically designed with multiple source and receiver lines to enable the acquisition of large volumes of seismic data. In other words, 3D surveys use closely spaced grids of shot lines such that the data are acquired at the surface over an area on which several seismic recording lines are located; moreover, the interpreted data is displayed as a 3D volume or cube which may then be sliced into multiple cross-sections (e.g. Figure 3b). Three-dimensional seismic intuitively provides a more precise and spatially continuous image of the subsurface geological structure.

Seismic attribute: a seismic attribute is a quantity derived from 2D or 3D seismic data that can be analyzed in order to extract or enhance information that might not otherwise be visible in the seismic images, leading to a better geophysical or geological interpretation of the seismic data [14]. Time, frequency, and amplitude are the examples of seismic attributes.

Seismic inversion: transformation of seismic reflection data into subsurface physical properties such as elastic parameters is called seismic inversion [15]. There are two classes of seismic inversion techniques [16]: (i) deterministic inversion, which gives one single realization of the physical properties of the rock (e.g., P-wave impedance); (ii) stochastic inversion, which generates multiple equiprobable realizations of the subsurface physical properties of the rock. Colored inversion [17], sparse-spike inversion [18], and simultaneous elastic inversion [19-20] are common examples of deterministic seismic inversion techniques. Geostatistical inversion [21] and Bayesian inversion [22] are two main types of stochastic seismic inversion.

3D Seismic Data Incorporation in Static Reservoir Modeling

3D seismic data, due to its excellent spatial resolution, play a key role not only in defining the reservoir structure and geometry, but also in constraining the reservoir property variations. To produce realistic models of reservoir lithofacies and corresponding petrophysical properties while avoiding non-physical results at the same time, seismic information should be actively incorporated into the reservoir modeling process. Three main frameworks for deriving reservoir properties from seismic data [23] are described below.

Deterministic relationships: First, deterministic relationships is established between seismic attributes and reservoir properties at borehole locations, then, these relationships are employed in generating a 3D cube of the reservoir property (Figure 4) [24].

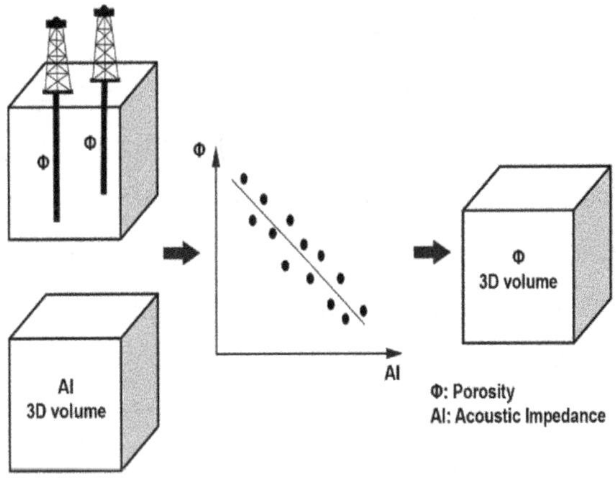

Figure 4: A schematic view of generating a static reservoir model based on deterministic relationships between seismic attributes and reservoir properties [1].

Geostatistics: Seismic data (e.g. seismic amplitude map and inverted seismic attributes) are incorporated as extra information for guiding geostatistical interpolation of the reservoir properties in areas far away from the wells, and multiple equiprobable realizations of the reservoir properties are produced (Figure 5).

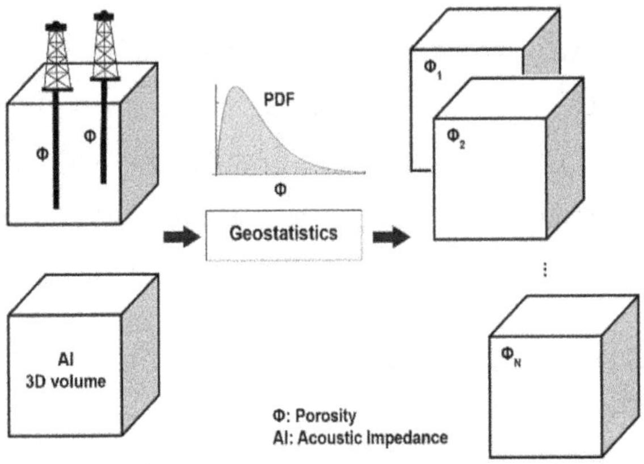

Figure 5: A schematic view of generating multiple equiprobable realizations of reservoir properties using geostatistical simulation [1].

Examples of geostatistical techniques for seismically-constrained reservoir property modeling include generalized regression methods such as cokriging [25-26], kriging with locally variable mean [27], stochastic simulation [28-30], and geostatistical inversion [21].

Seismic matching loop: The seismic response of the model is computed using a forward-modeling operator; moreover, the synthetic seismic response is compared with the real seismic data, and the best match is obtained by iteration (Figure 6). This process can be conducted manually or by using an optimization algorithm [31].

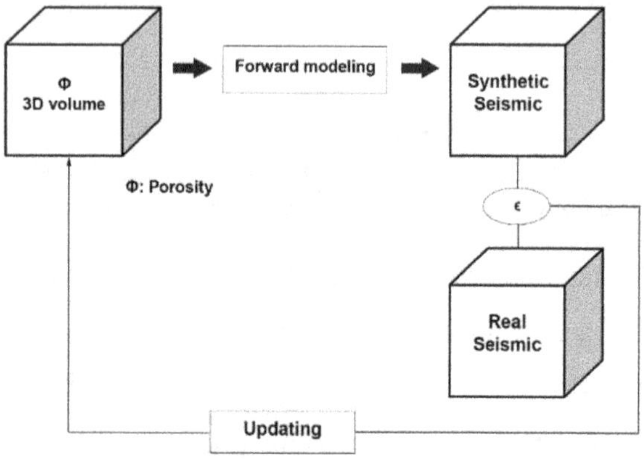

Figure 6: A schematic view of reservoir model building or updating by the seismic matching loop process [1].

Recent developments have combined geostatistics and a seismic matching loop (from either deterministic or statistical rock-physics relations) to estimate reservoir properties from seismic data. This process may be performed (i) by using a sequential or multistep inversion scheme [23,32-34], or (ii) based on a unified inversion scheme [35].

Seismic-constrained Static Reservoir Modeling

A typical workflow for constructing a seismically constrained static reservoir model is summarized in Figure 7. It consists of three main stages: data preparation, reservoir property modeling, and seismic matching loop. The details of each step are described below.

Data preparation: This is the first stage in the workflow where all the required input data, including the geology, geophysics, petrophysics, rock physics, and engineering data are transferred to a common framework. A 3D structural framework of the model in the depth domain is constructed. Moreover, this requires the interpreted horizons and faults to be converted from time to depth using an accurate velocity model derived from seismic velocities and calibrated to well logs. The zones between the horizons and the layers within each zone are defined. In addition, the cubes of the depth-converted seismic inverted attributes (if available) should be rescaled to the reservoir model grid size. Typically, multiple log datasets including measured depth, true vertical depth, lithology, porosity, resistivity, and so on are available at some wells, which also need to be rescaled to fit the cellular framework of the model. If core measurements, well test analyses (e.g., pressure transient), and other static and dynamic datasets are also available, they should be rescaled appropriately [36].

Reservoir property modeling: The second stage in the workflow is reservoir property modeling. This is usually carried out in a geostatistical framework and is based on two sequential steps:

First, reservoir lithofacies/flow units are drawn from the input data. There are two main classes of geostatistical technique for lithofacies modeling: object-based and pixel-based. Pixel-based methods offer more effective (or natural) ways of incorporating seismic data into a reservoir lithofacies model. Lithofacies modeling using a Bayesian classification technique is an example of a pixel-based simulation ([37-40]. Other seismic-constrained lithofacies modeling techniques are sequential indicator simulation (SIS) with seismic constraints [41,34], the truncated Gaussian simulation (TGS) method [7], and pluri-Gaussian simulation (PGS) [42].

Second, petrophysical properties such as porosity and net to gross are simulated facies-by-facies by utilizing some other class of geostatistical methods such as sequential Gaussian simulation (SGS) [43]. In practice, SGS coupled with collocated cokriging is widely used in the sense that a seismic attribute such as impedance is used as a secondary variable in the cokriging part of the SGS [6,44].

Seismic matching loop: The main objective of the seismic matching loop is to update the reservoir models to replicate the real seismic data as closely as possible. A key step in a successful seismic matching loop is the definition of an accurate (deterministic or statistical) petro-elastic model [34] that enables the reservoir properties to be inferred from the seismic data and vice versa. In the workflow, the seismic response of the reservoir models is iteratively determined using a well-designed petro-elastic model, and is then compared with the real seismic data to evaluate the match. Any resulting discrepancy is then used to guide the updating process of the reservoir models. As shown in Figure 7, the seismic matching loop can be performed at different levels: level 1, the seismic time domain [31], and level 2, the elastic depth domain [23, 45].

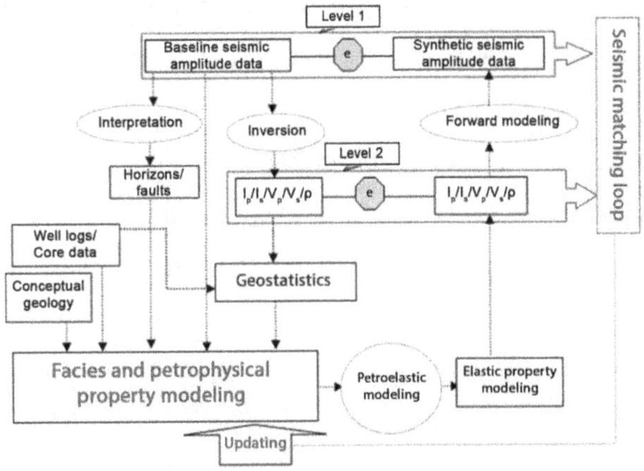

Figure 7: A typical workflow for seismic-constrained static reservoir modeling [1].

Challenges in 3D Seismic Data Incorporation into Reservoir Models

Integrating 3D and 4D seismic data into reservoir models presents several challenges and difficulties, some are still at the forefront of research and development (R&D) specialists. In this section, some of the challenges in the incorporation of 3D seismic data in static reservoir modeling are reviewed.

Reservoir modeling at seismic scale or at a finer grid scale: To model a reservoir, it is important to find a common grid scale that merges different data types on different scales. A common approach is either to scale up or to scale down; however, even the most sophisticated (upscaling/downscaling) techniques necessarily eliminate some parts of the data [46]. As an example, a reservoir containing highly porous and permeable layers about 1 m thick requires a fine-scale reservoir model. To integrate seismic data into such a model, the data must be scaled down accordingly. Different seismic downscaling techniques have been proposed, but these are generally not physics-based and suffer from simplifying assumptions about the seismic averaging process [47].

A promising strategy to define an optimum grid scaling system can be described as follows. When the lateral grid scale of a reservoir model is chosen, the seismic bin spacing is the best option because it preserves all the potential information that can possibly be extracted from the seismic data. However, the challenge is the choice of the vertical- or z-scale of the model. Seismic data typically have a lower resolution in the vertical direction than the other sources of information (e.g. well log data and core measurements). For modeling reservoir properties, the vertical scale should be related to the scale of the geological heterogeneities in the volume to be modeled. This scale is highly dependent upon the types of rock in the reservoir. It is essential to note that, while it is not the purpose of a reservoir model to include every small geological detail extracted from the geological analyses, neither large-scale nor regional information of interest is in this context. The purpose of the model is to simulate a specific volume of reservoir rocks that allow fluid flow through it. Thus, the optimal vertical scale of a reservoir model depends on the scale of the reservoir flow unit, which may vary from one or two centimeters to several meters.

Deterministic or stochastic seismic inversion: Deterministic inversion gives just one realization of any rock physical property, whereas stochastic inversion generates multiple equiprobable realizations. No single inversion technique is guaranteed to solve every kind of problems; moreover, every implementation of the deterministic or stochastic inversion might be suitable, provided it is applied to the right case.

Cascaded seismic and petrophysical inversion or direct petrophysical inversion of seismic data: A two-step sequential approach is often used to infer reservoir properties from seismic data: first, a seismic inversion is performed to estimate elastic properties; second, a petrophysical inversion is used to predict reservoir properties such as lithology and porosity from the inverted seismic data. Some other techniques have been proposed for direct petrophysical inversion of seismic data [31]. Although these techniques improve coherence between the reservoir property model and seismic data, the well calibration and quality control steps tend to be more complicated [44].

Reservoir model-to-seismic matching loop in amplitude time domain or in elastic depth domain: The reservoir model-to-seismic matching loop allows the validation of the reservoir properties introduced into the reservoir model by generating synthetic seismic impedance or amplitude data from the model and by attempting to match them to real seismic data. To perform reservoir model-to-seismic in the best domain in which matching loop is problematic, it is necessary to choose between the amplitude versus elastic property domain and between the time and depth domain.

The petro-elastic model gives absolute values of the elastic properties, whereas seismic inversion provides both absolute and relative values of the elastic properties. Therefore, in the elastic domain, it is necessary to first choose which form of the elastic parameters should be used for comparison: absolute or relative. Also, it should be noted that the low-frequency model used in the seismic inversion and the background trend used in the petro-elastic model are not the same. In addition, there is always uncertainty and instability in the seismic inversion process [48]. All of these challenging issues imply that the reservoir model-to-seismic matching loop in the elastic domain should be cautiously applied in order to reduce the risk of misinterpretation.

In the amplitude domain, a key challenge is the correct positioning of the modeled 3D seismic data in time versus real seismic amplitude data. A possible solution is to stretch the reservoir model in time between two interpreted horizons, for example, by using petro-elastic-derived velocities. In addition, since the modeled data usually does not match the real seismic data, using purely mathematical algorithms for comparison (e.g. subtraction and correlation) may not always work [48].

Manual or assisted reservoir model-to-seismic matching loop: A key question in the reservoir model-to-seismic matching loop process is the possibility of employing an assisted process for reservoir model estimation. A reservoir model matched only with seismic data is not a comprehensive solution; moreover, this only provides a set of reservoir properties that are consistent with the seismic data but not necessarily with well log data and geological information. Similarly, a reservoir model that takes account of geological knowledge is not only necessarily consistent with seismic data. The ultimate goal of an assisted reservoir modeling process is to generate a set of reservoir models that best match all the available static data (well logs, geological, and pre-production seismic data) to better initialize the subsequent history-matching process. Emami Niri and Lumley [1,34] proposed a solution to this problem by developing a multi-objective optimization approach for the reservoir model estimation process. From this approach, the resulting reservoir property models not only satisfy the seismic response of the models, but also match the well log data and the geological information defined in the parallel objective functions. An ensemble-based stochastic optimization algorithm simultaneously reduces the mismatch in all objective functions and produces optimal reservoir models that are the best compromise solutions to the defined objective functions.

Dynamic Reservoir Modeling

Dynamic reservoir modeling simulates the reservoir fluid flow behavior over production calendar time and contributes to the optimization of reservoir future production in terms of the recovery and economics. In the context of dynamic reservoir modeling, the term "history matching" is frequently used. History matching is the process of adjusting the dynamic reservoir flow model to match the production data measured at wells. In other words, it evaluates the past and present behavior of the reservoir and forecasts its future performance on that basis. Conventionally, a static model is initialized to be consistent with the geological, geophysical, and well log data for the reservoir, and then it is adjusted to match the historical production data in order to predict the future behavior of the reservoir as flow stimulation proceeds [49,50]. History matching process may be carried out either manually [51] or automatically by using optimization algorithms [52]. The main characteristics of these two approaches are summarized in Figure 8, which is adapted from elsewhere [49].

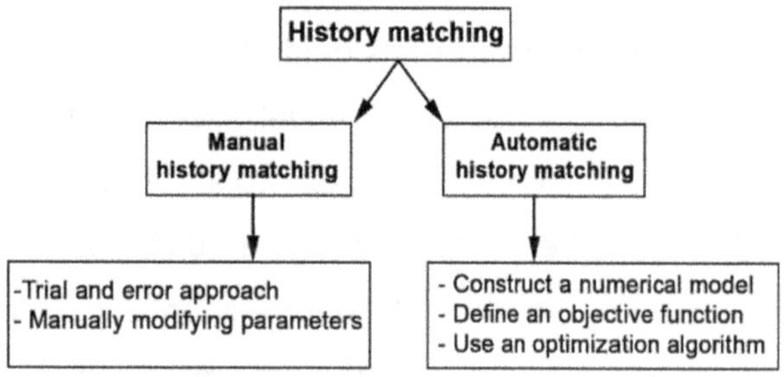

Figure 8: Manual versus automatic history matching.

Several researches such as single fluid phase history matching in a gas reservoir using adjoint methods [53], history matching in a two-phase oil-water reservoir [54], and 3D, three-phase oil-water-gas history matching studies [55] have been conducted on addressing history matching problems in hydrocarbon energy applications.

The quality of business decisions in reservoir management is highly dependent upon the quality of the history matching. However, it is notable that history matching is a time-consuming, costly, and ill-posed inverse problem and requires a large number of iteration runs, leading to non-unique predictions which make reservoir management decisions difficult [56].

4D Seismic Fundamentals

Four-dimensional seismic data is a series of 3D seismic datasets acquired over a producing reservoir in a calendar time period. It is used for imaging fluid movements, fluid contacts, monitoring flow paths and barriers, locating bypassed oil, placing new wells, identifying sealing faults, and mapping pressure compartmentalization and heat effects [57-60]. In general, there are two categories of reservoir properties which affect 3D seismic images: (i) time-independent geological properties such as lithology and porosity, and (ii) time-varying fluid-flow properties such as fluid saturation and pore pressure. The difference between two consecutive 3D seismic survey images approximately cancels out the static properties since, in effect, these do not change in that period of time; thus, the resulting image difference is caused only by dynamic property changes [46].

Figure 9 is a schematic outline of the physical principle of time-lapse seismic reservoir monitoring. If survey-1 data are acquired before production or injection in an oil and gas field, and survey-2 data are acquired afterwards, then production/injection has resulted in changes to the reservoir fluid saturations and pressures (Figure 10), and hence to the seismic compressional and shear wave velocities and density; therefore, ideally, no change has occurred in the upper and lower bounding formations during this time.

Changes in the elastic properties of the reservoir interval may then be inferred from the observed changes in the timing and amplitude of the seismic traces of monitor survey (δt and δA). Thus, each subsequent time-lapse seismic survey captures a new snapshot of the changes that have occurred in the reservoir in a given production time period. This simplified description explains how 4D seismic surveys help identify the production-induced changes.

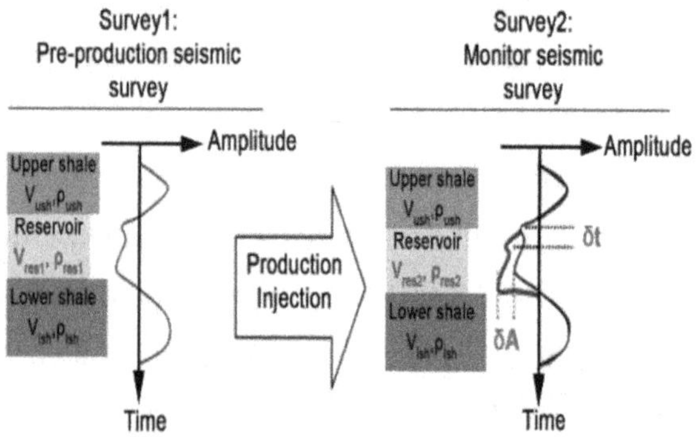

Figure 9: The physical principle of time-lapse seismic reservoir monitoring; production or injection results in changes in P- and S-wave velocities and density, which appear as changes in amplitude and timing in the monitor seismic survey (adapted from [59]).

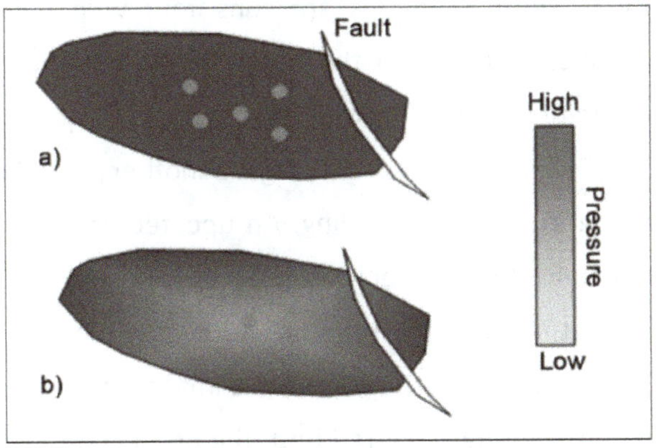

Figure 10: A schematic map of the reservoir pressure (a) before and (b) after production; green circles are the producer locations (adapted from [60]).

4D Seismic Data Incorporation in Dynamic Reservoir Modeling

Most of the data for reservoir surveillance and management studies come from well measurements. In the hydrocarbon energy applications, this includes well production data (e.g. oil, gas, and water flow rates), well-head and bottom-hole pressure measurements, production logging tools (PLT's), repeated logs (e.g. thermal decay time logs for water saturation changes), and so on. There are usually many reservoir model parameters, but often there are only a few wells. Obviously, there can be many non-unique models that fit the sparse data at the wells in part due to the huge challenge that is involved in extrapolating the well data to the entire volume of a reservoir. Other complementary dynamic datasets such as pressure interference tests and chemical tracers can provide useful information on the volumetric measurement of reservoir connectivities; however, such datasets cannot provide information on the location of reservoir heterogeneities [60]. Four-dimensional seismic data are considered to complement conventional reservoir surveillance techniques, and to reduce the level of uncertainty associated with the monitoring process [58,59]. Time-lapse seismic data in conjunction with geological and fluid-flow history matching yield a reservoir description which is more efficient and produces a better prediction of reservoir performance. The basic philosophy of this process, which is usually known as "4D seismic history matching", is illustrated in Figure 11.

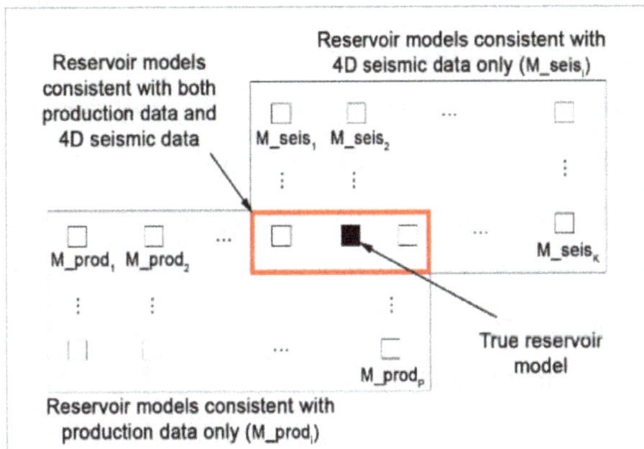

Figure 11: The basic philosophy of integrating 4D seismic data into the reservoir history matching process (adapted from [1]).

RESULTS AND DISCUSSION

The reservoir history matching process itself may produce several reservoir models which are consistent with production data only (M_prodi). There are also several other reservoir models which match 4D seismic data only (M_seisi). Constraining the reservoir models to be consistent with both production data and time-lapse seismic data reduces the uncertainties which are associated with the history matching process by narrowing the range of possible outcomes (which ideally bracket the true reservoir model).

Several authors have pointed out the benefits of incorporating 4D seismic data into a reservoir history matching procedure [61-66]. Most of the published work in this area used either synthetic data, or included 4D seismic data in a qualitative way. However, quantitative incorporation of 4D seismic in reservoir dynamic flow modeling has received considerable attention and is currently an active area of research in both academia and industries.

Qualitative approach: A perfect example of a qualitative integration of 4D seismic data in reservoir dynamic flow modelling is at the Draugen oil field, offshore Norway, where a 4D seismic image of the reservoir water flooding has given an outlook into what has been happing beyond the wells. Based on Pre-4D history matching, three possible dynamic flow model scenarios (depending on aquifer support, permeability assumptions, and fault transmissibility) are shown in Figure 12.

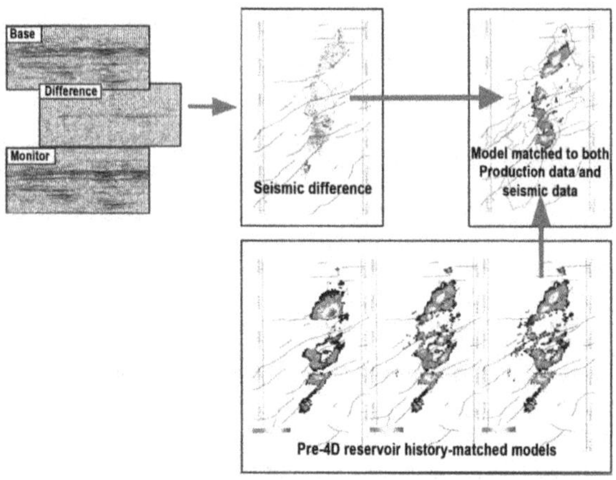

Figure 12: reservoir model updating at the Draugen oil field, offshore Norway (source: [59]). Based on Pre-4D history matching, three possible dynamic flow model scenarios are shown. All of these scenarios are consistent with only production data. On the other hand, the amplitude difference map of 4D seismic surveys gives another possible scenario of water flooding. An updated reservoir dynamic flow model has been built by selecting the correct features from the pre-4D history matched flow models and the 4D seismic images.

All of these scenarios are consistent with only production data. On the other hand, amplitude difference map of two seismic surveys acquired before and during production gave another possible scenario of water flooding. An updated reservoir dynamic flow model was built by selecting the correct features from the pre-4D history matched flow models and the 4D seismic images, resulting in a model that honored both the production and 4D seismic data [67].

Quantitative approach: The Workflow of the quantitative integration of 4D seismic data in reservoir dynamic fluid-flow simulation process is given in Figure 13.

This process, sometimes known as "parameter estimation using 4D history matching [56]," is essentially an underdetermined and nonlinear inverse problem. In this workflow, reservoir hydrodynamic simulator and time-lapse seismic simulator are inter-related. Using a reservoir flow simulator, flow rates and changes in saturations and pressure are modeled. A seismic simulator is then employed to map the changes in saturations and pressure into changes in P- and S-wave velocities and density, and consequently into changes in amplitude and timing of the seismic response of the reservoir model. The difference between the observed and simulated production data and the discrepancy between the observed and simulated seismic data are simultaneously used to update the reservoir parameters using an optimization algorithm. This iterative process helps improve the consistency of the dynamic behavior of the reservoir with both 4D seismic and production data.

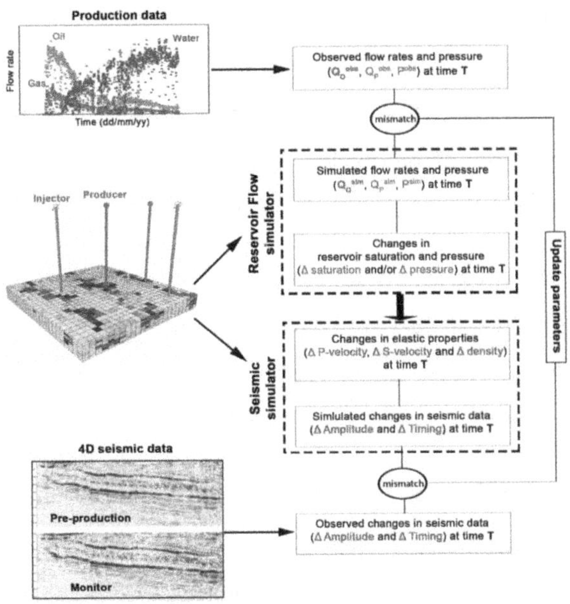

Figure 13: The Workflow of the quantitative integration of 4D seismic data in reservoir simulation process.

This joint quantitative matching of production and 4D seismic data usually consists of minimizing an objective function with two terms (e.g. Equation 1 from [68]): the first term accounts for the fit between the flow-simulated and the measured production fluid-flow data, and the second term measures the fit between the real (observed) and the synthetic (simulated) seismic data.

$$F = \sum_{j=1}^{nseries} \frac{w_j^P}{ntime_j} \sum_{i=1}^{ntime_j} \left(\frac{d_{i,j}^{obs} - d_{i,j}^{cal}(\alpha)}{\sigma_{i,j}^P} \right) + \sum_{j=1}^{ndata} \frac{w_j^P}{nvalues_j} \sum_{i=1}^{nvalues_j} \left(\frac{S_{i,j}^{obs} - S_{i,j}^{cal}(\alpha)}{\sigma_{i,j}^S} \right) \quad (1)$$

where, nseries and ndata are the number of production and seismic data series to be matched respectively; ntimej is the number of measurement times, and $nvalues_j$ is the number of observed seismic data; σ^P and σ^S represent standard deviations for production and seismic data errors respectively, and w is the weighting factor.

Challenges in 4D Seismic Data Incorporation into Reservoir Models

Integrating time-lapse 4D seismic data into the reservoir modeling and history matching processes poses a significant challenge due to the frequent mismatch among the initial reservoir model, the reservoir geology, and the pre-production seismic data. It is essential that synthetic 3D seismic data modeled from the initial static reservoir model closely match the real baseline 3D seismic data. Without such initial 3D seismic match, risks are introduced into the subsequent use of 4D seismic datasets to update reservoir simulation models [46,23].

Another key challenge is to determine which type of seismic data is the most suitable for comparison, and at what level it can be best incorporated into the history matching process [69]. According to Figure 14, the joint seismic history matching problem can be performed at any one of three different levels [70]:

Level 1: Amplitude level, in which the seismic amplitude or waveform data (forward-modeled from flow simulator outputs using a petro-elastic model and wavelet convolution) are compared with the real 4D amplitudes or waveforms [71,72]. In this level, seismic data have not been inverted and do not suffer from the errors related to seismic inversion; however, an additional seismic forward modeling step at each iteration step of the reservoir history matching loop is compulsory. Landa and Kumar [72] showed that although there are some difficulties and challenges in the seismic modeling step, the associated uncertainties are better recognized in the seismic forward modeling in comparison with the seismic inversion.

Level 2: Impedance level, in which the elastic

impedances (forward-modeled from a flow simulator using a petro-elastic model) are compared with the inverted seismic impedance values [73,74]. This is the most popular level of 4D seismic history matching due to less computational cost, less time of the process, and more convenience in comparison with the production data; therefore, it is considered by some of the authors as the optimum level of the 4D seismic history matching. Level 3: Pressure or saturation level, in which flow simulator outputs are directly compared with saturation and pressure values computed from the seismic inversion by using a petro-elastic model [75]. This level is a more intuitive level of comparison for the reservoir specialists, but it contains more uncertainties due to an additional step of the direct inversion of the seismic data into reservoir properties.

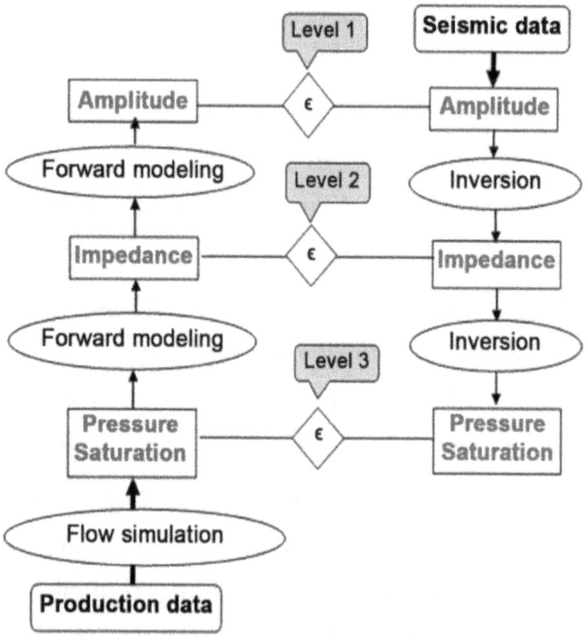

Figure 14: Joint seismic history matching can be performed at three different levels: amplitude, impedance, or pressure/saturation

Many other challenges are posed by the quantitative incorporation of 4D seismic data into the reservoir dynamic flow modeling process, including:

Weighting misfits: since each of reservoir production and 4D seismic datasets has a different sample density and accuracy, weighting the corresponding misfits in the global objective function is a challenging issue.

Data calibration: usually, the observed time-lapse seismic attributes are derived as a relative property, so they require calibration to be compared with reservoir properties.

Scaling issues: the observed data and simulated outputs have different scales; consequently, scaling either down or up is compulsory for a correct comparison.

Discrepancy in optimum parameters suggested by 4D seismic compared to production data: this can be related to calibration or rescaling, or to the seismic modeling process itself [76].

CONCLUSIONS

Static and dynamic reservoir modeling plays a central role in the energy industry. The successful integration of 3D and 4D seismic data in reservoir modeling process not only adds tremendous information on the spatial distributions and variations of the reservoir properties, but also minimizes the risks and uncertainties of the reservoir management decisions. This paper demonstrates the key roles of 3D and 4D seismic data in static and dynamic reservoir modeling processes, followed by a review of the workflows, algorithms, and associated challenges of the incorporation of 3D and 4D seismic data in reservoir model building and updating processes.

AKNOWLEDGMENTS

The main part of this paper comes from author's

Ph.D. Thesis. I would like to thank University of Western Australia for providing a great environment and Professor David Lumley for his supervision and support during this research.

NOMENCLATURES

DST	: Drill Stem Test
PGS	: Pluri-Gaussian Simulation
RFT	: Repeat Formation Test
SGS	: Sequential Gaussian Simulation
SIS	: Sequential Indicator Simulation
TGS	: Truncated Gaussian Simulation

REFERENCES

1. Emami Niri M., "Seismic Data Integration and Multi-objective Optimization for 3D Reservoir Characterization and Model Building," PhD Thesis, University of Western Australia, **2015**, 1-87.
2. Mallet J. L., "Numerical Earth Models," *Journal of society of Petroleum Engineers*, **2008**, *3*, 1-7.
3. Zakrevsky K., "Geological 3D Modeling," EAGE Sustainable Earth Sciences Publications, **2011**, 1-6.
4. Pyrcz M. J. and Deutsch C. V., "Geostatistical Reservoir Modeling (2nd ed.)," Oxford University Press, **2014**, 1-429.
5. Dubrule O., Basire C., Bombarde S., Samson P., Segonds D., and et al., "Reservoir Geology Using 3-D Modeling Tools," *SPE Annual Technical Conference and Exhibition*, Society of Petroleum Engineers (SPE-38659-MS), **1997**, 1-16.
6. Caers J., "Petroleum Geostatistics," Society of Petroleum Engineers, Richardson (1st ed.), **2005**, 1-107.
7. Doligez B., Beucher H., Lerat O., and Souza O., "Use of a Seismic Derived Constraint: Different Steps and Joined Uncertainties in the Construction of a Realistic Geological Model," *Oil and Gas Science and Technology-Revue de l'IFP*, **2007**, *62*(2), 237-248.
8. Dubrule O., "Geostatistics for Seismic Data Integration in Earth models (1st ed.)," *SEG/EAGE Distinguished instructor short course*, **2003**, *6*, 1-167.
9. Dutton D., "A New Technique to Improve 3-D Fault Geometry Interpretation from a Seismic Dataset: an Example from a Welded Growth Fault Family, Offshore Angola," AAPG Annual Meeting, **2003**.
10. Eaton T. T., "On the Importance of Geological Heterogeneity for Flow Simulation," *Sedimentary Geology Journal*, **2006**, *184*(3), 187-201.
11. Sheriff R. E. and L. P. Geldart, "Exploration Seismology (2nd ed.)," Cambridge University Press, **1995**, 1-426.
12. Onajite E., "Seismic Data Analysis Techniques in Hydrocarbon Exploration," Elsevier, **2013**.
13. Yilmaz Ö., "Seismic Data Analysis," Tusla: Society of Exploration Geophysicists (1st ed.), **2001**, 1-647.
14. Chopra S. and Marfurt K. J., "Seismic Attributes-A historical perspective," *Geophysics Journal*, **2005**, *70*(5), 3-28.
15. Tarantola A., "Inverse Problem Theory and Methods for Model Parameter Estimation," Society for Industrial and Applied Mathematics (SIAM), **2005**, 1-26.
16. Kemper M., "Rock Physics Driven Inversion: the Importance of Workflow," *First Break*, **2010**, *28*(10), 69-81.
17. Lancaster S. and Whitcombe D., "Fast Track

Colored Inversion," 69th Annual International Meeting, SEG (Society of Exploration Geophysicists), Extended Abstracts, **2000**, 1572-1575.

18. Debeye H. and Riel V. P. "Lp-norm Deconvolution 1," *Geophysical Prospecting*, **1990**, *38*(4), 381-403.
19. Fatti J. L., Smith G. C., Vail P. J., Strauss P. J., and et al., "Detection of Gas in Sandstone Reservoirs Using AVO Analysis: A 3-D Seismic Case History Using the Geostack Technique," *Geophysics*, **1994**, *59*(9), 1362-1376.
20. Hampson D., Russell B., and Bankhead B., "Simultaneous Inversion of Pre-stack Seismic Data," 75th Annual International Meeting, SEG, Extended Abstracts, **2005**, 1633-1637.
21. Haas A. and Dubrule O., "Geostatistical Inversion-a Sequential Method of Stochastic Reservoir Modeling Constrained by Seismic Data," *First Break Journal*, **1994**, *12*(11), 561-569.
22. Gunning J. and Glinsky M. E., "Delivery: an Open-source Model-based Bayesian Seismic Inversion Program," *Computers and Geosciences*, **2004**, *30*(6), 619-636.
23. Ravalec-Dupin L., Enchery G., Baroni A., and Da Veiga S., "Preselection of Reservoir Models from a Geostatistics-based Petrophysical Seismic Inversion," *SPE Reservoir Evaluation and Engineering*, **2011**, *14*(05), 612-620.
24. Angeleri G. and Carpi R., "Porosity Prediction from Seismic Data," *Geophysical Prospecting*, **1982**, *30*(5) 580-607.
25. Doyen P. M., "Porosity from Seismic Data A Geostatistical Approach," *Geophysics*, **1988**, *53*(10), 1263-1275.
26. De Buyl M., "Optimum Field Development with Seismic Reflection Data," *The Leading Edge*, **1989**, *8*(4), 14-20.
27. Cole J., Nebrija E. L., Saggaf M. M., Al-Shabeeb, et al., "Integrated 3D Reservoir Modeling for Permian Khuff Gas Development in Ghawar Field, Saudi Arabia," *The Leading Edge*, **2003**, *22*(7), 666-669.
28. Emami Niri M. and Lumley D., "Initializing Dynamic Reservoir Models for History Matching Using Pre-production 3D Seismic Data," 6th International Petroleum Technology Conference (IPTC-16650-MS), **2013**.
29. Emami Niri M. and Lumley D., "An Uncertainty Analysis in Quantitative Integration of Inverted 3D Seismic Data for Static Reservoir Modeling," 13th International Congress of the Brazilian Geophysical Society, **2013**, 993-997.
30. Emami Niri M. and Lumley D., "Initializing Reservoir Models for History Matching Using Pre-production 3D Seismic Data-constraining Methods and Uncertainties," *Exploration Geophysics Journal*, **2015**, 1-12.
31. Bornard R., Allo F., Coleou T., Freudenreich F., and et al., "Petrophysical Seismic Inversion to Determine More Accurate and Precise Reservoir Properties," SPE Europec/EAGE Annual Conference (SPE-94144-MS), **2005.**
32. Emami Niri M. and Lumley D., "A Multi-Objective Stochastic Optimization Approach for Estimation of Subsurface Geomodels," ASEG Extended Abstracts, **2013**, 1-4.
33. Emami Niri M. and Lumley D., "Reservoir Model Estimation by Simultaneous Optimization of Multiple Objective Functions," 76th EAGE Conference and Exhibition, **2014**, Extended Abstracts.
34. Emami Niri M. and Lumley D., "A Multi-

Objective Stochastic Optimization Method for Estimation of Subsurface Geomodels," *Journal of Applied Geophysics*, **2016**, *129*, 187-199.

35. Bosch M., Mukerji T., and Gonzalez E. F., "Seismic Inversion for Reservoir Properties Combining Statistical Rock Physics and Geostatistics: A Review," *Geophysics Journal*, **2010**, *75*(5), 165-176.
36. Saussus D. and Sams M., "Facies as the Key to Using Seismic Inversion for Modeling Reservoir Properties," *Journal of First Break*, **2012**, *30*(7), 45-52.
37. Avseth P., Mukerji T., and Mavko G., "Quantitative Seismic Interpretation: Applying Rock Physics Tools to Reduce Interpretation Risk (1st ed.)," Cambridge University Press, **2005**, 313-319.
38. Coulon J., Lafet Y., Deschizeaux B., Doyen P., and et al., "Stratigraphic Elastic Inversion for Seismic Lithology Discrimination in a Turbiditic Reservoir," 76th Annual International Meeting, SEG, Extended Abstracts, **2006**, 2092-2096.
39. Emami Niri M. and Lumley D., "Probabilistic Reservoir Property Modeling Jointly Constrained by 3D Seismic Data and Hydraulic Unit Analysis," SPE Asia Pacific Oil and Gas Conference and Exhibition, Society of Petroleum Engineers,, SPE-171444-MS, **2014**.
40. Emami Niri M. and Lumley D., "Probabilistic Reservoir Property Modeling Jointly Constrained by 3D Seismic Data and Hydraulic Unit Analysis," *SPE Reservoir Engineering and Evaluation*, **2016**, *19*(2), 253-264.
41. Pyrcz M., "Geostatistical Assignment of Reservoir Properties on Unstructured Grids," SPE Annual Technical Conference and Exhibition, Society of Petroleum Engineers (SPE-77427-MS), **2002**.
42. Chilès J. and Delfiner P., "Geostatistics: Modeling Spatial Uncertainty," Wiley, **1999**, 1-186.
43. Deutsch C. V. and Journel A. G., "Geostatistical Software Lbrary and User's Guide," Oxford University Press, **1992**.
44. Doyen P. M., "Seismic Reservoir characterization: An Earth Modeling Perspective," *EAGE Publications Bv*, **2007**, *2*, 1-147.
45. Emami Niri M. and Lumley D., "Simultaneous Optimization of Multiple Objective Functions for Reservoir Modeling and Seismic Data Matching," *Geophysics*, **2015**, *80*(2), M53-M67.
46. Lumley D. and Behrens R., "Practical Issues of 4D Seismic Reservoir Monitoring: What an Engineer Needs to Know," SPE Reservoir Evaluation and Engineering, **1998**, *1*(06), 528-538.
47. Bahar A., Abdel-Aal O., Ghani A., Silva F., and et al., "Seismic Integration for Better Modeling of Rock Type Based Reservoir Characterization: A Field Case Example," Abu Dhabi International Conference and Exhibition (SPE-88793-MS), **2004.**
48. Modin D. and Guilloux Y., "Geomodel to Seismic Feedback Loop-Method and Case Study," 71st EAGE Conference and Exhibition, Extended Abstracts, **2009**.
49. Rwechungura R. W., Dadashpour M., and Kleppe J., "Advanced History Matching Techniques Reviewed," *SPE Middle East Oil and Gas Show and Conference,* Society of Petroleum Engineers, SPE-142497-MS, **2011**, 1-19.

50. Hajizadeh Y., "Population-based Algorithms for Improved History Matching and Uncertainty Quantification of Petroleum Reservoirs," PhD Dissertation, Heriot-Watt University, **2011**, 1-83.
51. Saleri N. and Toronyi R., "Engineering Control in Reservoir Simulation: Part I," *SPE Annual Technical Conference and Exhibition, Society of Petroleum Engineers*, SPE-18305-MS, **1988.**
52. Gao G., Li G., and Reynolds A. C., "A Stochastic Optimization Algorithm for Automatic History Matching," *SPE Journal*, **2007**, *12* (2), 196-208.
53. Fasanino G., Molinard J., de Marsily G., and Pelce V., "Inverse Modeling in Gas Reservoirs," *SPE Annual Technical Conference and Exhibition*, Society of Petroleum Engineers, SPE-15592-MS, **1986**, 1-15..
54. Watson A., Wade J., and Ewing R., "Parameter and System Identification for Fluid Flow in Underground Reservoirs," Proceedings of the Conference Inverse Problems and Optimal Design in Industry, Springer, **1994**, 81-108.
55. Li R., Reynolds A., and Oliver D., "History Matching of Three-phase Flow Production Data," *SPE Journal*, **2003**, *8*(4), 328-340.
56. Dadashpour M., "Reservoir Characterization Using Production Data and Time-lapse Seismic Data," PhD Dissertation, Norwegian University of Science and Technology, **2009**, 1-137.
57. Lumley D., "Seismic Time-lapse Monitoring of Subsurface Fluid Flow," PhD Thesis, Stanford University, **1995**, 1-79.
58. Lumley D., "Time-lapse Seismic Reservoir Monitoring," *Geophysics*, **2001**, *66*(1), 50-53.
59. Calvert R., "Insights and Methods for 4D Reservoir Monitoring and Characterization (1st ed.)," *SEG/EAGE Distinguished Instructor Short Course*, **2005**, *8*, 1-217.
60. Johnston D. H., "Practical Applications of Time-lapse Seismic Data," *SEG Distinguished Instructor Short Course*, **2013**, *16*, 1-27.
61. Anderson R. N., Boulanger A., He W., Xu L., and et al., "4D Time-lapse Seismic Monitoring in the South Timbalier 295 Field Gulf of Mexico," Offshore Technology Conference, OTC-8312-MS, **1997**, 868-871.
62. Burkhart T., Hoover A. R., and Flemings P. B., "Time-lapse (4-D) Seismic Monitoring of Primary Production of Turbid Reservoirs at South Timbalier Block 295, offshore Louisiana, Gulf of Mexico," *Geophysics*, **2000**, *65*(2), 351-367.
63. Lygren M., Fagervik K., Valen T., Hetlelid A., and et al., "A Method for Performing History Matching of Reservoir Flow Models Using 4D Seismic Data," *Petroleum Geoscience*, **2003**, *9*(1), 83-90.
64. Vasco D. W., "Seismic Imaging of Reservoir Flow Properties: Time-lapse Pressure Changes," *Geophysics Journal*, **2004**, *69*(2), 511-521.
65. Jourdan J., Lefeuvre F., and Dubucq D, "Integration of 4D Seismic into the Dynamic Model: Girassol, Deep Offshore Angola," *First Break Journal*, **2006**, *24*(4), 51-58.
66. Castro S. A., "A Probabilistic Approach to Jointly Integrate 3D/4D Seismic, Production Data and Geological Information for Building Reservoir Models," PhD Thesis, Stanford University, **2007**, 1-103.
67. Guderian K., Kleemeyer M., Kjeldstad A, Pettersson S., and et al., "Draugen Field-successful Reservoir Management Using 4D

Seismic," 65th EAGE Conference and Exhibition, Extended Abstracts, **2003**.

68. Roggero F., Ding D. Y., Berthet P., Lerat O., and et al., "Matching of Production History and 4D Seismic Data-Application to the Girassol Field Offshore Angola," *SPE Annual Technical Conference and Exhibition*, Society of Petroleum Engineers, SPE-109929-MS, **2007**, 1-20.
69. Fahimuddin A., Aanonsen S., and Skjervheim J. A., "Ensemble Based 4D Seismic History Matching-integration of Different Levels and Types of Seismic Data," *SPE Europe/ EAGE Annual Conference and Exhibition*, SPE-131453-MS, **2010**, 1-15.
70. Waggoner J., "Integrating Time-Lapse 3D (4D) Seismic Data with Reservoir Simulators," 63rd EAGE Conference and Exhibition, Extended Abstracts, **2001**.
71. Huang X., Meister L., and Workman R., "Reservoir Characterization by Integration of Time-lapse Seismic and Production Data," *SPE Annual Technical Conference and Exhibition,* Society of Petroleum Engineers, SPE-38695-MS, **1997**, 1-9.
72. Landa J. L. and Kumar D., "Joint Inversion of 4D Seismic and Production Data," SPE Annual Technical Conference and Exhibition, *Society of Petroleum Engineers*, SPE-146771-MS, **2011**, 1-7.
73. Gosselin O., Aanonsen S., Aavatsmark I., Cominelli A., and et al., "History Matching Using Time-lapse Seismic," *Journal of Petroleum Technology*, **2004**, *56*(4), 66-67.
74. Dong Y. and Oliver D. S., "Reservoir Simulation Model Updates via Automatic History Matching with Integration of Seismic Impedance Change and Production Data," International Petroleum Technology Conference, IPTC-12550-MS, **2008**.
75. Landa J. L. and Horne R. N., "A Procedure to Integrate Well Test Data, Reservoir Performance History and 4-D Seismic Information into a Reservoir Description," *SPE Annual Technical Conference and Exhibition*, Society of Petroleum Engineers, SPE-38653-MS, **1997**, 1-16..
76. Stephen K. D., "Seismic History Matching with Saturation Indicators Combined with Multiple Objective Function Optimization," *SPE Europe/ EAGE Annual Conference and Exhibition*, SPE-164857-MS, **2013**, 1-10.

3

Critical Parameters Affecting Water Alternating Gas (WAG) Injection in an Iranian Fractured Reservoir

Karim Akbari Aghdam[1] and Seyyed Saeed Ghorashi*[2]

[1] Petroleum and Gas Engineering Department, Pars Oil and Gas Company (POGC), Asaluyeh, Bushehr, Iran
[2] Research Institute of Petroleum Industry (RIPI), Tehran, Iran

ABSTRACT

Microscopic oil displacement of water flooding and sweep efficiency of continuous gas injection could be improved by water alternating gas (WAG) injection. The WAG injection process aims to squeeze more oil out of the reservoirs; in this method, water and gas are alternatively injected into the reservoir. Also, availability of hydrocarbon or CO_2 gases in the field makes it attractive for gas-based enhanced oil recovery (EOR) methods such as water alternating gas (WAG) injection. Conducting some simulations are required to optimize EOR methods for investigating the effect of parameters affecting WAG injection. Reducing and controlling the mobility ratio, creating stable front, and preventing early fingering of gas are the advantages of water alternating gas injection, which have promoted extensive applications throughout the world. Critical parameters, including WAG ratio, injection rates, gas composition variation, cycle times and some others which affect the WAG injection as an enhanced oil recovery method are studied thoroughly in this paper. Because of higher mobility of water relative to gas, injected water has more efficiency, but the excess use of water will cause early breakthrough. This study suggests that injecting proper volume at suitable times with different rates during injection time provides a 10 -15 % improvement in the recovery factor for one pore volume which is injected by using commercial reservoir simulator ECLIPSE 300. The best rate variation during a cycle time of WAG injection and choosing of first injection phase are discussed in this paper.

Keywords: Water Alternating Gas (WAG), Simulation, Mobility Ratio, WAG Ratio, Enhanced Oil Recovery (EOR), Permeability

INTRODUCTION

WAG process is one of the enhanced oil recovery methods that has the advantage of increasing microscopic efficiency by using a miscible displacement, while it is maintaining a good macroscopic efficiency by using alternating water. Many successful results have been reported in this EOR process, and it is widely used in the North Sea [1]. An increasing demand for energy has forced oil companies to think about secondary and tertiary oil recovery methods to increase the recovery from the existing reservoirs [2]. To increase the extent of

***Corresponding author**

Seyyed Saeed Ghorashi
Email: ghorashis@ripi.ir

the reservoir contact by the injected gas, it must be injected alternatively with water. This mode of injection is called water alternating gas (WAG) it's the popularity of which has been increasing since 1950 [3]. This method is being widely practiced in the oil fields, e.g. in Gulfaks fields [4]. This EOR method also has been used in gas/condensate reservoirs [5]. Microscopic oil displacement of water flooding and consequently the sweep efficiency of continuous gas injection could be improved by water alternating gas injection. About 55% of the total oil productions by enhanced oil recovery (EOR) methods in the United States are resulted from gas-injection methods, most of which are WAG processes [6]. The WAG process improves the macroscopic and microscopic sweep efficiency of water and gas injection process at the same time. The cyclic nature of the WAG process causes (1) an increase in water saturation during the water injection half cycle and (2) a decrease in water saturation during the gas injection half cycle. This process of inducing cycles of imbibition and drainage causes the residual oil saturation to be typically lower than that of water flooding and similar to those of gas flooding. The key parameter, which could lead to frontal instability and by passing of oil by injection fluids are the heterogeneity of reservoir and the mobility ratio of the fluids. Gravity segregation is another factor that impacts the recovery [6]. The injected volume for each phase is an important factor to achieve a good sweep efficiency and economical process. Another important parameter is the composition of the injected fluid and its effect on model to predict the process so that operating strategy design can be performed [7]. Continuous slug injection performs better than WAG when the largest permeability layers are at the bottom of the aquifer, richer gases are used, and the vertical to horizontal permeability ratio is small [8]. Christensen et al. presented an extensive review of 59 field cases [9, 10]. WAG injection is used with immiscible and miscible gases [11, 12].

In this process, if the reservoir rock is water-wet, water fills the small pore spaces due to their high capillary pressure, and gas acts in an opposite fashion, which means that it goes through the larger pores. This shows that, in this process, gas acts as a complementary factor for the water flooding [13]. The WAG displacement will be optimized if the mobility ratio is favorable (less than 1). Increasing the viscosity of the gas or reducing the relative permeability of the gas can result in a reduction of the mobility ratio. The reduced mobility of the gas phase can be achieved by injecting water and gas alternately. It is important to adjust the amount of water and gas so that the best possible displacement efficiency will be achieved [14]. Too much water will result in poor microscopic displacement, and too much gas will result in poor vertical and maybe also horizontal sweep [15].

Layered reservoirs may represent favorable geological conditions for gas injection. For instance, if a high permeability layer is situated below a low permeability layer, it prevents quick gravity-segregated tonguing in the top zone towards the production intervals. The low vertical permeability of the layers also contributes to a better WAG sweep-out [16]. Furthermore, the studies indicate that the improved recovery by WAG method over regular water flood is possible mainly because of the trapped gas effect and better conformance. WAG flooding is a successful method for improving oil recovery. More than 80% of the WAG-flooded projects in the world have been reported profitable [17]. In recent decades, approximately 40% of big gas injection projects all over the world such as Canada, Russia, Turkey, and Norway have been performed as WAG [18]. Based on the results of an Iranian study on an Asmari reservoir portion of Mansoori oil field, it was shown that the WAG process can be optimized in order to have more efficiency than either water or gas injection. In addition, lower costs and the production conditions are kept in more desirable conditions [19].

Kharrat et al. in 2010 studied the parametric investigation of WAG injection process in naturally fractured reservoirs in Iran, and they concluded that by the optimization of significant parameters in WAG injection process more enhanced oil recovery can be expected than either water or gas injection alone [20].

Jafari et al. in 2008 conduct a series of experiments with the numerical simulation of different EOR techniques in a non-fractured carbonate core from an Iranian offshore oil reservoir. The results showed that the implementation of the WAG process at an optimal injection volume, optimum rate of injection fluids, and optimum WAG ratio can lead to a higher oil recovery in comparison with the other alternating scenarios such as gas injection and water injection [21].

The main purpose of this work is an investigation of the effect of parameters affecting WAG injection, and consequently the optimization of the crucial parameters of WAG process. In this study, the simulation studies are run on a synthetic model with real PVT data of the reservoir fluid in order to perform a qualitative study of the parameters which are crucial for a WAG process design by using a commercial reservoir simulator ECLIPSE 300. Firstly, a synthetic model was simulated, and then reservoir fluid was simulated by using the real PVT data. As the main procedure of this paper, 8 critical parameters affecting the incremental oil recovery, namely injected gas variation, cycle time of injection, first phase to inject, WAG ratio, water rate, gas rate, best period, and time of injection were investigated. The oil recovery was considered as the main indicator for choosing the optimized scenario.

FIELD DESCRIPTION

The investigated oil reservoir in this paper is located in an Iranian central oil field. For a better evaluation of the effect of several parameters, it is better to use a synthetic geological model. This synthetic model has a length of 2000 ft., a width of 1000 ft., and a thickness of 150 ft. The real model fluid and rock data and reservoir initial conditions were used. The oil was light oil with a gravity of 35° API and an oil formation volume factor of 1.39 RB/STB. Rock and reservoir fluid characteristics of this study are illustrated in Table 1.

Table 1: Rock and reservoir fluid characteristics.

Datum depth (ft.)	6100-6210	Total pore volume (MM Reservoir barrel)	21.373
Average reservoir pressure @ datum depth (psia)	3035	Reservoir temperature (°F)	210
Reservoir oil °API	35	Initial gas saturation (%) Two rock types	0
Initial oil saturation (%) (water wet rock)	0.85	Initial oil saturation (%) (oil wet rock)	0.807

MODEL DESCRIPTION

In this study, different injection scenarios were simulated through a commercial simulator, and a gridding network of reservoir was designed using geological data (Figure 1). In this network, the reservoir was divided longitudinally and latitudinally into 20 and 10 grid blocks. 5 grid blocks were defined for the reservoir in vertical direction (Table 2). Simplicity Block Center method was used to grade the reservoir. Geological indications and the information related to well drilling showed no fracture in the reservoir rock; therefore, single porosity model was used for the simulation of reservoir.

For better investigating the critical parameters affecting oil production, two rock types with different wettability conditions were chosen. The oil wet rock, which was from one of Iranian central oil fields, and water wet rock were chosen from one of Iranian southern fields. Reservoir rock has the same rock type in the model with rock compressibility of 3.4×10-7 (1/psi). The relative permeability of two types of rocks is shown in Figures 2 and 3. PVT data were prepared using the results of experiments on reservoir fluid. The reservoir fluid model was simulated by a commercial simulator PVTI © 2009 Schlumberger.

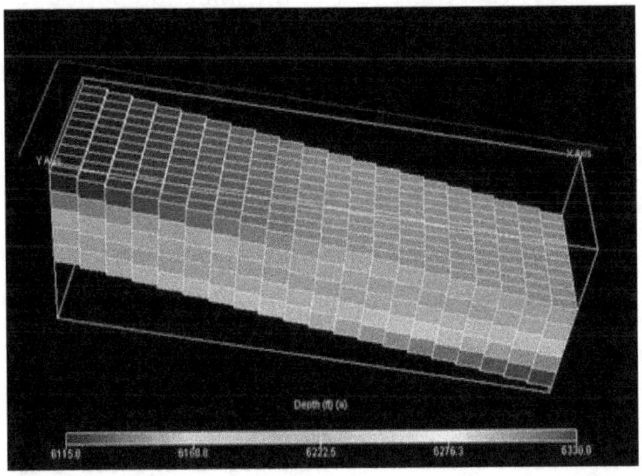

Figure 1: A three dimensional view of a synthetic model.

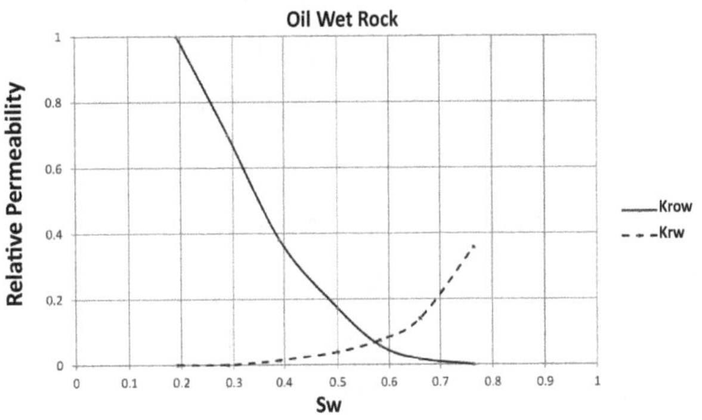

Figure 2: Relative permeability curves for oil wet rock (from one of Iranian central oil field).

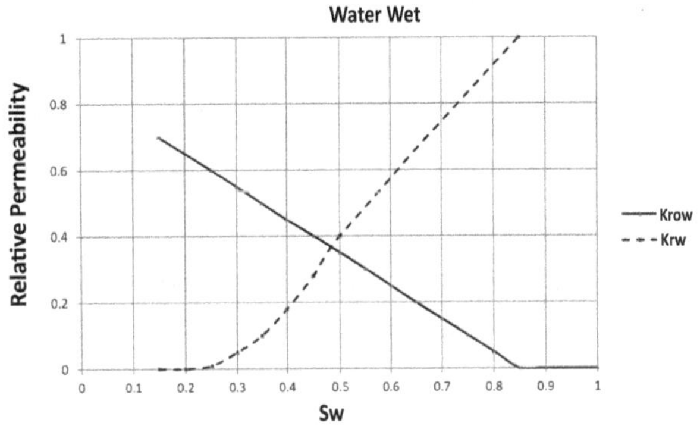

Figure 3: Relative permeability curves for water wet rock (from one of Iranian southern field).

Table 2: Model characteristics.

Type of porous medium	conventional	*x* grid block size (ft.)	100
Number of cell in *x*-direction (Nx)	20	*y* grid block size (ft.)	100
Number of cell in *y*-direction (Ny)	10	*z* grid block size (ft.)	30
Number of cell in *z*-direction(Nz)	5	Permeability in x, y, and z direction (m/day)	80
Number of cell	1000	Porosity(%)	16

RESULTS AND DISCUSSIONS

Gas variation effect on oil recovery

In this case, water and gas were injected at a rate of 5000 STB/day and 3000 Mscf/day respectively. In WAG simulation, alternatively, the injected pressure is set at 3050 psia, which is equal to average reservoir pressure; the total amount of one pore volume was also injected. The results are listed in Table 4. Water wet rock has a higher incremental recovery compared to oil wet rock due to WAG injection process. In oil wet rocks, water cannot produce the oil which is wetted the walls of rock, but in water-wet rock, wetting phase is water, thereby pushing non-wetting oil through the production well so easier. Since the initial oil saturation for water-wet rock is different from oil-wet rock, for the correct comparison of the recovery for both rock types, the incremental recovery, which is equal to the difference between the recoveries due to WAG injection and the recovery obtained by natural depletion at the same time, was used.

Since minimum miscibility pressure (MMP) for carbon dioxide is less than other gases, so this gas can make miscibility at a low injected pressure, whereas other gas cannot do the same. Thus, by using this gas at a low injected pressure, high oil recovery can be obtained. In this part of study, first contact miscibility pressure (FCMP) and MMP for four different possible injecting gases were simulated by PVTI and ECLIPSE300 software (© 2009) Schlumberger respectively. It is well worth mentioning that for obtaining the more accurate MMP results, the dynamic slim tube simulator was chosen. In Table 3, the FCMP and MMP results for CO_2, solvent (40% C_2 and 60% C_1), N_2, and C_1 are presented.

Table 3: The amount of MMP and FCMP for different gases.

Component	CO_2	Solvent (40% C_2 and 60% C_1)	N_2	C_1
First Contact Miscibility Pressure (psia) **Simulated by PVTI software**	3400	3980	14600	6400
Minimum Miscibility Pressure (psia) **Simulated by dynamic Slim tube**	3000	3970	13200	5500

Table 4: Gas variation effect on oil recovery.

Component	CO_2	Solvent (40%C_2 and 60% C_1)	N_2	C_1
Incremental Recovery for Water Wet rock	37.45	37	33.07	22.1
Incremental Recovery for Oil Wet rock	29.53	26.40	22.91	16.94

Effect of Cycle Time of Injection

To investigate the effect of the injection period for each fluid (gas and water) on the efficiency of oil production, several cycle times were used. It was assumed that the cycle time for each injection fluid is the same. For example, if cycle time for water was one year, this value would be inserted into the software. The results of these simulations are shown in Figures 4 to 6. As the cycle time of injection increases, the recovery of producing oil increases (Figure 5). When gas is injected after water injection, water retains the easier flow of gas in the reservoir, so it makes higher sharing of gas in production. As the cycle time period increases, recovery rises, but increasing this parameter would cause the high production of water in production wells. When water reaches its breakthrough point, the effectiveness of water injection for gas phase will be reduced.

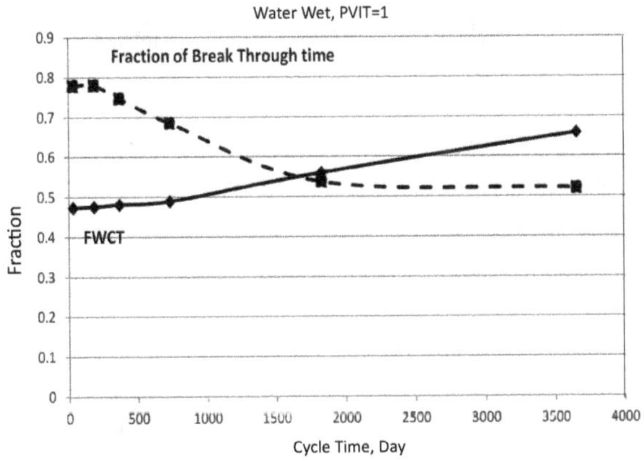

Figure 4: Fraction of break through time and FWCT (total field water cut) at different cycle times in water wet rock, total pore volume of injection (PVIT) =1; vertical axes is total field water cut (FWCT) based on the number of fractions, and it is dimensionless.

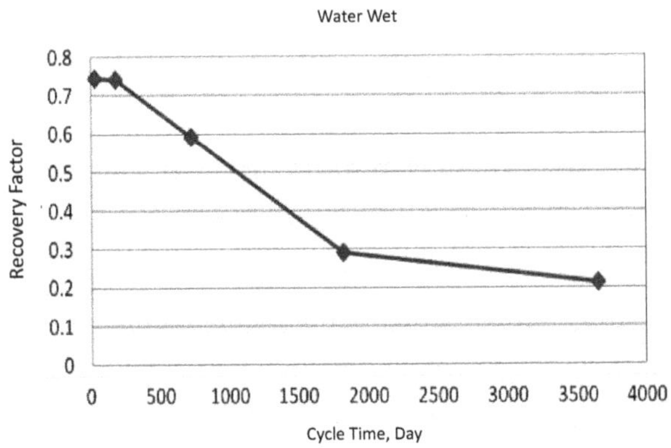

Figure 5: Recovery of oil production at different cycle times in water wet rock (PVIT=1).

Cycle time has an optimum value for each reservoir which must be obtained in order to minimize the amount of total field water cut (FWCT) and increase the breakthrough time for each injected phase. By increasing cycle time, one of the facilities (water or g=as) should be hung off, so it is unfavorable from economical points of view. To make breakthrough time dimensionless, each breakthrough time was divided by the total time of injection, which is a number between zero and one i.e.:

Fraction of breakthrough time = Breakthrough time / total time of injection (1)

To include economic dimension in the recovery for choosing the best cycle time, a new value was introduced as:

A= (B×C/D) (2)

where, *A* is a dimensionless value for the recovery of oil production (from economical view); *B* is the fraction of breakthrough time; *C* is recovery, and *D* is the dimensionless field water cut (FWCT).

By plotting *A* in terms of the logarithm of cycle time, the best cycle time could be determined, which is shown in Figure 6. When *A* value decreases, production is not economical. Since *A* has a constant value for a year cycle time, it must be concluded that the optimal value is one year cycle time.

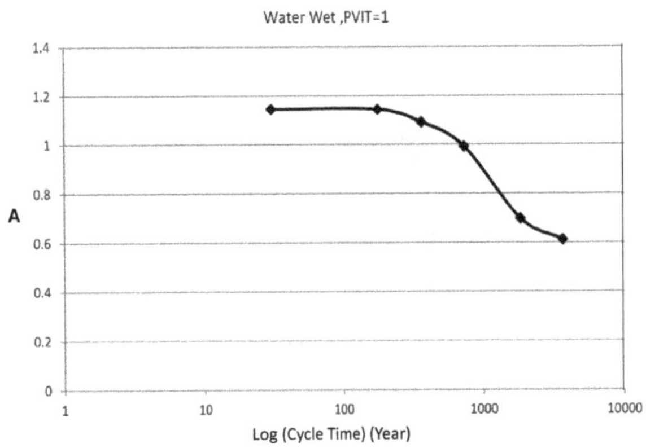

Figure 6: A value (recovery from economical view) at different cycle times in water-wet rock (PVIT=1).

Selecting the First Phase to Inject

Selecting the first phase to inject in the WAG injection process is one of key parameters which must be determined. Water and gas rates were selected 2000 STB/day and 1500 MSCF/day respectively. The total injected pore volume was 0.7 PV, and water-wet rock data were chosen as the input to the simulator. Injecting water phase as a first phase produces higher recovery value than the gas phase. At the beginning of injection, the reservoir oil saturation (also the permeability of oil phase) is high enough, and this value decreases over times of injection, so it is better to use a fluid (like water) which has higher mobility than gas and can produce more oil at the early times of injection. If gas phase was chosen as the first injected fluid, then the gas breakthroughs very soon and cannot produce oil like before the breakthrough time; thus, if water was injected after gas phase, the saturation and also permeability of oil phase would be decreased, and water could not produce high oil volume in comparison with the time when water was injected as the first fluid. In the case of injecting gas as the first fluid, if the cycle time of injection was increased, it would result in low oil production after early breakthrough times.

The results of these simulation scenarios are illustrated in Figures 7 and 8. Figure 7 shows the recovery comparison for several cycle times for the two conditions (W-G) and (G-W). (W-G) means injecting the water phase first, while (G-W) means injecting the gas phase first. Figure 8 shows the fraction of breakthrough time (FBT) of injection. As mentioned before, when water was first injected, FBT would be decreased, which means water very quickly reaches the producing well. When the gas was the first injected fluid, water FBT would then be increased. In this kind of injection, for the highest cycle time of injection, FWCT would be the minimum.

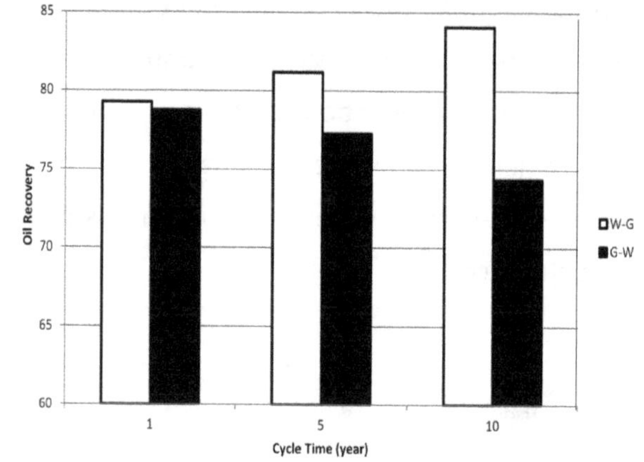

Figure 7: Recovery comparison for several cycle times for two conditions, namely W-G and G-W in water wet rock.

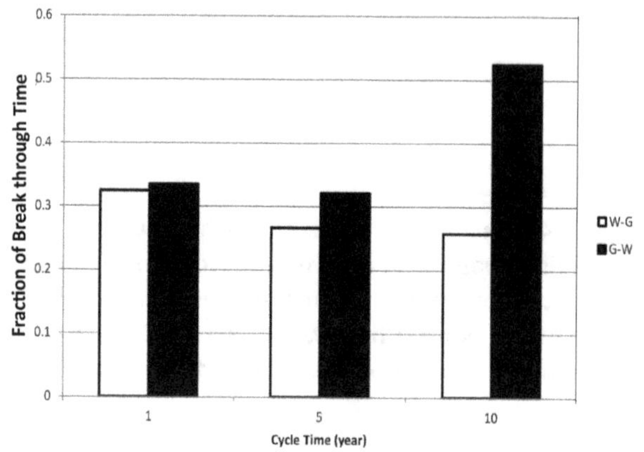

Figure 8: FBT comparison at several cycle times for two conditions, namely W-G and G-W in water wet rock.

Selecting the Best WAG Ratio

WAG ratio is defined as the total volume of water injection divided by the total volume of gas injection, i.e.:

$$WAG\,Ratio = \frac{Total\,Volume\,of\ \ Water\,injection}{Total\,Volume\,of\,Gas\,injection} \tag{3}$$

A new parameter was defined to better investigate the simulation scenarios. This parameter is TWAG ratio which is defined as the time of water injection in one cycle divided by the time of gas injection in one cycle, i.e.:

$$\text{TWAG ratio} = \frac{\text{Time of water injection in one cycle}}{\text{Time of gas injection in one cycle}} \tag{4}$$

Although by increasing WAG, ratio recovery increases, this is not the general rule, and, in addition to the WAG ratio, recovery depends on the water and gas rates. Two scenarios were defined, first water rate was considered constant, and to obtain the same amount of fluid injection at different TWAG ratios, the rate of gas injection was varied; in the second scenario, the gas rate was kept constant, and the water rate varied to have the same amount of fluid injection at several TWAG ratios. To investigate both total field water cut (FWCT) and breakthrough time, a dimensionless parameter was defined as:

$$FWCTT = \left[\frac{t_f}{t_o} (FWCT) \right] \tag{5}$$

where, t_f is equal to the total time of injection, and t_o is the time at which breakthrough is started. From the mathematical point of view, FWCTT is between zero and infinite. The brief review of the simulation process is presented below.

Constant Water Rate

In this part of the simulation, four water injection rates were selected: 7000 STB/day, 4000 STB/day, 3000 STB/day, and 1500 STB/day. At water injection rates, in order to inject one pore volume (1 PV) of the reservoir model, by choosing different TWAG ratios and using Equation 10, the appropriate gas injection rate could be chosen.

Total pore volume injected to the reservoir (PVIT) is equal to the sum of total water injected (PVIW) in the reservoir conditions plus the total gas injection in the reservoir (PVIG) conditions.

$$PVIT=PVIW+PVIG \tag{6}$$

$$PVIT= q_w \times t_w \times B_w + q_w \times t_w \times B_w \tag{7}$$

$$t_{total} = t_w + t_g \tag{8}$$

$$TWAG\,ratio = TWAG = \frac{t_w}{t_g} \tag{9}$$

$$q_g = \frac{PVIT(1+TWAG)}{t_{total}B_g} - q_w \frac{B_w}{B_g} TWAG \tag{10}$$

where, t_{total} is the total time of injection. When the water injection rate is constant, we perform simulation for a constant period of time in all the conditions. Hence, by knowing TWAG, gas rate can be obtained. When the water rate is low, gas must be injected at a high rate, and this could be uneconomical to prepare the facilities to inject gas at high rates. To involve the economic dimensions, a dimensionless parameter was introduced for WAG ratio as:

$$WAGG\,Ratio = WAG\,Ratio \times \frac{q_w}{q_g} \tag{11}$$

$$Economic\,desire \propto low\,gas\,rates \propto 1/q_g. \tag{12}$$

The results are shown in Figures 9 and 10. By increasing the WAG ratio at all water rates, the recovery would be increased. Also, at a constant WAGG ratio, by increasing the water rate the recovery and FWCTT increases are not desirable, and in order to maximize recovery and minimize FWCTT, the optimized WAGG ratio must be determined.

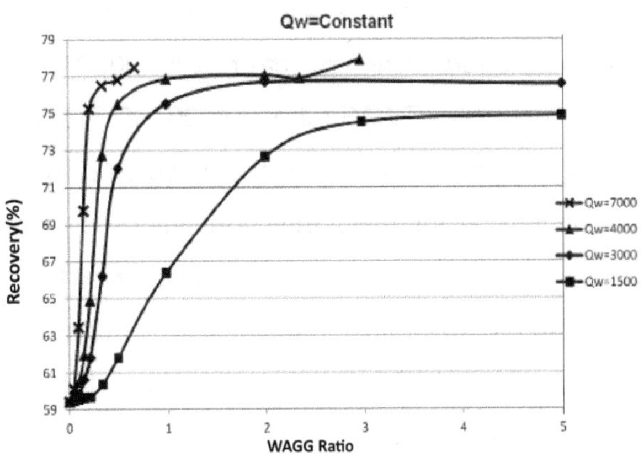

Figure 9: Recovery at different WAGG ratios; water rate is kept constant for each curve.

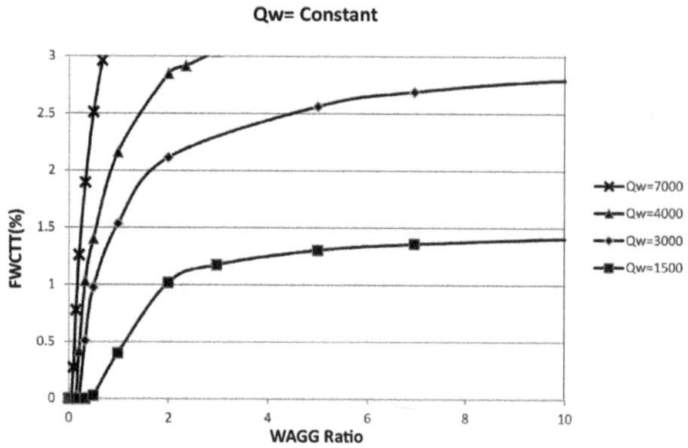

Figure 10: FWCTT at different WAGG ratios; water rate is kept constant for each curve.

Constant Gas Rate

For this type of simulation, the gas rates were remained constant. For having a constant volume of fluid injection, the water rates varied at different TWAG ratios. Two gas rates, namely 4000 MSCF/day and 1000 MSCF/day, were chosen for this purpose. The results of this simulation are shown in Figure 11. When gas was injected at a rate of 4000 MSCF/day, it resulted in a sharp increase in recovery from zero WAG ratios, and this reflects the water rate effect on production; after a quick increase in recovery, it reaches the optimized value and increasing WAG ratio after this stage does not affect the recovery; thus, it is very important to choose the optimized WAG ratio when gas rates remain constant. When the gas rate is 1000 MSCF/day for the same TWAG ratio of 4000 MSCF/day, a greater volume of water must be injected to have the same total volume of injection, and because of this, at low rates (like 1000 MSCF/day), recovery is higher in the range of low WAG ratios. However, by increasing the WAG ratio, it reaches the optimized value. FWCT for both gas rates has the same trend. Total pore volume injected to the reservoir is equal to the sum of total water injected in the reservoir conditions plus the total gas injected in the reservoir conditions, i.e:

$$PVIT = PVIW + PVIG \tag{13}$$

$$q_w = \frac{PVIT(1+TWAG)}{t_{total} B_w TWAG} - \frac{q_g B_g}{B_w (TWAG)} \tag{14}$$

where, B_g and B_w are gas and water formation volume factors respectively. When the gas injection rate is constant and the simulations are held for a constant period of time in all the conditions, water rate could be obtained by knowing TWAG and using Equation 14.

Selecting the Optimum Period and Time of Injection

In order to obtain the optimum time to inject after starting oil production and to select the best period of injection, three periods of 10 years, 20 years, and 40 years of injection were selected for a total of 40 years of production. For ten years of injection period, there were four possible statuses, namely an injection in the first decade, the second decade, the third decade, or the fourth decade of the production period. For a twenty-year injection period, there were two statuses of injections in the first half of the production or in the second period of the production time period. The best time to start injection is when the reservoir has not produced much oil and there is sufficient oil in the reservoir. This technique will be successful, if the saturation of oil is not reduced severely. This point is important to prevent the reservoir from reducing its productivity, and this is possible when the reservoir pressure and oil saturation does not reduce too much; in fact, when oil saturation and reservoir pressure reduces, the injection could not produce much oil from the reservoir, so for investigating the effect of these two parameters in oil recovery in different scenarios, a new value was introduced as defined by:

$$X=\Delta S_o \times P_{\text{start time of injection}} \quad (15)$$

$$X=(S_{wf}-S_{wi}) \times P_{\text{start time of injection}} \quad (16)$$

The results are depicted in Figures 11-13. It is important to note that when the productivity of oil reduces, the injection does not play an important role in production.

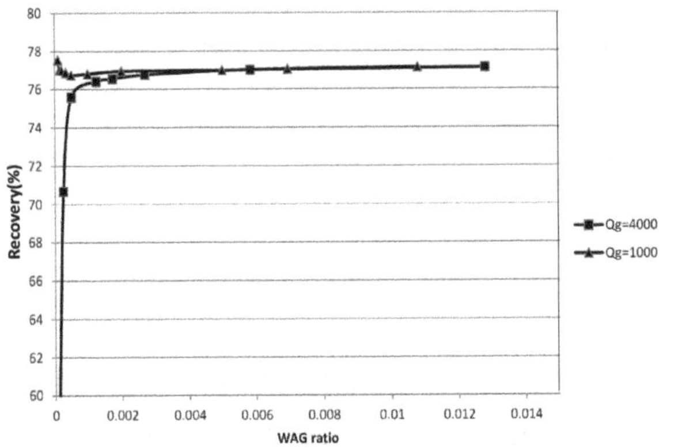

Figure 11: Recovery at different WAG ratios; gas rate is kept constant for each curve.

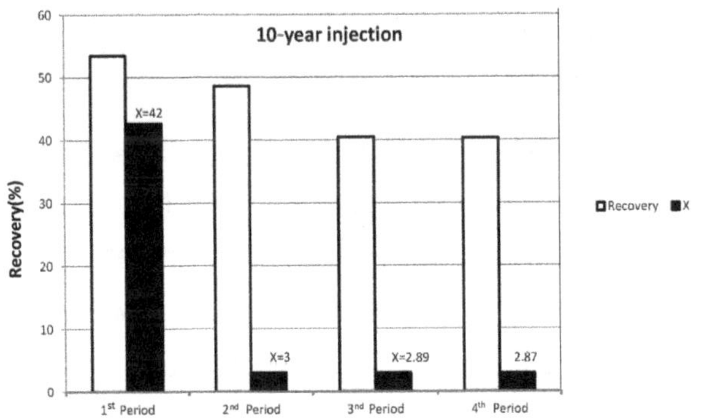

Figure 12: Recovery and X value for four possible statuses in a ten-year injection period.

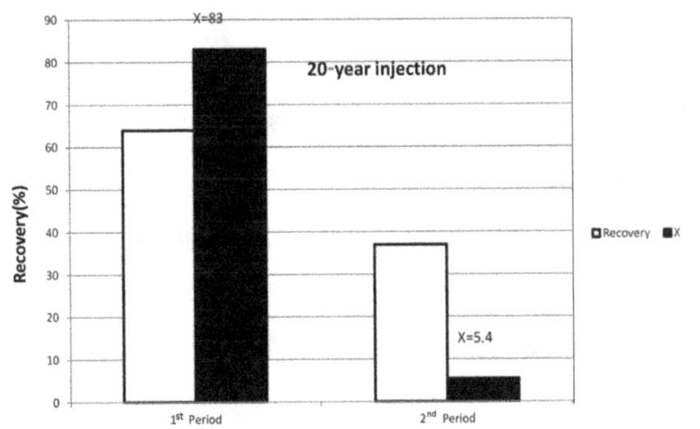

Figure 13: Recovery and X value for two possible statuses in a twenty-year injection period.

Figure 14 shows the amount of oil production based on injection time. For ten and twenty years of injections, the first injection modes were chosen; it is clear that increasing the time of injection does not mean higher oil production. For up to twenty years of injection, we could expect almost the same production, but when the injection process continues for forty years, recovery rate decreases, and, from an economical point of view, it is not an optimized period for the injection.

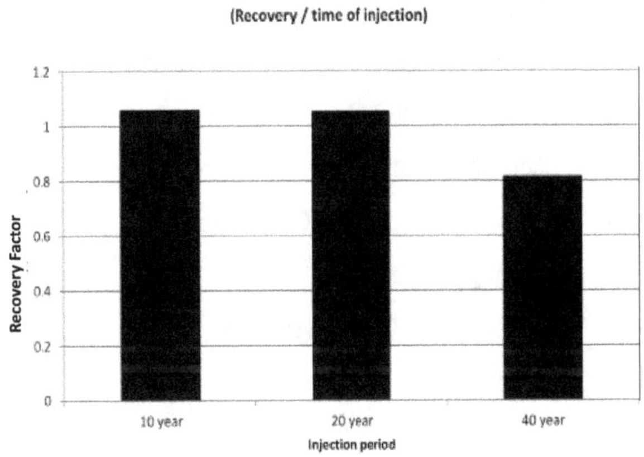

Figure 14: Recovery per time of injection for three different injection periods.

CONCLUSIONS

Based on the studies of different parameters affecting the WAG process, it is concluded that:

1- Carbone dioxide is miscible with oil at low pressures, so by using this gas in WAG injection process, the highest recovery will be achieved.

2- By increasing the cycle time of injection, recovery increases; however, increasing cycle time causes high water production, so the optimum value of cycle time must be found.

3- If water is first injected, recovery will be higher compared to when gas is first injected.

4- By increasing WAG ratio from zero, recovery increases dramatically; however, recovery increases very slowly by further increasing WAG ratio.

5- From an economical point of view, the optimum period of injection must be determined to have the maximum oil production at a minimum time of injection, and this depends on the reservoir pressure and oil saturation.

Abbreviation

API: American Petroleum Institute
EOR: Enhanced Oil Recovery
FCMP: First Contact Miscibility Pressure
FWCT: Total Field Water Cut
IOR: Improved Oil Recovery
K_{row}: Oil Relative permeability in presence of water
K_{rw}: Water Relative Permeability
MMP: Minimum Miscibility Pressure
Mscf: Thousand Standard Cubic feet
PVIG: Total gas injected
PVIT: Total Injected Pore Volume
PVIW: Total Water Injected
RB: Real Barrel
STB: Stock Tank Barrel
S_w: Water Saturation
WAG: Water Alternative Gas injection

REFERENCES

1. Esmaiel T. E., Fallah S., and Kruijsdijk C. V., "Determination of WAG Ratios and Slug Sizes Under Uncertainty in a Smart Wells Environment," SPE 93569, Presented at *the 14th, SPE Middle East Oil & Gas Show and Conference*, Bahrain Exhibition Center, Bahrain, **2005**.
2. Willhit G. P., "Enhanced Oil Recovery," *Society of Petroleum Engineering*, Texas University, United States, **1998**.
3. Esmaiel T. E., Fallah S., and Kruijsdijk C. V., "Determination of WAG ratios and slug sizes under uncertainty in a smart wells environment," Paper SPE 93569 Presented at *the 14th SPE Middle East Oil and Gas Show and Conference*, Bahrain, **2005**.
4. Dalen V., Instefjord R., and Kristensen R., "A WAG Injection Pilot in the Lower Brent Formation at the Gullfaks Field," *7th EAPG IOR Europe*, Moscow, Russia, **1993**, 460-468.
5. Udosen E. O., Ahiaba O., and Aderemi S. B., "Optimization of Gas Condensate Reservoir Using Compositional Reservoir Simulator," SPE Paper 136964, **2010**.
6. Chen S., LI H., Yang D., and Tontiwachwuthikul P., "Optimal Parametric Design for Water-Alternating-Gas (WAG) Process in a CO_2 Miscible Flooding Reservoir," PETSOC, **2009**.
7. Haajizadeh M., Narayanan R., and Waldren D., "Modeling Miscible WAG Injection EOR in the Magnus Field," Paper SPE 66378 Presented at *the SPE Reservoir Simulation Symposium*, Houston, **2001.**
8. Bermudez L., Johns R. T., and Parakh H., "Parametric Investigation of WAG Floods above the MME," SPE Journal 84366, **2007**.
9. Tehrani D. H., Danesh A., and Sohrabi M., "Water Alternating Gas Injection Studies-Phase 1," Department of Trade and Industry, Improved Oil Recovery Research Dissemination Seminar, **2001.**
10. Christensen J. R., Stenby E. H. and Skauge A ., "Review of WAG Field Experience," SPE 39883, **1998.**
11. Hallam R. J., Ma T. D., and Reinbold E. W., "Performance Evaluation and Optimization of the Kuparuk Hydrocarbon Miscible Water-Alternating-Gas Flood," *7th EAPG IOR Europe*, Moscow, Russia, **1993**, 403-415.
12. Ma T. D. and Youngren G. K., "Performance of Immiscible - Water- Alternating-Gas Injection at Kuparuk River Unit, North Slope, Alaska," **1994.**
13. Danesh A., "PVT and Phase Behavior of Petroleum

Reservoir Fluids," Department of Petroleum Engineering, Heriot Watt University, Edinburgh, Scotland, **1998**, 215-230.

14. Sulaiman W. R., "A Numerical Simulation Study to Evaluate the Use of Water Alternating Gas (WAG) As a Pilot EOR Method in Dulang Unit, Malaysia," Ph. D. Thesis, Imperial College, London, **2003**, 93-99.
15. Larsen J. A. and Skauge A., "Simulation of the Immiscible WAG Process Using Cycle- Dependent Three Phase Relative Permeabilities," Paper SPE 56475 Presented at *the ATCE*, Houston, **1999**.
16. Van Poollen H. K., "Enhanced Oil Recovery," Pennwell Publishing Company, **1980**, 118-119.
17. Alvarado V., Ranson A. and Hernández K., "Selection of EOR/IOR Opportunities Based on Machine Learning," Paper SPE 78332, Presented at *the SPE 13th European Petroleum Conference held in Aberdeen*, Scotland, U. K., **2002**.
18. Rehman T., "A Techno Economical Evaluation of Miscible Flooding," Dalhousie University, **2008**, 12.
19. Safdarian M., "Parametric Study of WAG Process its Performance over the Synthetic Fractured Model," Master of Science Thesis, Petroleum University of Technology, **2009**.
20. Farzaneh S., "Parametric Investigation of WAG Injection Process in Naturally Fractured Reservoirs," *6th International Congress on Chemical Engineering*, Iran, **2010**.
21. Jafari M., Badakhshan A., Taghikhani V., Rashtchian D., et al., "Experimental Study and Simulation of Different EOR Techniques in a Non-Fractured Carbonate Core from an Iranian Offshore Oil Reservoir," *Iranian Journal of Chemical Engineering*, *27*(2), **2008**, 7-9.

4

An Injection Rate Optimization in a Water Flooding Case Study with an Adaptive Simulated Annealing Techniques

Vahid Azamipour and Mehdi Assareh*

Thermodynamics Research Laboratory, School of Chemical Engineering, Iran University of Science and Technology, Tehran, Iran

ABSTRACT

This paper introduces an effective production optimization and a water injection allocation method for oil reservoirs with water injection. In this method, a two-stage adaptive simulated annealing (ASA) is used. A coarse-grid model is made based on average horizon permeability at the beginning iterations of the optimization to search quickly. In the second stage, the fine-grid model is used to provide the accuracy of the final solution. A constrained optimization problem to maximize an objective function based on net present value is implemented. Allocation factors from the streamline simulation are used to help for the appropriate estimation of initial water injection rates. The proposed optimization scheme is used for a field sector simulation model. The results show that the optimized rates confirm the increment of total oil production. Optimized oil production and total water injection rates lead to an increase in the total oil production from 385.983 (initial guess) to 440.656 Msm^3. This means a recovery factor increment by 14.16%, while the initial rates were much higher than the optimized rates. Moreover, the recovery factor of optimized production schedule with an optimized total injection rate is 2.20% higher than the initial production schedule with an optimized total water injection rate. The allocation of the water injection rates and the revision of allocation rates result in 446.383 and 450.164 Msm^3. The revision of the water rates allocation provides a reduction of water cut during production.

Keywords: Life-Cycle Optimization, Adaptive Simulated Annealing, Water flooding, Streamline Simulation.

INTRODUCTION

In the context of oilfield development, the dynamic reservoir modeling and the production optimization have been attractive research areas. The complexity of geological models, facilities and the operational constraints complicate the formulation and solution of production optimization problems. Especially in large fields, because of the costs and complexities of the full simulation of a reservoir model, finding the optimized production strategies becomes very time-consuming, and it is practically impossible to derive all possible solutions. In fact, for the optimization of such problems, the efficiency and viability of the method are important subjects. The determination of a production scenario is one of the significant tasks in the reservoir management

***Corresponding author**
Mehdi Assareh
Email: assarehm@iust.ac.ir

because it affects the total oil production and project benefit. Many studies have been carried out to improve the reservoir management by optimizing the economic life-cycle performance of the reservoir models [1-4]. The focus on such problems is related to the maximization of the recovery factor and economic parameters (e.g. net present value). Due to many non-unique possible solutions for a certain objective function, the optimization techniques are needed to achieve the improved one when initial oil rates have a considerable impact on the total oil production. There are many researches addressing different aspects of production scheduling. Van Essen et al. maximized short-term gains with a high discount rate by selecting the net present value (NPV) criteria as short-term gains [5]. According to the literature, NPV, which is a cumulative objective function, is maximized in life-cycle optimization techniques. In similar approaches used by Bailey et al. and Yasari et al., the goal was to minimize the variance of the NPV distribution and maximize the average NPV [6,7]. Siraj et al. used the explicit uncertainty handling in the model-based economic optimization to examine the balance between short-term and long-term gains [8]. In these life-cycle optimization techniques, NPV is maximized. Van Essen et al. proposed the robust optimization method including a model uncertainty in the optimization by considering an average NPV over an ensemble of geological model realizations [9]. They showed that their method led to a reduction of the variance of the optimized strategy and an increment of the expected value when applying to the different geological realization. Van Essen et al. used the adjoint formulation to attain robust optimization [5]; however when the adjoint approach is worked with commercial simulation, it needs to access to the simulation source code. Ikewun et al. has applied a decline curve for optimizing workflow. They exerted simulation models and the decline curve analysis (DCA) as a production optimization and forecasting tools for shale gas reservoirs [10]. Salam et al. introduced their approach to optimize production strategy which used the NPV; also, they considered the capital expenditure [11].

Proxy models and upscaling are two possible common solutions for fast objective function evaluation in the reservoir optimization. Proxy modeling is implemented in many field development and production optimization problems. The usage of a proxy function or a surrogate model is very effective to reduce the high computational cost of function evaluation for the numerical simulator. The neural network is used as an approximation optimization to reduce the number of simulation runs and represent the strongly non-linear behavior of a complex petroleum production system [12,13]. Onwunalu et al. combined an optimization algorithm with an enhanced statistical proxy led to speed up field development optimizations [14]. However, the neural network presents some practical difficulties where there are too many degrees of freedom or too many constraints, where the input space and the output space cannot be correlated at all, or where the learning costs of the neural network for difficult problems is high [15]. There are several approaches towards using proxy applications in the literature, including Neuro-Fuzzy [16], kriging [17,18], and statistical proxies [19].

Upscaling is interesting in the problems with high-resolution models. High-resolution geological models have been built to capture the heterogeneities of the reservoir. For typical history matching and prediction simulation studies, these models increase the computational cost and are time-consuming. Upscaling a reservoir model from the fine grid to a coarser grid has been performed for the simplification of reservoir models. By Usage of a quite accurate approach in which the cells of the fine model are merged together to form a single cell in the coarse model, an appropriate coarsening is achieved from a fine grid model that preserves a good estimate of the value of objective function (for more detailed information see King et al. [20]). Jessen et al. proposed a technique to simulate water flooding with coarse-scale dual-porosity [21]. In the early time, the performance of their approach could be acceptable in comparison with the fine model; however, at later times, due to the simplistic representation of flow and transport between active and passive porosities, it could not perform accurately. Chen et al. developed an upscaling model for near-well flow in heterogeneous reservoirs and improved the precision of standard upscaling methods [22].

Moreover, the need for fast objective function evaluation, selecting an appropriate automated optimization technique can provide a better production scenario. Ekkawong et al. used linear programming and decline curve analysis in a gas field with small compartmentalized reservoirs with several constraints [23]. Kritsadativud et al. used a gradient-based optimization algorithm with both linear and non-linear constraints in a single time-step [24]. The stochastic optimization algorithms are more effective for escaping from the local optima, due to continuing the search quite randomly in areas beyond attractive zones. There are also numbers of field applications of stochastic optimization techniques. The genetic algorithm (GA) has been applied to the optimization of well placement [14,25,26]. Morales et al. introduced a risk-constrained algorithm based on the GA to allow the user to input the risk factor desired and individual weights for each realization [27]. Their model provided different optimum well placement, which depended on the risk level wanted by the user to take. Guyaguler et al. used hybrid genetic algorithms based on the GA, polytope search, and Kriging Proxy to reduce the computational effort in the well place location optimization of injection wells under uncertainties [17]. A field development optimization procedure based on the GA and other optimization tools (neural networks and evolution strategies) has been proposed and applied to a number of reservoir models [11]. Other optimization algorithms, including simultaneous perturbation stochastic approximation, simulated annealing [28,29, 30], and particle swarm optimization have been applied to optimization problems [31, 32].

Managing water flooding is one of the most commonly used approaches toward improving oil recovery and reducing the costs, which requires to know about the amount of water injection and how to allocate it through injection wells. Indeed, the success of a water flooding project depends on the ability to sweep the remaining oil towards oil production wells. However, an improper design may result in increasing the costs associated with the water cycling and poor sweeping the remaining oil. Thiele and Batycky introduced well allocation factors (WAF) or injection efficiencies for each injector to optimize the amount of water injection and produced fluids between well pairs in a water flooding reservoir; then, the rate of injection was reallocated to improve the water flooding performance [33]. Also, Ghori et al. reported the application of this approach in the real field [34]. Alhuthali et al.

presented a streamline-based workflow to equalize the arrival time of the front of water flooding at all producers within selected sub-regions of a water flooding project [35]. Furthermore, Alhuthali et al. extended their study to optimize a water flooding under geological uncertainty [36]. Ambia proposed using the GA and the particle swarm optimization in water flooding problems to determine an optimum pattern design and production and injection strategy to maximize NPV of the project [31]. Van den Hoek et al. presented a new methodology combining fluid-flow and fracture-growth within the context of an existing standard reservoir simulator which was coupled with induced fractures via special connections to eliminate most of the numerical instabilities [37].

In the present work, the objective function that is NPV with a discount rate to consider the long-term and the short-term gains is optimized through the optimization of oil production rates and total water injection in a field sector in which the water injection is started from the beginning of the oil production. Initial guesses for oil production wells are related to the capacity of the production of each well. The adaptive simulated annealing (ASA) is combined with a polytope search to provide the optimization with the ability of escaping from the local minima. In the first iterations, using a coarse grid block model in ASA causes the time length of the life-cycle optimization strikingly to be reduced. In later iterations, the fine grid model is substituted to confirm the preciseness of the NPV value and fine-tuning of the optimization variables. In addition, the total water injection is allocated to each well by this workflow. The goal was to optimize the allocation of water injection well rates. The initial guesses for injection well rate allocation are determined by the injection efficiency or allocation factor of each well from the streamline simulation. Revision of all allocation water rates by considering a limited water cut led to control the water cut and the water production. The offered workflow is a new approach combined with two stages optimization (coarse grid-fine grid model) to allocate water injection rates and the usage of water allocation factor for the initial guess. In the next section, the approach is described in detail. Next, the case study results are presented to demonstrate the improvement in reducing the time length of the life-cycle optimization, when it keeps the precision of the final optimized solution.

EXPRIMENTAL PROCEDURES

Methodology

In this approach, the optimization of oil production rates of each well results in the optimization of total oil production. Furthermore, the reliability of the approach is improved by including practical constraints for the rates of each well. Later on, the optimized water injection rates allocation and revision of allocated water injection in each time step are led to an increase in the final objective function. The NPV is considered as the objective function in order to include both the long-term and the short-term gains. The objective function J is NPV for the cumulative oil and water production over a fixed time horizon which is introduced by Van Essen et al. [38] and can be expressed by the following mathematical formulation:

$$J = \sum_{K=1}^{K} \left[\frac{CWI^K + CWP^K - COP^K}{(1+b)^{\frac{t_k}{t_{ref}}}} \times \Delta t_k \right] \quad (1)$$

where, *K* shows the total number of time steps *k*, and Δt_k is the time interval of time step *k* in days. *b* is the discount rate for a certain reference time t_{ref}.

In the above objective function, the water injection cost is defined as:

$$CWI^K = \sum_{j=1}^{N_{inj}} r_{winj} \times q_{winj.k} \quad (2)$$

The produced water production cost is defined by:

$$CWP^K = \sum_{j=1}^{N_{prod}} r_{wp} \times q_{wp.k} \quad (3)$$

The produced oil production benefit is calculated by:

$$COP^K = \sum_{j=1}^{N_{prod}} r_o \times q_{o.k} \quad (4)$$

where, r_o , r_{wp}, and r_{winj} are the oil price, the water production cost, and the water injection cost in $/m^3 respectively. $q_{winj.k}$, $q_{wp.k}$, and $q_{o.k}$ represent the total flow rate of injected water, produced water, and produced oil at time step *k* in m^3. N_{prod} and N_{inj} are the number of production and injection wells respectively. The optimization problem with the objective function defined in Equation 1 can be formulated as follows:

$$\max_{u_{1:K}} J(u_k), \quad (5)$$

Which is subject to:

$$g_{k+1}(u_k x_k x_{k+1}) = 0.\ k = 0.\dots.K-1 x_0 = \overline{x_0} \quad (6)$$

and

$$c_{k+1}(u_{k+1} \cdot x_{k+1}) \le 0. \quad (7)$$

where, u is the control vector (input vector), and x is the state vector (grid block pressures and saturations); g is a vector-valued function presenting the system equations, and x_0 stands for a vector of the initial conditions of the reservoir. A colon in a subscript indicates a range, e.g. $u_{1:K} = \{u_1, u_2, \dots, u_k\}$. The vector of inequality constraints g is related to the wells limitations.

The core optimization algorithm in this work is simulated annealing (SA) which has been introduced by Kirkpatrick [39]; in addition, it has been used as an optimization technique for the combinatorial optimization. SA is originated from the physical process of metallurgical annealing, when at first, a metal is at high temperature in a heating bath and is slowly cooled. Molecules in the hot metal are randomly dispersed. While the temperature is falling, the molecules are organized in a way, which is very near to the global minimum energy. The algorithm can escape the trap of local optimum by a metropolis algorithm using a random number generation method. The random search causes the algorithm to store variations which lead to decreasing the energy (objective function), and some variations increasing it [39]. If the new objective function is decreased, the set of optimization parameters will be accepted automatically. Otherwise, a Boltzmann probability function is called, and a random number is created. If the random number is less than the parameter set probability, it will be accepted as the next iteration:

$$P(accept) = \begin{cases} \Delta J < 0 & accept \\ \Delta J > 0 \rightarrow P = Exp\left(\frac{-\Delta J}{T}\right) & \end{cases} \quad (8)$$

Boltzmann probability function depends on T being the temperature as a global time-varying parameter of the objective functions, i.e. J(uk$_1$) and J(uk$_2$). The SA algorithm can escape from a local minimum which is worse than the global one because the probability function P should be positive even when J(uk$_2$) is larger than J(uk$_1$). Adaptive simulated annealing (ASA) is an improved version of simulated annealing. In the first step of adaptive simulated annealing, i.e. the initial temperature, the initial solution should be entered, and the objective function should be initialized. Then, y^i, a random parameter introduced by Ingber [1] is defined in Equation 12 and helps create a random solution, α^i_{k+1} (in the problem, this is the initial rates for wells) which is defined in Equation 9. Also, α^i_k is the current solution in the dimension i generated at annealing-time *k* within the range, and $\boldsymbol{B}_i$ and $\boldsymbol{A}_i$ are the upper and lower bounds for α^i_k.

$$\alpha^{i}_{(k+1)} = \alpha^{i}_{k} + y^{i}(B_i - A_i) \tag{9}$$

$$y^{i} \in [-1,1] \tag{10}$$

y^{i} is generated from a u^{i} from the uniform distribution:

$$u^{i} \in U[0,1] \tag{11}$$

$$y^{i} = \operatorname{sgn}\left(u^{i} - \frac{1}{2}\right)T_i\left[\left(1 + \frac{1}{T_i}\right)^{|2u^{i}-1|} - 1\right] \tag{12}$$

Finally, the value of objective function is calculated based on the new solution. The difference between two objective functions (delta) distinguishes whether the new solution is accepted or not. When the delta is equal or lower than zero, the new solution would be accepted, and the temperature based on a formulation proposed by Ingber [40] would be adjusted which is shown in Equation 13 and Equation 14, where T_{0i} is the initial temperature. Also, n_i and m_i can be considered free parameters to help tune ASA for specific problems. D is the dimension of the space problem. As it is represented in Figure 1, if the delta is greater than zero, the metropolis function is called. When the metropolis function is accepted, the new solution will be stored and the temperature will be adjusted by Equation 13. Otherwise, if the temperature is not reached the final temperature and rejected by the Boltzmann probability function, a new solution is generated in the algorithm. At the End, when the final temperature is reached, the algorithm will be terminated.

$$T_i(k) = T_{0i}(\exp\left(-c_i k^{\frac{1}{D}}\right), \quad 0 < c < 1 \tag{13}$$

$$c_i = m_i \exp(-n_i / D) \tag{14}$$

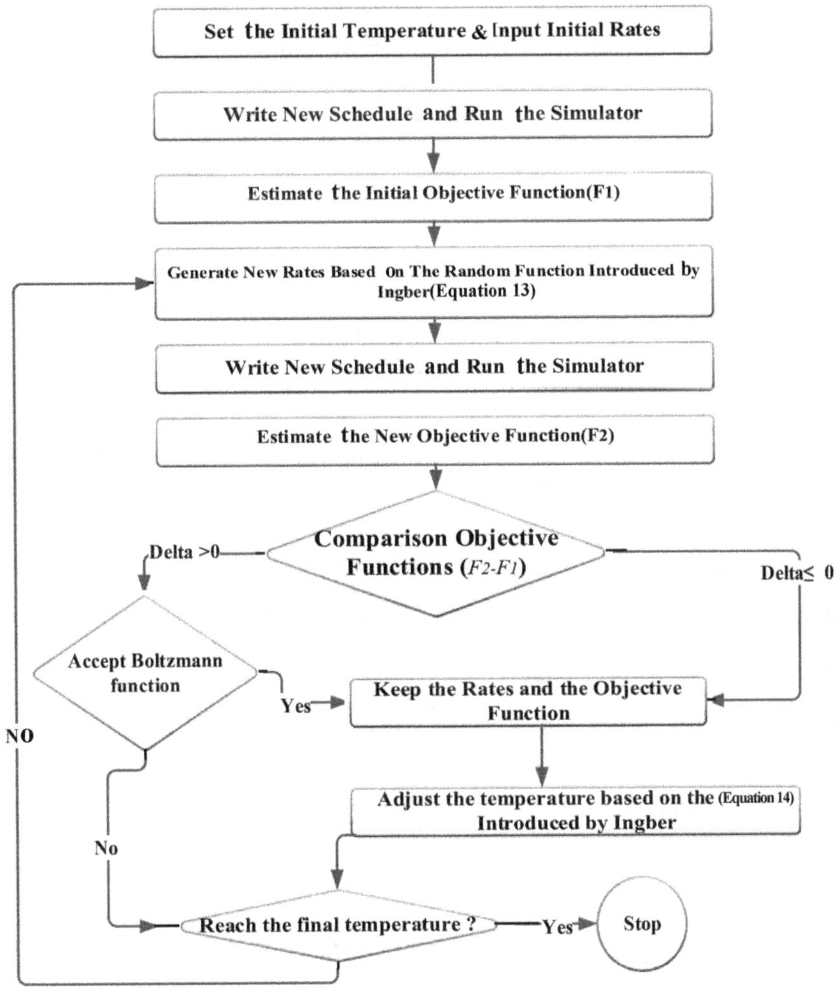

Figure 1: An adaptive simulated annealing flow diagram.

In the initial steps of adaptive simulated annealing (ASA), a coarse grid model is used which is created based on the horizon permeability map. This coarse grid model is used to reduce the length time needed to find the optimized solution. The optimization is performed in two sections; in the first section, the coarse grid block model is applied and does not have a substantial impact on the final solution because the solution varies in a large search space. In the second section, the fine grid block model is used to keep the preciseness of the final solution. Therefore, by dividing ASA in two sections of coarse and fine grid block models, the method can find an approximate optimized solution rapidly. By switching to fine grid section, the precise optimized solution can be guaranteed. Figure 2 shows the flowchart of ASA using fine grid and coarse grid models. The first guesses for the oil production rates in ASA come from their capacity of production, i.e. $NTGkhS_o$ where *NTG, k, h*, and S_o are net to gross, permeability, height, and oil saturation respectively. This makes a reliable guess for all rates of production wells and helps accelerate the optimization by improving the next guess. The initial oil rates and the total water injection are the input and the objective function calculated through the amount of oil and water production. Until the temperature is got to a specified temperature, the coarse grid block model is applied. In addition, a polytope search, a hill-climbing algorithm, enhances the speed of the method by finding the best solution between the previous iterations and picking the best one for the next guess in the ASA. The polytope algorithm introduced by Nelder and Mead tries to find the direction that increases the value of the objective function. In the next step, the fine grid block model is used when the temperature does not reach the final temperature, i.e. the stopping criteria.

To simulate the water flooding process, the coupled system of non-linear equations system for oil and water material balance for each simulation cell should be solved in ECLIPSE.

In the presented method, water injection rate allocation is performed after the optimized total amount of water injection has been determined. This allocation is done by using the same workflow for finding the optimized oil production rates. This means a two-stage ASA (coarse and fine grid block models) is used to be coupled with a polytope search to speed up the process. The speed of the usage of the coarse grid block model and the accuracy of the solution in the section of fine grid block model help optimize the allocation of water injection by an efficient workflow. For finding a trustworthy initial allocated injection water rate, water well allocation factor is an acceptable clue to assign an appropriate water injection rate to each well. Well allocation factor is the ratio of injected water to the oil produced at offset wells, which allows us to calculate the efficiency of injection wells calculated by the streamline based on flow simulation. Injection efficiency introduced by Thiele and Batycky [33] is defined as follows:

$$I_{eff} = \frac{\text{off} - \text{set oil production}\left[\text{rb/day}\right]}{\text{water injection}\left[\text{rb/day}\right]} \tag{15}$$

At optimized oil production rate, water injection rates are allocated. The well allocation factor comes from streamline simulation in each time step, and then the average of the total time steps for each injection well is calculated. By the allocation of the water injection rate to each well, the more remaining oil is swept towards oil production wells. The improved sweep efficiency provides a higher objective function and satisfies the economic consideration such as the reduction of the total amount of water production and the total oil production increment. A revision of the allocation rates during oil production and water injection results in better total water production controls, and the breakthrough time by specifying the economical operation limits of the wells such as water cut. If a well violates the limitation value such as specified water cut, the worst offending connection, which has the highest water cut in the well, will be closed. Even the entire well will be shut down in a case of very high water cut. This remedy has a considerable effect on the reduction of the total water production and the water cut; moreover, this remedy postpones the breakthrough of water injected front. Because of decreasing the total water production, the cost of water production is declined significantly. Therefore, it increases the value of the objective function.

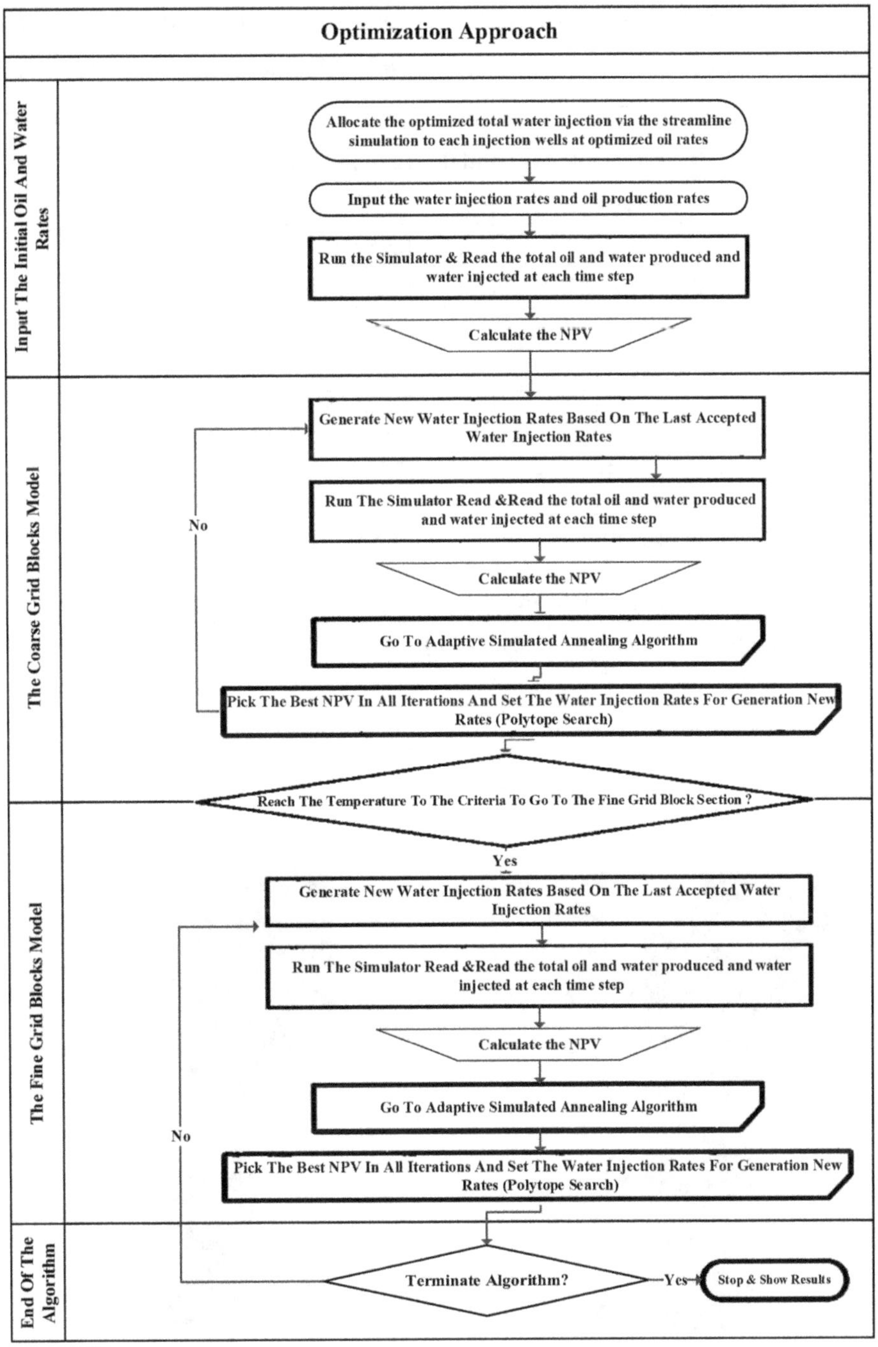

Figure 2: Workflow of the Optimization Algorithm.

RESULTS AND DISCUSSION

The optimization procedure was applied to a sector of a real three-dimensional oil reservoir model with two phases (water and oil) in Libya. The mentioned field sector named Mudarko was given to the authors for this research work. The reservoir model consists of 47310 (83×114×5) grid blocks, 14 injection wells, and 33 production wells. The control mode of production was on the liquid production rate. The life cycle of the reservoir covered a period of 24820 days. The initial and bubble point pressure were 641.5 and 267.93 psi respectively. The distribution of the porosity and the permeability is shown in Figure 3.

The reservoir porosity is ranging from zero to 0.15. According to the porosity map, the lower part of the figure exhibits very low porosity (very low quality rock) which shows a barrier-like behavior. This makes the average porosity in the reservoir to be 0.056.

It should be noted that the permeability in the all directions is the same, and it changes from 4.2 to 1028 mD. All remaining geological and fluid properties used in the model are presented in Table 1 and Table 2.

Table 1: Geological and fluid properties of the model

Property	Value	Units
ϕ	0.05759	-
ρ_o(at 14.7 psi)	847.376534	kg/m^3
ρ_w (at 14.7 psi)	1117.316	kg/m^3
c_o (at $_{Pint}$)	1.459E-05	1/psi
c_w (at $_{Pint}$)	4.579e-006	1/psi
μ_o (at $_{Pint}$)	0.3289	Pa.s
μ_w (at $_{Pint}$)	0.5	Pa.s
Temperature	87	°C

Table 2: Statistical information of simulation grid blocks dimensions

	Min (m)	Max (m)	Delta (m)	Mean (m)	Std. dev.	Variance
DX	9.74	17.12	7.37	13.14	1.93	3.72
DY	501.75	501.81	0.06	501.75	0.02	0
DZ	498.75	498.88	0.13	498.8	0.06	0

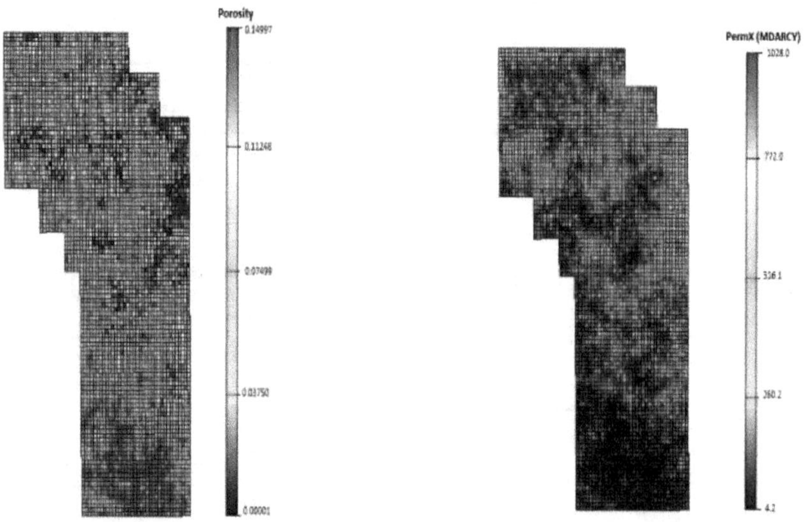

Figure 3: The map of the porosity (left) and the permeability map (right).

The coarse grid model contains 1150 grid blocks, so the number of grid blocks is significantly decreased. The upscaling procedure is performed in a very careful procedure based on the visual contrast in the permeability as can be seen from Figure 4. This produces a suitable coarse scale model with a production behavior such as the field pressure decline and the oil production rate close to fine scale one as shown in Figure 5. However, everything is not exactly the same, especially if it is investigated well by well. The time elapsed for the fine grid model and coarse grid one are 413.32 and 54.94 seconds respectively. A substantial speed-up is achieved by this short run time.

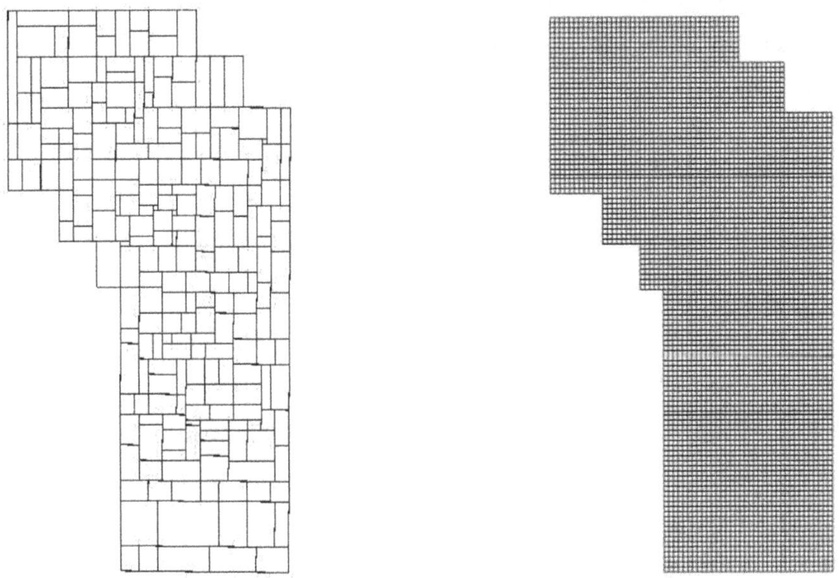

Figure 4: The coarse grid model (left) and the fine grid model (right).

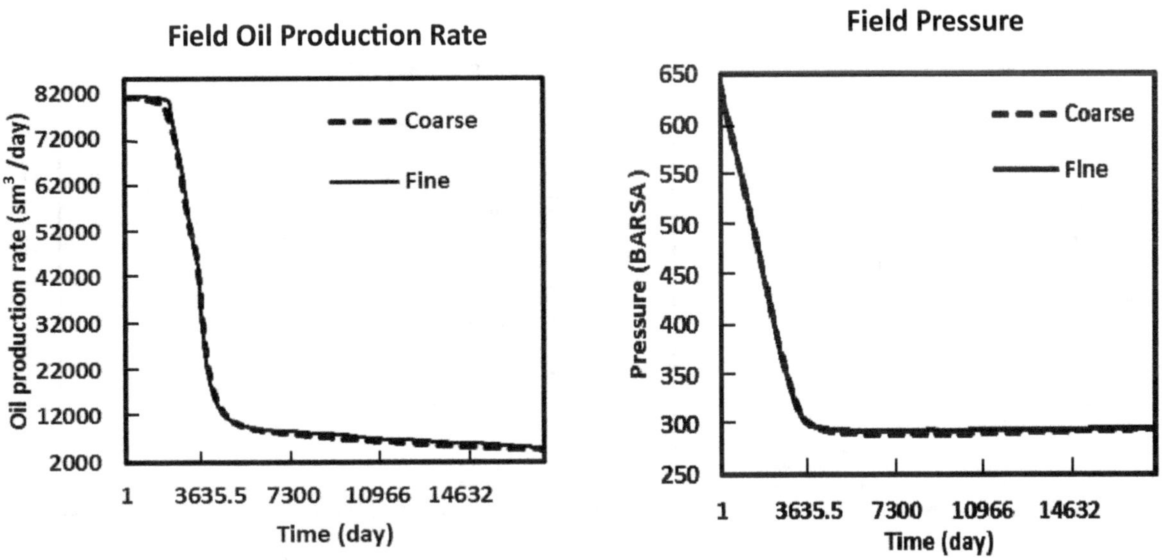

Figure 5: Field oil production rate versus time for both model coarse and fine models (left), and field pressure versus time for both model coarse and fine models (right).

Through implementing the optimization approach, the optimized total oil production can be found. NPV, the objective function defined in Equation 1, is codded in C# programming language and coupled with ECLIPSE for simulation runs where r_o = 128 $/m^3 (20.33 $/bbl), r_w=19 $/m^3 (3.02 $/bbl), and r_{inj}=6 $/m^3(0.95 $/bbl). The minimum initial oil rate production for all production wells is defined 500 m^3/day, and the maximum water injection rate for all injection wells is defined to be 10000 m^3/day as practical constraints in the optimization method. The optimized scenario was selected among 567 scenarios created in the ASA algorithm. In the context of optimization, there are two words that those are used sometimes instead of each other: optimized solution and optimal solution. However, some authors distinguish between the two words. The optimal solution is the global maximum; however, the optimized solution is a solution improved from the starting point and closer to the global maximum. The algorithm, which has been used for optimization, is a stochastic optimization procedure. Such procedure can investigate the variables space with several different directions due to Metropolis algorithm. The outcome of the stochastic procedure is different for two different runs. The optimization repetition also helped us to select among the best results.

As the result shows in Figure 6, although the initial oil production rate of optimized scenario was substantially lower than the first guess for the oil production rate, the total field oil production was increased from 385.983 *Msm³* (the initial guess) to 440.656 *Msm³*, i.e. 14.16% increment. Due to increasing water injection from 18100 m^3/day to 25073 m^3/day in the optimized solution, more oil was swept to the oil production wells. Therefore, another scenario was performed with the optimized total water injection rate and the initial oil production rates, which is simply called initial scenario with an optimized injection. This means that the oil production rates and total water injection rate are subject to optimization. The result revealed that the total optimized oil production is still 2.20% better than this scenario (see Figure 6). In addition, a recovery factor comparison is displayed in Figure 7; the recovery factor for the optimized scenario is 32.24%, which is higher than the recovery factors of the initial scenario (28.22%) and the initial scenario with the optimized injection (31.51%).

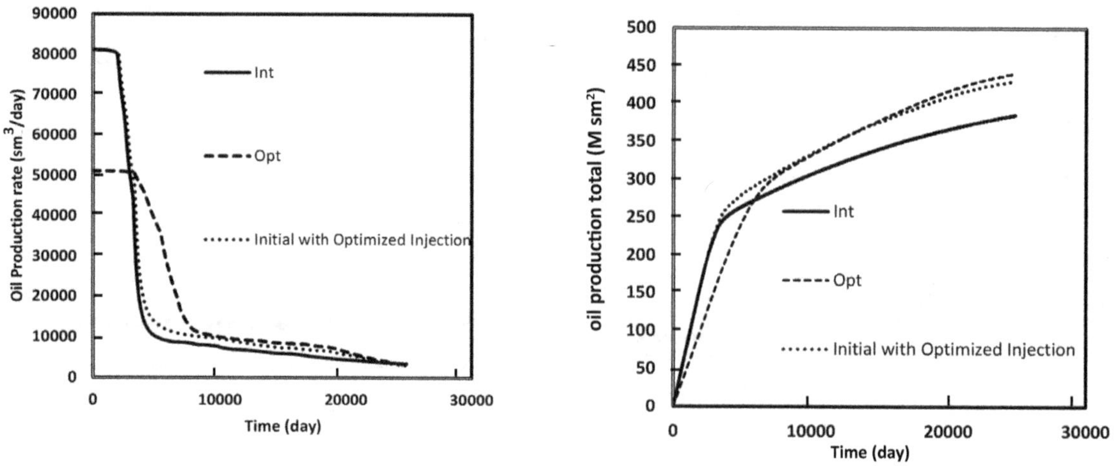

Figure 6: Field oil production rate, comparison between the initial scenario and Initial with optimized injection (dotted line) (left); total field oil production comparison between the initial scenario (solid line) and Initial with optimized injection (dotted line) (right).

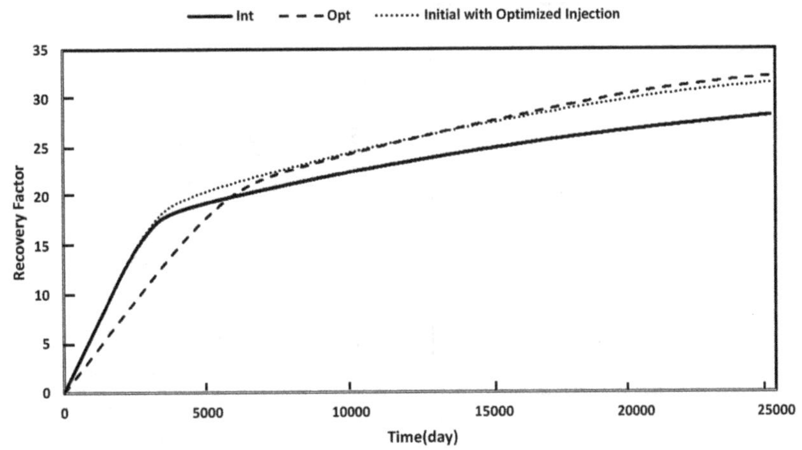

Figure 7: Recovery factor comparison between the initial scenario (solid line) and the initial scenario with optimized injection (dotted line).

In Figure 8 to Figure 9, oil production rates and the total oil production for two wells in the optimized scenario, initial scenario, and initial scenario with optimized injection are presented. There are cases for which, the optimized total oil production is higher than other scenarios; however, the initial oil rate of the optimized wells is lower than that of others. Furthermore, it should be noted that if the plots, one by one, for all wells are check, not all the total oil production of wells in the optimized scenario have a better result than the first scenario. Indeed, the general effect of all of the wells in the optimized scenario has a higher outcome. Figure 10 shows the objective function in each iteration. The polytope search helps the objective function to have an ascending profile.

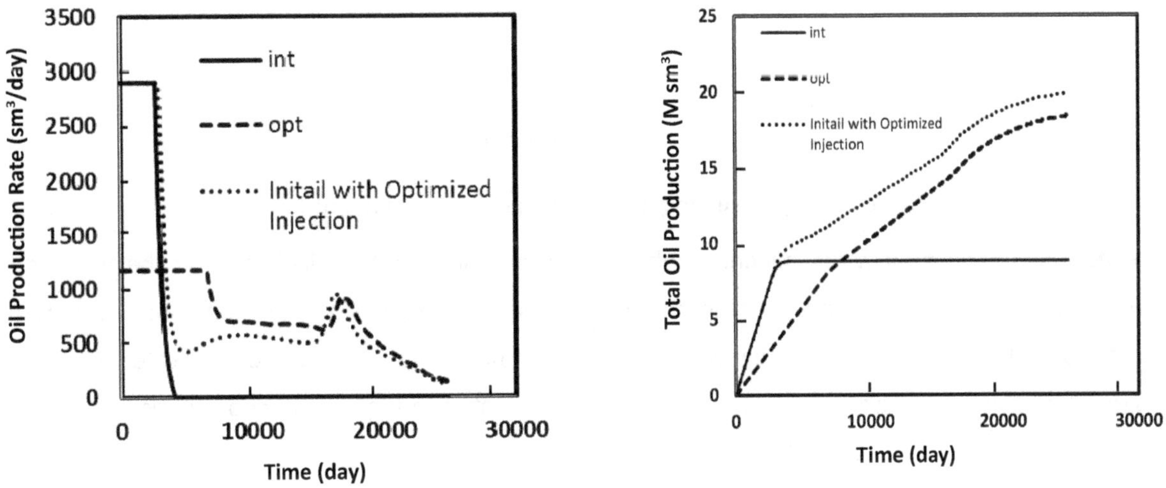

Figure 8: Oil production rates comparison in well M_P12-3 (left); the total oil production comparison in well M_ P12-3, among optimum scenario, initial scenario, and initial scenario with optimized water injection (optimized the total water injection rate and the initial oil production rates) (right).

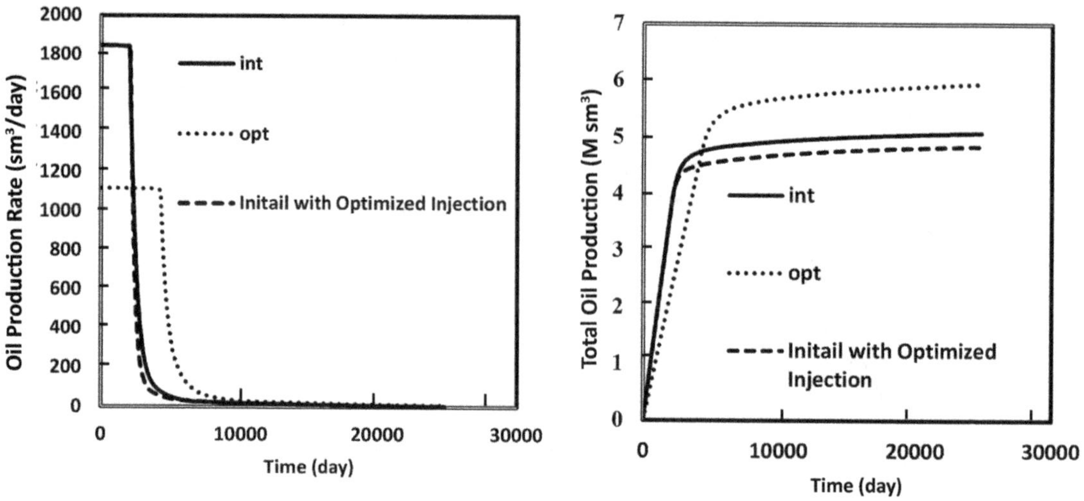

Figure 9: Oil production rates comparison in well M_P13-4 (left); the total oil production comparison in well M_P13-4, among optimum scenario, initial scenario, and initial scenario with optimized water injection (optimized the total water injection rate and the initial oil production rates) (right).

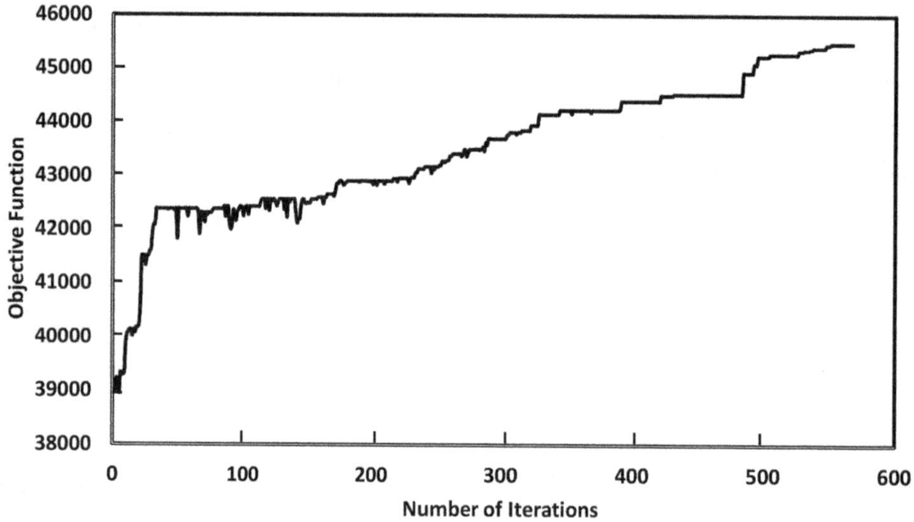

Figure 10: Objective function versus the number of iterations.

In the next step, the total water injection, which was calculated from the optimized scenario, is allocated to each injection well by a new workflow. For simplicity, this optimization of water injected allocation (water injection rate allocation in the injection well) is called the allocation scenario. For finding the optimized water allocation scenario, 567 scenarios are created in the ASA algorithm by being codded in C# programming language and being coupled with ECLIPSE for simulation runs at the same oil production rates calculated from the optimized oil production scenario. The mentioned optimization workflow was used to allocate water injection at the optimized oil production rates, which were calculated from previous optimization. For the first guess to allocate water injection, the water allocation factor calculated by streamline simulation was used in order to allocate total optimized water injection. There is not too much difference between the initial water allocation rates which come from streamline simulation and the final result for water injection wells which come from the presented method. Hence, the first allocation factors are an acceptable optimized water allocation. Moreover, the revision of water allocation rates considered some economic constraints for water cut and minimum oil production rate at the optimized oil production scenario. The allocation of water injection can have a significant effect on the improvement of the water flooding efficiency. In fact, the optimized total water injection rate should be distributed over water injection wells in a way that most of the remaining oil can be produced. As we can see in Figure 11, by making a comparison between the optimized oil production scenario, which is achieved from the previous optimization of oil production rates, and total water injection (the optimized oil scenario), the optimization of the allocation of water injection rates scenario (allocation scenario) and the total oil production are represented as 450.164 Msm^3 and 440.565 Msm^3 respectively. In addition, a recovery factor comparison is presented in Figure 12; moreover, the recovery factor is improved from 32.24% (the optimized oil scenario) to 32.91% (the allocation scenario). Although the total oil production does not have a remarkable improvement, the total water production in the allocation and revision scenario is strikingly decreased (Figure 13). The total water produced in the optimized oil scenario and the allocation scenario was 201.117 Msm^3 and 132.427 Msm^3 respectively; in other words, the total water produced was decreased by about 35%. However, the total water injection in the allocation scenario was increased by 4%. Water injection rates allocation helped decrease the water produced; therefore, the cost of water produced was decreased, and the total benefit was improved. Water cut shown in the Figure 14 also demonstrates that during the production the water cut in the allocation scenario is lower than the optimized oil. Considering all of the mentioned information such as the increment in the total oil production, the reduction in the total water production, and the improvement in controlling water cut during the life cycle of the reservoir leads to the enhancement of the objective function in the revision of water rate allocation scenario. This means gaining more profit and a reduction in the costs of water production and injection as shown in Figure 15. The optimized objective function for the allocation scenario was 48800, while the optimized objective function of the optimized oil scenario was 45400.

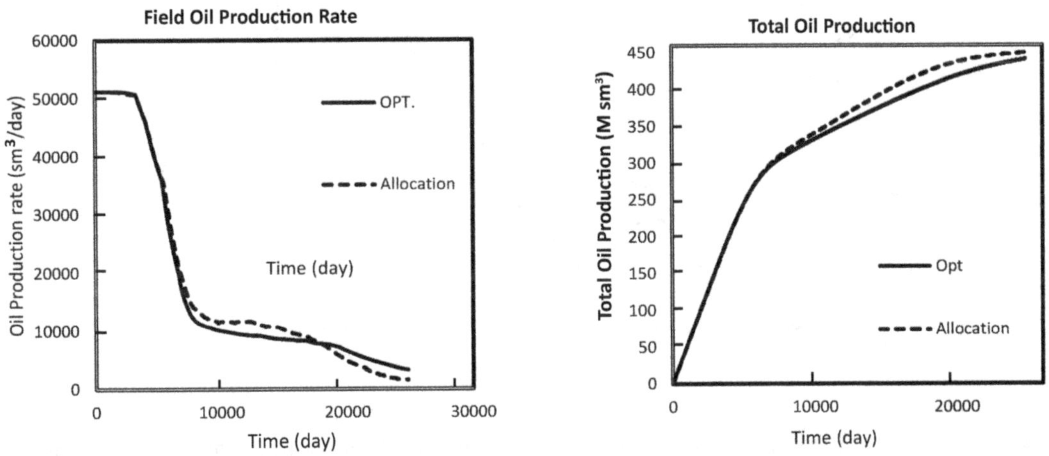

Figure 11: Field oil production rates comparison between the optimized oil scenario and the allocation scenario (left); total oil production comparison between the optimized oil scenario and the allocation scenario (right).

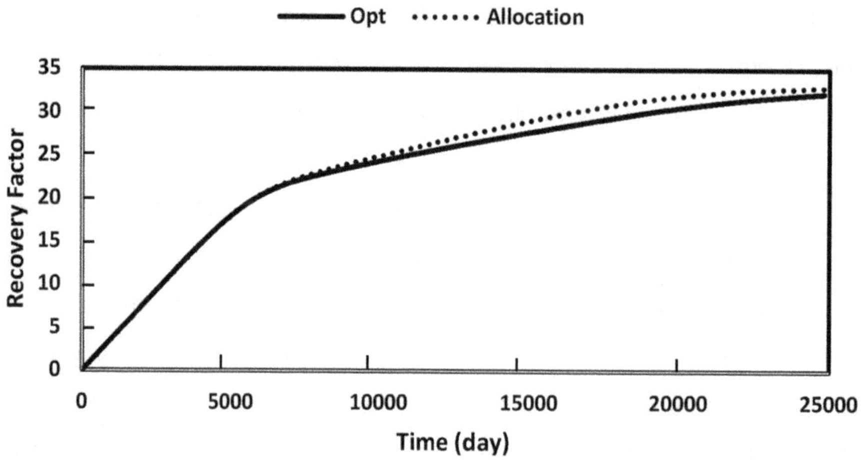

Figure 12: Recovery factor between the optimized oil scenario and allocation scenario.

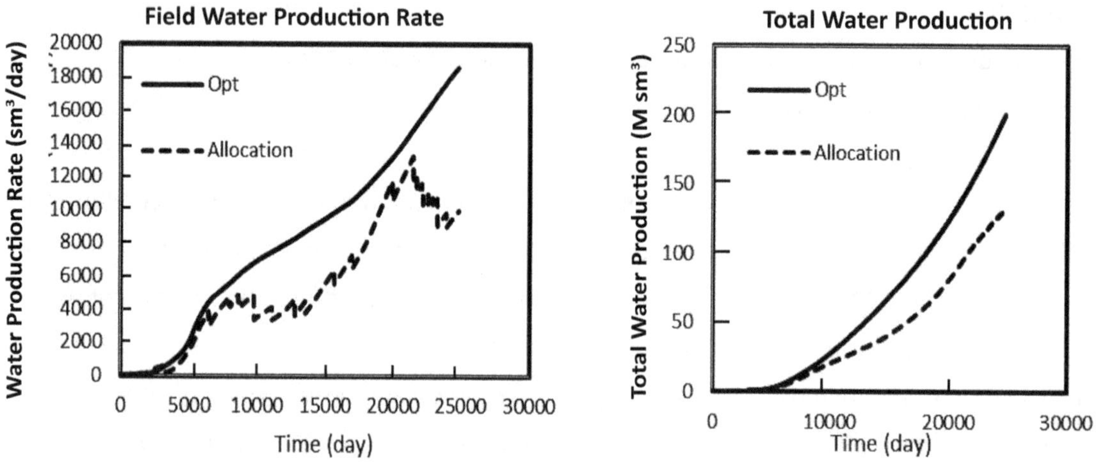

Figure 13: Field water production rates comparison between the optimized oil scenario and the allocation scenario (left); total water production comparison between the optimized oil scenario and the allocation scenario (right).

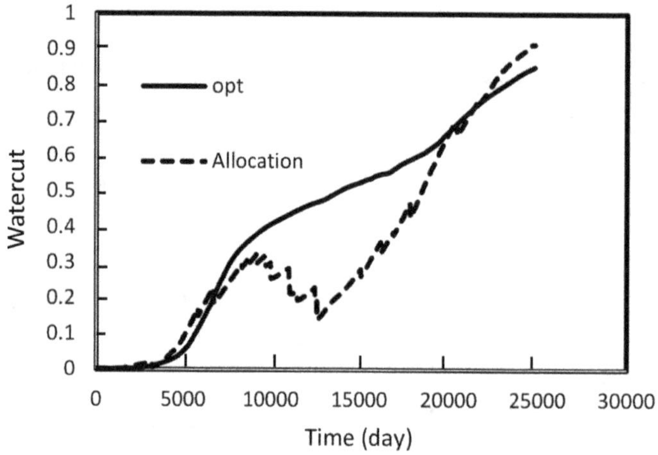

Figure 14: Water cut comparison between the optimized oil scenario and the allocation scenario.

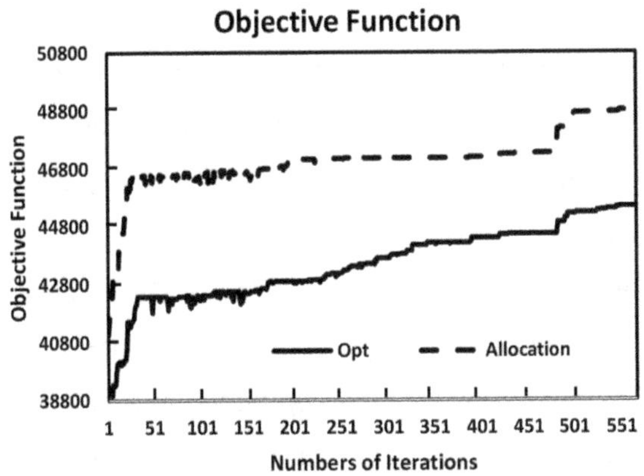

Figure 15: objective functions comparison between the optimized oil scenario and the allocation scenario.

The uncertainty in NPV comes from both the modeling process of water flooding and varying economic conditions, which include economic variables such as the interest rate, oil price etc. Including these uncertainties can be a step further in the future research to improve the optimization objective function definition, and to create a more reliable solution.

CONCLUSIONS

In this work, an effective scheme was offered to use a two-level adaptive simulated annealing technique for the optimization of initial oil production rates and total water injection in a reservoir undergoing water flooding and the allocation of water injection rates to each injection well. This method noticeably causes the run time of the optimization process to be decreased relative to the direct optimization. The objective function is based on NPV to consider the long-term and short-term gains. Based on the numerical simulation, it is concluded that:

- The optimization of oil well rates is an important issue in the total oil recovery. Consequently, the initial oil rate selection for wells should be accurate. To escape from the local minima during multi-variables optimization workflow, the usage of stochastic optimization algorithms such as ASA including Metropolis mechanism aids the optimization procedure.
- Using the coarse grid model for the early temperature cooling in ASA algorithm provides the optimization approach with reducing the run time of the optimization. The coarsening of the grid blocks based on the permeability map is performed, and reducing the number of grid blocks has a considerable effect on the performance of the proposed optimization for the first stage of optimization. For the final steps, the fine grid blocks model was used to keep the accuracy and reliability of the solution. The comparable trends of reservoir behavior such as field pressure decline and oil production rates in both of the fine and coarse grid models are necessary factors to generate a reliable coarse grid model.
- The optimized total water injection can be allocated through the proposed workflow for each injection well. The allocated water injection rates not only increase the total oil production, but also decrease the total water production.
- The revision of allocated water helps the method to consider some economic constraints, e.g. for water cut. This revision leads to better control on the water cut.
- The water allocation factors calculated from streamline simulation provide the acceptable initial water injection rates for the allocation of total water injection.

NOMENCLATURES

Δt_k	: time interval of time step k (y)
ASA	: Adaptive simulated annealing
b	: discount rate
c	: compression (1/psi)

GA	: Genetic algorithm
J	: objective function
k	: time step counter
NPV	: Net present value ($)
NTG	: net to gross
o	: oil
p	: pressure (bar)
q	: rate production/injection
r	: revenues/costs
S	: saturation
SA	: Simulated annealing Subscripts
t_k	: time at time step k (date)
t_{ref}	: reference time (date)
u	: input vector
w	: water
winj	: Injected water
wp	: produced water
x	: state vector
x_0	initial conditions
μ	: viscosity (cp)
ρ	: density (kg/m^3)
φ	: porosity

REFERENCES

1. Asheim H., "Maximization of Water Sweep Efficiency by Controlling Production and Injection Rates," *European Petroleum Conference*, London, United Kingdom, **1988**.
2. Sudaryanto B. and Yortsos Y. C., "Optimization of Fluid front Dynamics in Porous Media Using Rate Control. I. Equal Mobility Fluids," *Physics of Fluids* (1994-present), **2000**, *12*, 1656-1670.
3. Brouwe D. and Jansen J., "Dynamic Optimization of Water Flooding with Smart Wells Using Optimal Control Theory," *European Petroleum Conference*, Aberdeen, United Kingdom, **2002**.
4. Sarma P., Aziz K., and Durlofsky L. J., "Implementation of Adjoint Solution for Optimal Control of Smart Wells," *SPE Reservoir Simulation Symposium*, The Woodlands, Texas, **2005**.
5. Van Essen G., Van den Hof P., and Jansen J. D., "Hierarchical Long-term and Short-term Production Optimization," *SPE Journal*, **2011**, *16*, 191-199.
6. Bailey W. J. and Couet B., "Field Optimization Tool for Maximizing Asset Value," *SPE Reservoir Evaluation & Engineering*, **2005**, *8*, 7-21.
7. Yasari E., Pishvaie M. R., Khorasheh F., Salahshoor K., and et al., "Application of Multi-criterion Robust Optimization in Water-flooding of Oil Reservoir," *Journal of Petroleum Science and Engineering*, **2013**, *109*, 1-11.
8. Siraj M. M., Van den HofP. M., and Jansen J. D., "Model and Economic Uncertainties in Balancing Short-Term and Long-Term Objectives in Water-Flooding Optimization," *SPE Reservoir Simulation Symposium*, Houston, Texas, **2015**.
9. van Essen G., Zandvliet M., Van den Hof P., Bosgra O., and et al., "Robust Water Flooding Optimization of Multiple Geological Scenarios," *SPE Journal*, **2009**, *14*, 202-210.
10. Ikewun P. O. and Ahmadi M., "Production Optimization and Forecasting of Shale Gas Wells Using Simulation Models and Decline Curve Analysis," *SPE Western Regional Meeting*, Bakersfield, California, USA, **2012**.
11. Salam D. D., Gunardi I., and Yasutra A., "Production Optimization Strategy Using Hybrid Genetic Algorithm," in *Abu Dhabi International Petroleum Exhibition and Conference*, Abu Dhabi, **2015**.
12. Yeten B., "Optimum Deployment of Nonconventional Wells," Stanford University, **2003**.
13. Zangl G., Graf T., and Al-Kinani A., "Proxy Modeling in Production Optimization," *SPE*

Europec/EAGE Annual Conference and Exhibition, **2006**.

14. Onwunalu J. E., Litvak M. L. , Durlofsky L. J., and Aziz K., "Application of Statistical Proxies to Speed up Field Development Optimization Procedures," *Abu Dhabi International Petroleum Exhibition and Conference*, Abu Dhabi, **2008**.
15. Carpenter W. C. and Barthelemy J. F., "A Comparison of Polynomial Approximations and Artificial Neural Nets as Response Surfaces," *Structural Optimization*, **1993**, *5*, 166-174.
16. Zarei F., Daliri A., and Alizadeh N., "The Use of Neuro-fuzzy Proxy in Well Placement Optimization," *Intelligent Energy Conference and Exhibition*, Amsterdam, The Netherlands, **2008**.
17. Guyaguler B. and Horne R. N., "Uncertainty Assessment of Well Placement Optimization," *SPE Annual Technical Conference and Exhibition*, New Orleans, Louisiana, **2001**.
18. Schuetter J., Mishra S., Zhong M., and LaFollette R., "Data Analytics for Production Optimization in Unconventional Reservoirs," *Unconventional Resources Technology Conference*, San Antonio, Texas, USA, **2015**.
19. Artus V., Durlofsky L. J., Onwunalu J., and Aziz K., "Optimization of Nonconventional Wells under Uncertainty Using Statistical Proxies," *Computational Geosciences*, **2006**, *10*, 389-404.
20. King M. J., Burn K. S., Wang P., Muralidharan V., et al., "Optimal Coarsening of 3D Reservoir Models for Flow Simulation," *SPE Annual Technical Conference and Exhibition*, Dallas, Texas, **2005**.
21. Jessen K. and Evazi Yadecuri M., "Simulation of Water Flooding with Coarse-scale Dual-porosity Representation of Highly Heterogeneous Reservoirs," *SPE Western Regional & AAPG Pacific Section Meeting 2013 Joint Technical Conference*, Monterey, California, USA, **2013**.
22. Chen Y., Bernath A., Rai H., and Muron P., "Development and Application of Upscaling Techniques for Modeling Near-well Flow in Heterogeneous Reservoirs," *SPE Annual Technical Conference and Exhibition*, Orleans, Louisiana, **2009**.
23. Ekkawong P., Kritsadativud P., Lerlertpakdee P., and Amornprabharwat A., "Innovative and Automated Workflow for Fast Production Optimization and Forecast in Gulf of Thailand Gas Fields Using Linear Programing Optimization," *SPE Digital Energy Conference and Exhibition*, The Woodlands, Texas, USA, **2015**.
24. Kritsadativud P., Jafarpour B., and Ekkawong P., "Fast Production Optimization with Decline Curve Analysis Under Facility Constraints: A Field Case Study," *SPE Western Regional Meeting*, Garden Grove, California, USA, **2015**.
25. Emerick A. A., Silva E., Messer B., Almeida L. F., et al., "Well Placement Optimization Using a Genetic Algorithm with Nonlinear Constraints," *SPE reservoir simulation symposium*, The Woodlands, Texas, **2009**.
26. Kazakov K. V. and Bravichev K. A., "Automated Way to Search for the Optimum Irregular Well Placement Using a Genetic Algorithm (Russian)," *SPE Russian Petroleum Technology Conference and Exhibition*, Russia, **2016**.
27. Morales A. N. , Nasrabadi H., and Zhu D., "A New Modified Genetic Algorithm for Well Placement Optimization under Geological Uncertainties," *SPE EUROPEC/EAGE Annual Conference and Exhibition*, Vienna, Austria, **2011**.
28. Castaño K. P. and Ojeda G., "Optimizing Thin-

layer Mapping through Spectral Inversion: Performance of Genetic Algorithms and Simulated Annealing," *2010 SEG Annual Meeting*, Denver, Colorado, **2010**.

29. Khan H., Saputelli L. A., Ranjan P., Carvajal G. A., et al., "Multi-Objectives Constrained Water flood Optimization in Tight Carbonates," *SPE Reservoir Characterization and Simulation Conference and Exhibition*, Abu Dhabi, UAE, **2013**.
30. Yang D., Zhang Q., and Gu Y., "Integrated Optimization and Control of the Production-injection Operation Systems for Hydrocarbon Reservoirs," *Journal of petroleum science and Engineering*, **2003**, *37*, 69-81.
31. Ambia F., "A Robust Optimization Tool Based on Stochastic Optimization Methods for Water Flooding Project," *SPE Annual Technical Conference and Exhibition*, San Antonio, Texas, USA, **2012**.
32. Lagos S. R., Sabbione J. I., and Velis D. R., "Very Fast Simulated Annealing and Particle Swarm Optimization for Microseismic Event Location," *2014 SEG Annual Meeting*, Denver, Colorado, USA, **2014**.
33. Thiele M. R. and Batycky R., "Water Injection Optimization Using a Streamline-based Workflow," *SPE Annual Technical Conference and Exhibition*, Denver, Colorado, **2003**.
34. Ghori S. G., Jilani S. Z., Alhuthali A., Krinis D., et al., "Improving Injector Efficiencies Using Streamline Simulation: A Case Study in a Giant Middle East Field," *SPE Middle East Oil and Gas Show and Conference*, Manama, Bahrain, **2007**.
35. Alhuthali A., Oyerinde A., and Datta-Gupta A., "Optimal Water Flood Management Using Rate Control," *SPE Annual Technical Conference and Exhibition*, San Antonio, Texas, USA, **2006**.
36. Alhuthali A. H. H., Datta-Gupta A., Yuen B. B. W., et al, "Optimal Rate Control under Geologic Uncertainty," *SPE Symposium on Improved Oil Recovery*, Oklahoma, USA, **2008**.
37. Van Den Hoek P. J., Al-Masfry R. A., Zwarts D., Jansen J. D., et al., "Waterflooding Under Dynamic Induced Fractures: Reservoir Management and Optimization of Fractured Water Floods," *SPE Symposium on Improved Oil Recovery*, Tulsa, Oklahoma, USA, **2008**.
38. Van Essen G., Jansen J. D., and Van den Hof P., "Hierarchical Long Term and Short Term Production Optimization," *Society of Petroleum Engineers*, New Orleans, Louisiana, **2009**.
39. Kirkpatrick S., "Optimization by Simulated Annealing: Quantitative Studies," *Journal of Statistical Physics*, **1984**, *34*, 975-986.
40. Ingber L., "Adaptive Simulated Annealing (ASA): Lessons learned," *Control and Cybernetics*, **1996**, *25*, 33-54.

5

An Experimental Investigation of Magnetized Water Effect on Formation Damage

Ehsan Khamehchi *, Seyed Arman Hosseini Kaldozakh, and Ali Alizadeh

Department of Petroleum Engineering, Amirkabir University of Technology, Tehran, Iran

ABSTRACT

In oil industries, water injection into oil reservoirs for pressure maintenance, oil displacement, and oil recovery is a common technique. Formation damage during water injection is a major problem in this process. Formation damage from the incompatibility of formation water (FW) and injection water (IW) causes a reduction in the permeability around the injection wells. Therefore, it is necessary that the formation damage be minimized using specific techniques such as the injection of scale inhibitors and water compatible with formation water. It has been proven that moving water through relatively weak magnetic field changes water properties. These changes involve density, electrical conductivity, salts dissolving ability, sedimentation rate of solid particles etc. This study was conducted to investigate the effect of magnetized water injection on the decline in rock permeability. Therefore, a magnetic field device was designed and combined with a formation damage setup. The results indicate that, in the presence of magnetic field, water injection causes less damage to rock, and the permeability reduction in this case is lower than when non-magnetized water is injected. In addition, the results show that a higher magnetic field flux reduces the permeability damage.

Keywords: Formation Damage, Water Injection, Magnetized Water, Scale Precipitation

INTRODUCTION

Experience in oil industries has indicated that many oil wells have flow limitation because of scale deposition in producing formation and equipment. Formation damage induced by oilfield scale is one of the difficult phenomena that occurs during water injection project. Mixing incompatible waters takes place in the reservoir during injection [1-7]. Formation damage due to oilfield scale is the result of precipitation and the accumulation of scale around the well bore [4,7]. This may influence reservoir performance, well bore performance, and deliverability of the reservoir system [8]. Because of the extensive use of water injection for oil displacement and pressure maintenance in oilfields, many reservoirs experience the problem of scale deposition when injection water begins to breakthrough.

In most cases, the scales formed in wells are caused by the formation of sulfate and the carbonate scales of calcium and strontium. Because of their proportionate hardness and low solubility, there are

***Corresponding author**
Ehsan Khamehchi
Email: khamehchi@aut.ac.ir

restricted processes available for their removal and preventive measures such as the squeeze inhibitor treatment which must be taken. Therefore, It is important to gain a proper understanding of the kinetics of scale formation and its detrimental effects on formation damage in both inhibited and uninhibited conditions [9].

Magnetic treatment methods have been studied for the past few decades as a new alternative for preventing scale, and the magnetic treatment technique is an alternative to the use of scale inhibition chemicals [10-12]. Magnetized water is water that passes through a magnetic field. Magnetizing water is an inexpensive and environmentally friendly method. Nevertheless, the effect of magnetic field on water is a controversial issue. Taking water from a relatively weak magnetized field causes many changes in water properties such as viscosity, pH, ability to dissolve salts, the rate of deposition of solid particles, and so on [12-18].

Many researchers have reported changes in the physical properties of water passed through the magnetic field. Joshi and Kamat [19] reported that the pH of distilled water changed up to 0.4 pH units. Also, Parsons et al. [20] recorded a reduction of 0.5 pH units after passing water through a magnetic field. Iwasaka and Ueno [21] found out that the size of the water clusters altered when they were exposed to a magnetic field. The dissolution rate into water of oxygen is significantly accelerated by the presence of a magnetic field, and the water vaporization rate, an essential process for all biological processes, is significantly affected by the application of a static magnetic field [22]. Chou and Lee [23] studied the effects of the amount of magnetic treatment by a permanent magnet on surface tension and showed that as the number of treatments increased, the surface tension of the sample decreased. Otsuka et al. [16] concluded that no changes in properties of pure water, distilled from ultra-pure water in a vacuum, were observed after magnetic treatment. However, when the same magnetic treatment was carried out after that, the distilled water was exposed to O_2; moreover, properties such as surface tension were changed. Sueda et al. [17] examined the maximum mass and diameter of a dripped water droplet on the tip of a glass capillary, and found out that both of which were affected strongly by magnetic fields.

Many researchers have studied the effects of magnetic field on reducing scales. Properly installed and configured magnetic treatment devices (MTD's) have had many successes in reducing the amount of scale build-up in pipes. In an experiment performed by Smith [24], permanent magnets reduced the formation of scale in six out of six hot-water storage tanks with an average of 34%. The maximum reduction was 70% and the minimum reduction was 17%. Brower [14] explains that magnetic systems treat water by passing it through a magnetic field. The dipolar movements of the molecules of dissolved solids and water molecules are affected in such a way that at the instant of crystal formation, the crystal form is divided into thin layers, and the ions align according to a single magnetic axis. The magnetic field then influences the production of a much greater number of nuclei. Hence, the solids precipitate as much finer crystals, which tend to remain separated because of the excess similar charge. Lipusa and Dobersekb [12] attained successful results with the scale on a heating copper-pipe spiral being 2.5 times thinner due to magnetized water treatment compared with untreated water. Busch [25] attained a 22%

reduction in scale, using artificially prepared hard water. Parsons et al. recorded a 48% reduction in scale in his experiment. Tai et al. [26] discovered in their research that the crystal growth rates of calcite were suppressed completely in the presence of the magnetic field at a low pH and in a supersaturating condition. By contrast, the growth rate seemed to increase at a high pH and relative supersaturating. According to Alimi et al. [10], the treatment-pH and the water flow rate of the MTD have an important impact on the nucleation type and on the amount of calcium carbonate.

The purpose of this study was to investigate whether or not magnetic field can affect permeability damage because of scale precipitation in a sandstone rock. A magnetic treatment device was designed using electromagnetic concepts and coupled with a formation damage setup. Finally, two different magnetic field fluxes were used for producing magnetized water, and the decline in rock permeability was examined by injecting magnetized water and non-magnetized water.

EXPERIMENTAL PROCEDURES

Magnetic Device

Magnetic treatment methods have been studied and have been available for the past few decades as an alternative to chemical methods to prevent and control scale formation. Magnetic treatment devices can be based on the electromagnets or permanent magnets. Although electromagnets can produce the magnetic fields of great intensity, many magnetic devices can be used to prevent scale formation. Magnetic devices can be designed to meet the specific requirements such as field strengths, field directions, and uniformity. Depending on the design lines of the magnetic field, it can be parallel or perpendicular to the flow direction [27]. Several authors claim that the important factors which promote magneto–hydrodynamic forces are the conductivity of solution, linear velocity of fluid, and flux density of magnetic field [10, 11, 28-30]. It was found out that even a weak magnetic field (B = 1000 G) influenced aragonite/calcite ratio in precipitated $CaCO_3$ [31]. Gabrielli et al. used a series of pairs of permanent magnets with a uniform magnetic field of 1600 G for investigating the scale reduction of magnetic water treatment [11]. Tombacz et al. have tested both flowing and static systems, and concluded that only in a flowing system, the magnetic effect is observed. The magnetic flux density ranges from 1000 G to 8000 G among those magnetic treatment experiments [32]. Kobe et al. took 5000 G as the magnetic flux density in their experiments to obtain successful treatment results [29].

Therefore, in this study for making magnetized water, a magnetic device was designed by using electromagnetism concepts. The device, which is designed to create a 4500 G magnetic field, consists of an inductor with an air gap and a coil. The magnetic field is created in the inductor and the air gap by applying an electric current to the coil.

Figure 1 shows a simple magnetic circuit with an air gap having in the middle of a length cut (lg) a leg. The winding provides Ni ampere-turn. The magnetic flux generated in the air gap is equal to the magnetomotive force Ni divided by the sum of the reluctances of the core and of the air gap. Supposing that the leakage flux is negligible, by applying the Ampère's circuital law, one may obtain[33]:

$$Ni = H_g l_g + H_c l_c \tag{1}$$

Hc and *Hg* can be written in terms of the magnetic flux as reads:

$$H_c = \frac{B_c}{\mu_c} = \frac{\phi_c}{\mu_c A_c} \tag{2}$$

$$H_g = \frac{B_g}{\mu_g} = \frac{\phi_g}{\mu_0 A_g} \tag{3}$$

According to Gauss's law of magnetism, the net outward flux of *B* through any closed surface must be equal to zero. Hence, the flux of *B* must be the same over any cross section of the magnetic circuit, which can be written as follows:

$$B_g A_g = B_c A_c \tag{4}$$

By combining the above equations, the magnetic flux is given by:

$$\phi = B_g A_g = \frac{Ni}{\frac{l_c}{\mu_c A} + \frac{l_g}{\mu_g A}} \tag{5}$$

If the reluctance of core and air gap is written as $R_c = \frac{l_c}{\mu_c A}$ and $R_g = \frac{l_g}{\mu_g A}$ respectively, Equation 5 can be written as follows:

$$Ni = (R_c + R_g)\phi \tag{6}$$

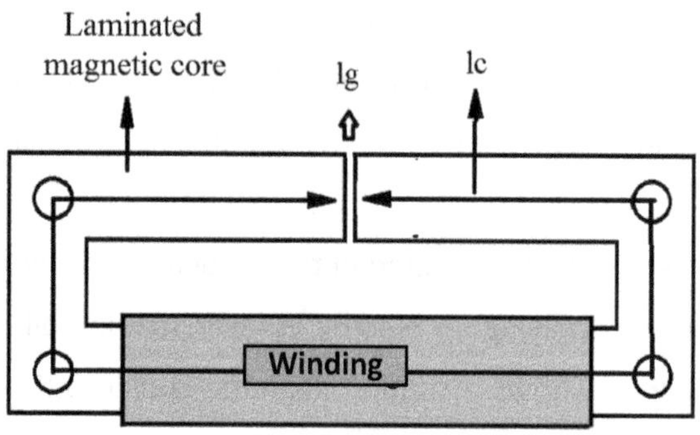

Figure 1: A simple magnetic circuit with an air gap having a length of l_g.

According to the above equations, the magnetic circuit with an air gap can be represented in a series electric circuit shown in Figure 2. The optimum number of winding turns and thickness of magnetic device to produce a magnetic flux of 4500 Gauss in the air gap can be determined by doing a reverse calculation. UI laminated magnetic cores (Figure 3) have been used for designing magnetic device. The calculations were performed for different sizes of UI laminated magnetic cores, and the optimum of winding turns and magnetic device thickness have been determined. Table 1 shows the calculations of different UI laminated magnetic cores. At last, UI-150 magnetic cores were chosen for magnetic device manufacturing, and for them, the calculated N and thickness are 440 turns and 11 cm respectively. This manufactured device is capable of creating a magnetic field with a peak flux density of 4500 Gauss. The air gap in the device is small (5 mm), and the magnetic field influenced the fluid properly. The magnetic field direction is perpendicular to the flow direction. The designed magnetic device is shown in Figure 4.

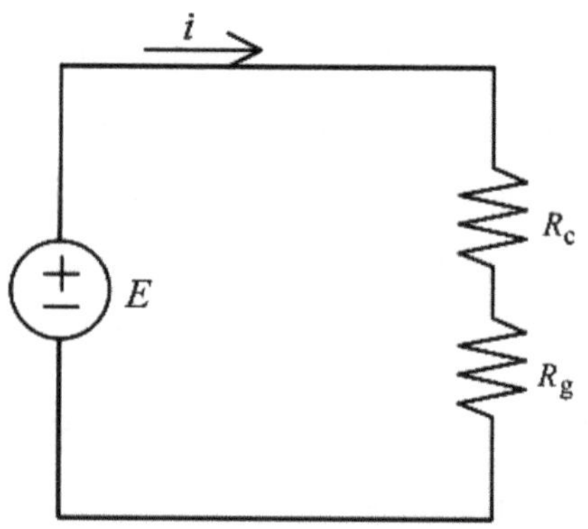

Figure 2: Series electric circuit.

Figure 3: UI laminated magnetic cores.

Figure 4: The magnetic device.

Experimental Section Setup

Here, water injection experiments were conducted by using a core flood system of the FDS350 apparatus [34]. For investigating the effect of magnetized water on rock permeability behavior, the designed magnetic treatment device was coupled with FDS350 apparatus. Figure 5 shows the magnetic treatment device coupled with FDS350.

Figure 5: Magnetic device coupled with FDS350.

Materials

In all of the experiments, three sandstone cores with an average porosity of 16% and an absolute permeability of 3.11 mD were used. The properties of these cores are reported in Table 2. All the cores were washed with toluene and methanol, separately in Soxhlet apparatus, and they were then dried using an oven at 100 °C for four hours before the test. The composition of formation and injected water used for the experiments are listed in Table 3. The formation water was used for the initial saturation of cores and permeability measurements, while the injection water was used for water injection processes in the experiments.

Table1: Calculations of different UI laminated magnetic cores.

Magnetic core	UI-75	UI-85.5	UI-96	UI-108	UI-125	UI-150	UI-180	UI-210	UI-240	UI-360
$l_c(m)$	0.445	0.508	0.571	0.643	0.746	0.895	1.075	1.255	1.435	2.155
$l_g(m)$	0.005	0.005	0.005	0.005	0.005	0.005	0.005	0.005	0.005	0.005
Ni_{min}	649.2	651	652.8	654.8	657.7	661.9	667	672.1	677.2	697.6
Ni_{max}	1623	1627	1632	1637	1644	1655	1668	1680	1693	1744
x(cm)	22	19.3	17.19	15.28	13.25	11	9.167	7.857	6.875	4.583
N	440	440	440	440	440	440	440	440	440	440
i_{min}	1.465	1.47	1.474	1.478	1.485	1.494	1.506	1.517	1.529	1.575
i_{max}	3.664	3.674	3.684	3.695	3.712	3.736	3.764	3.793	3.822	3.937

Table 2: Core properties.

Core	Length (cm)	Diameter (cm)	Porosity (%)	Absolute Permeability (md)
A	6.39	3.8	12.59	3.1028
B	6.48	3.8	11.62	3.1465
C	6.83	3.8	11.85	3.0484

Table 3: Formation and Injection Water Compositions.

Ion	Formation Water (ppm)	Injection Water (ppm)
Ca^{+2}	6998.41	480.99
K^{+}	1966.68	468.99
Na^{+}	24996.86	12229.97
Mg^{+2}	758.17	1773.99
Cl^{-}	54600.39	23009.02
HCO^{-3}	421.27	0.122
SO_4^{-2}	108.24	3169.24
TDS	82893.28	40872.69
pH	6.7	7.2
Viscosity at 50°C (cP)	1.24	1.18

At first, a dry core was saturated with formation water, and it was then loaded into the core holder of the FDS350 apparatus. The confining pressure was set at 500 psi above the pore pressure during the test. For initial permeability measurement, formation water was injected to the cores at three rates, namely 0.2, 0.3, and 0.4 mL/min. Apparatus records automatically temperature, confining and the pore pressure, flow rates, and differential pressures. Then, it calculates the permeability using the Darcy's law. The pressure differences in the initial permeability measurements at different rates are presented in Figure 6. After permeability measurements, formation water was again injected into the core at the rate of 0.3 mL/min to stabilize the differential pressure proportional to this rate. Then, non-magnetized or magnetized water injection process was started at the same rate (0.3 mL/min), and differential pressure and temperature data were collected by FDS350 program during the test. In this step, damage permeability is calculated by Darcy's law.

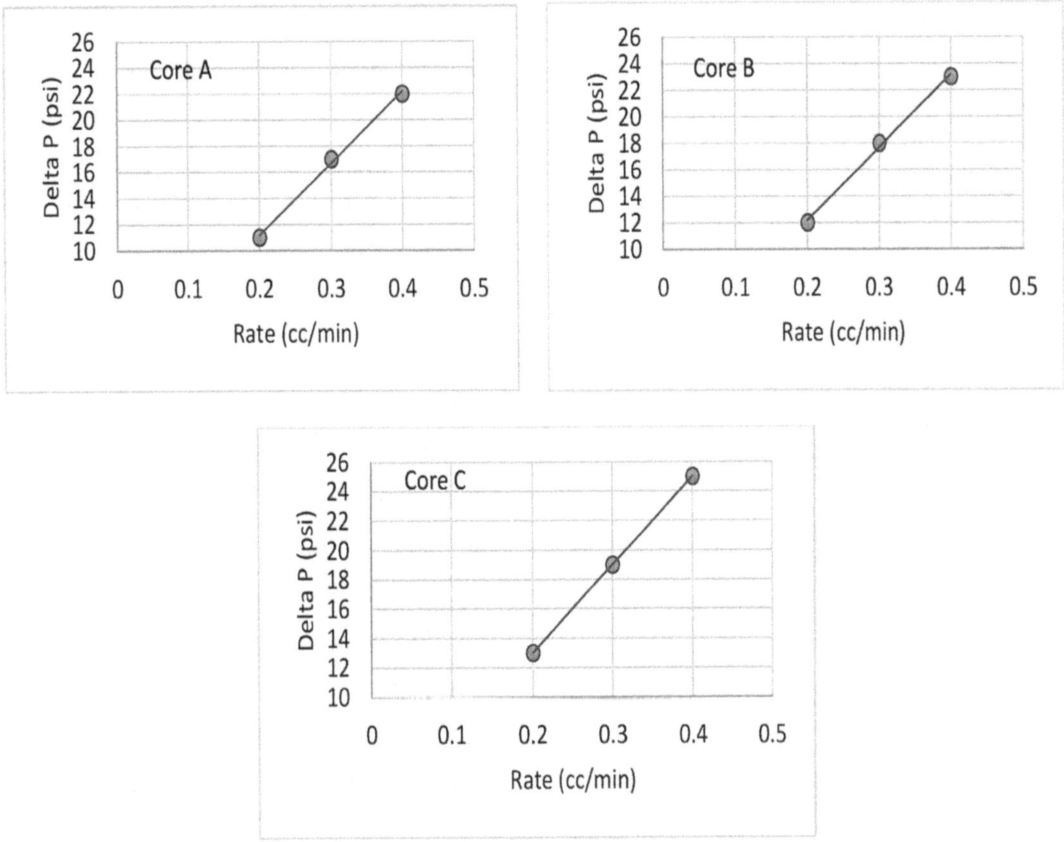

Figure 6: Differential pressure of cores at initial permeability measurement.

For producing magnetized water, a pump of FDS350 was coupled with the magnetic treatment device. Injection water was circulated through the magnetic treatment device by using the pump about one hour for each test.

In this study, three main experiments were considered, as listed below:

1. Injecting non magnetized injection water;
2. Injecting magnetized formation water (magnetic field =3000 G);
3. And injecting magnetized injection water (magnetic field = 4500 G).

The experiments were performed at a temperature of 50 °C, and the confining pressure during each test was 500 psi above inlet pressure (i.e. the effective stress is 500 psi).

RESULTS AND DISCUSSION

The main purpose of this part of investigation is to study rock permeability decline caused by scale deposition during magnetized water injection. As described before, three experiments conducted include non-magnetized water and two different magnetized water injections. All the experiments were conducted by using sandstone cores, which were initially saturated with formation water.

Figures 7-9 present the differential pressure variations versus injection time. These figures were plotted just after injection water was started to be injected at the flow rate of 0.3 mL/min into the core samples. The differential pressure was stabilized to a constant pressure before injecting injection water by the injection of formation water at a rate of 0.3 mL/min.

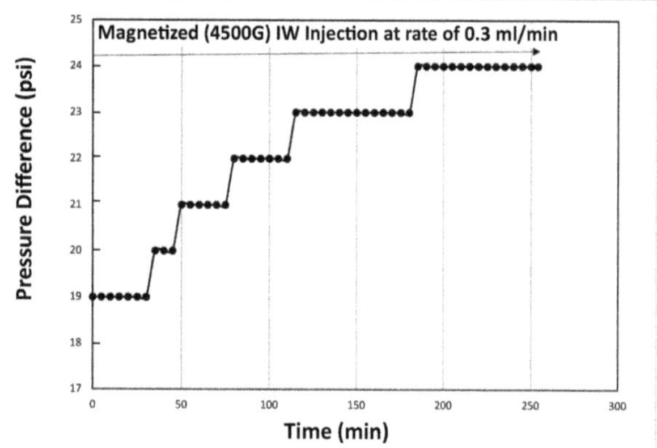

Figure 7: Pressure difference of non-magnetized water injection.

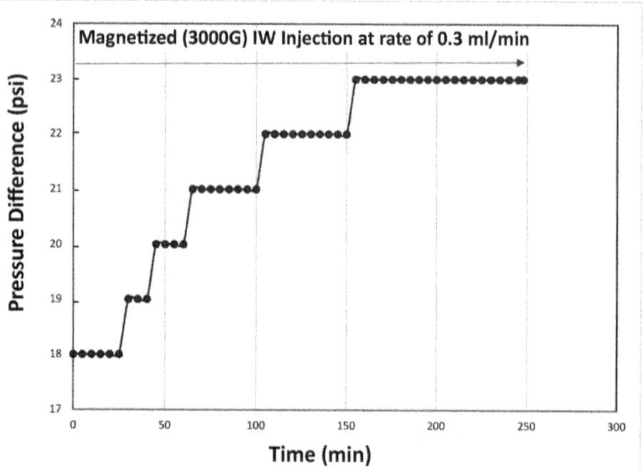

Figure 8: Pressure difference of magnetized water (magnetic field=3000 G) injection.

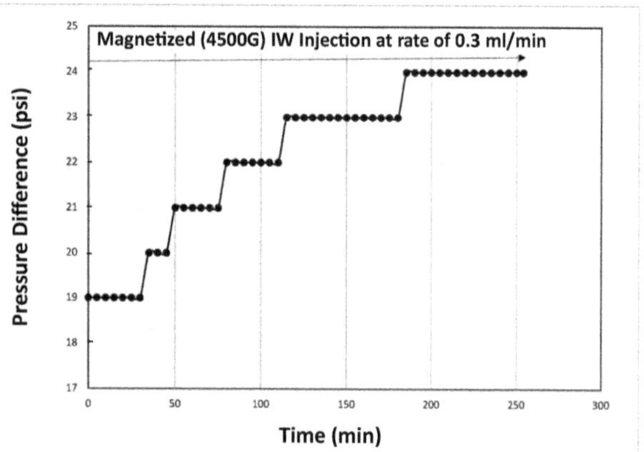

Figure 9: Pressure difference of magnetized water (magnetic field=4500 G) injection.

As shown in Figure 7, at the beginning of non-magnetized water injection into the core samples, the differential pressure was 17 psi; as the injection was consumed, the differential pressure across the core started to build up due to formation water and injection water interaction. This pressure build-up was because of the scale precipitation in the core. As injection process continued, the formation water tends to exist from the core. The interaction of formation water and non-magnetized water in the core was also reduced. Therefore, at the end of the process, the differential pressure was slowly increased. Magnetized water injection also caused damage due to scale precipitation. As shown in Figures 7 and 9, the differential pressure of injection water which was treated with 3000 and 4500 G magnetic field was increased during the process. The effect of magnetized water injection on formation damage cannot be explained by pressure drop curves properly; therefore, for its explanation, permeability damage curves are used in the next part.

The Effect of Magnetized Water Injection on Formation Damage

To investigate the effect of injection water injection on permeability reduction, the variation of permeability ratio was plotted as a function of time. Figure 10 shows the permeability declining trend versus the injection time of the 3000 G magnetized against non-magnetized water injection. At the beginning of flowing period, because of incompatibility of formation water and injection water, a sharp decline in permeability value was observed. The permeability decline rate was decreased during time. The reason for this behavior may be due to the fact that rate of plugging increases as the interaction of more injection water with formation water. This phenomenon was observed in both tests; however, in the presence of a magnetic field, the permeability decline rate was less than the other one. The growth rate of scale crystals is suppressed completely in the presence of a magnetic field. In addition, when water is treated with magnetic fields, the available dipole molecules affect crystals, and crystal growth is limited. When scales cannot grow enough, they may not flocculate and can pass the pore throats, which results in reducing formation damage.

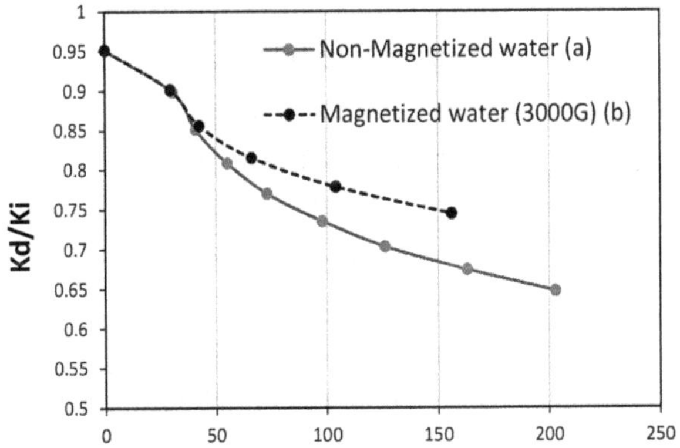

Figure 10: Effect of 3000 Guess magnetized water on formation damage.

The Effect of Magnetic Field Strengths

Herein, the magnetic field flux was increased from 3000 G to 4500 G, while the other conditions were kept constant to eliminate the effect of other parameters. The permeability decline trend was similar to the other tests as shown in Figure 11. Curve "a" in Figure 11 illustrates the permeability decline for 4500 G magnetized water. A decrease in permeability was also observed in this case, which is because of the scale precipitation in core samples due to the incompatibility of formation water and injection water, but this reduction is less than other cases. This result shows that by increasing magnetic field strength, the permeability damage due to scale precipitation is preferentially decreased. The effect of magnetic field strength was obvious in these experiments, and by injecting magnetized injection water to a core, which was saturated with formation water, the permeability improvement was about 10%. As discussed in literature, magnetic field has an effect on the scale precipitation, and numerous mechanisms have been investigated by researchers in this context. For example, Chibowski et al. or Barrett and Parsons have observed that a magnetic treatment applied to hard water decreased the quantity of scale deposited in the well [31, 35]. The principle of the phenomenon is still not well understood, and various contradictory hypotheses have been proposed. Two different approaches, namely magnetohydrodynamic (MHD) phenomena or hydration effects, were reported by Knez and Pohar [36]. MHD phenomena depend on the flow of the magnetized treated water. Busch et al. have assumed that the Lorenz forces $\vec{F} = q.\vec{v} \times \vec{B}$ exerted on charged species induce local convection movements in the liquid which could contribute to accelerating associations among ions [25]. MHD phenomena could also concern the electrical double layer near the charged surface of particles. This interpretation seems receivable for many experimental results of the literature where highly supersaturated waters were treated by a magnetic field. The aggregation of the colloidal particles under the influence of electrostatic phenomena would contribute to the acceleration of the crystal growth and the precipitation process [35].

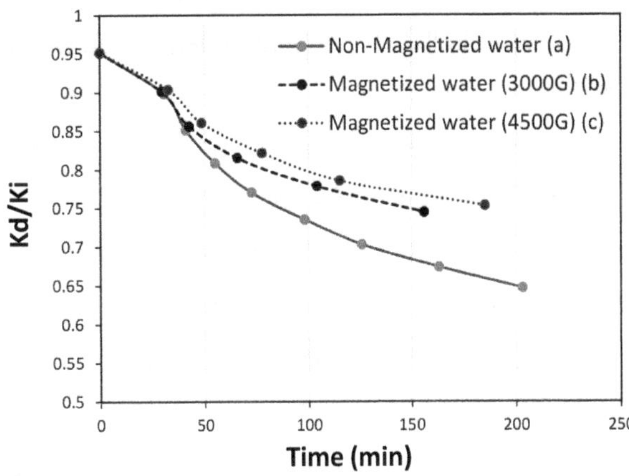

Figure 11: Effect of increasing the magnetic field fluxon formation damage.

There are many researches on the effect of magnetic field on water and scale precipitation, and some theories have been considered for interpreting this phenomenon as discussed. Therefore, the only goal of the researcher was to distribute and investigate this phenomenon in porous media. The results of the experiments confirm the positive effect of magnetized water injection on the permeability improvement when incompatible waters are contacted to each other. However, such research needs more investigation to align this phenomenon with considered theories.

CONCLUSIONS

A magnetic treatment device was used for the laboratory determination of the effect of magnetized water injection on formation damage. The designed magnetic treatment device was easily coupled with the formation damage setup. This device can be used to treat water with different magnetic field fluxes up to 4500 G. Laboratory core flooding experiments were conducted in which two different magnetized water (treated with 3000 and 4500 G magnetic field flux) and non-magnetized water were injected into sandstone core samples. The experiment results confirm that:

• The permeability decline trend of magnetized water injection was similar to that of non-magnetized water injection.

• The permeability reduction of core samples was lower in the presence of a magnetic field.

• It was observed that the injected water, which was treated at a higher magnetic field flux, produced a less decline in permeability, but it was not so significant.

• Increasing magnetic field flux from 3000 G to 4500 G improved permeability by about 2%, while this improvement for 3000 G magnetized water was about 10%.

NOMENCLATURES

μ	: Permeability of magnet (H/m)
$\mu_0 = 4\pi \times 10^{-7}$	: Permeability of free space (H/m)
A	: Cross sectional area of the core (m^2)
B	: Magnetic flux density (Gauss)
c	: Subscript to the core
FW	: Formation Water
g	: Subscript to the air gap
H	: Magnetic field strength (Ampere/m)
i	: Eclectic current (Ampere)
IW	: Injection water
Kd/ Ki	: Formation damage (damaged permeability to Initial permeability)
l	: Length (m)
ϕ	: Magnetic flux (Wb)
N	: Number of turns in winding
R	: Reluctance (ohm)
x	: Device thickness (cm)

REFERENCES

1. Adesina F. and Omole O., “Improved Model for Predicting Formation Damage Induced by Oilfield Scale,” *Advances in Sustainable Petroleum Engineering and Science 2009*, **2009**, *1*(3), 233-246.
2. Atkinson G., Raju K., and Howell R., “The Thermodynamics of Scale Prediction,” *SPE International Symposium on Oilfield Chemistry*, Society of Petroleum Engineers, **1991**.
3. Civan F., “Modeling Well Performance under Nonequilibrium Deposition Conditions,” *SPE Production and Operations Symposium*, Society of Petroleum Engineers, **2001**.
4. Fadairo A. and Falode O. “Predictive Tool for Predicting Sulfate Build up Rate around the Wellbore,” *International Journal of Oil, Gas and*

Coal Technology, **2009**, *2*(4), 347-364

5. Fadairo A., Omole O., and FalodeO., "A Modified Model for Predicting Permeability Damage due to Oilfield Scale Deposition," *Petroleum Science and Technology*, **2009**, *27*(13), 1454-1465.
6. Fadairo A. S., Omole O., and Falode O., "Effect of Oilfield Scale Deposition on Mobility Ratio," *CIPC/SPE Gas Technology Symposium 2008 Joint Conferenc*, Society of Petroleum Engineers, **2008**.
7. Moghadasi J., Kalantari-Dahaghi A. M., and Gholami V., "A New Model to Describe Particle Movement and Deposition in Porous Media," *SPE 99391, Presented at 15th SPE Europe Conference and Exhibition*, Vienna, Austria. **2006**.
8. Oddo J. and Tomson M., "Why Scale Forms in the Oil Field and Methods to Predict It," *SPE Production & Facilities*, **1994**, *9*(01), 47-54.
9. Moghadasi J., Jamialahmadi M., Müller-Steinhagen H., Sharif A., and et al., "Scale Formation in Iranian Oil Reservoir and Production Equipment during Water Injection," *International Symposium on Oilfield Scale*, **2003**.
10. Alimi F., Tlili M., Ben Amor M., and Gabrielli M. B. A., and et al., "Influence of Magnetic Field on Calcium Carbonate Precipitation," *Desalination*, **2007**, *206*(1), 163-168.
11. Gabrielli C., Jaouhari R., Maurin G., and Keddam M., "Magnetic Water Treatment for Scale Prevention," *Water Research*, **2001**, *35*(13), 3249-3259.
12. Lipus L. and Dobersek D., "Influence of Magnetic Field on the Aragonite Precipitation," *Chemical Engineering Science*, **2007**, *62*(7), 2089-2095.
13. Amiri M. and Dadkhah A. A., "On Reduction in the Surface Tension of Water due to Magnetic Treatment," *Colloids and Surfaces A: Physicochemical and Engineering Aspects*, **2006**, *278*(1), 252-255.
14. Brower J., "Magnetic Water Treatment," *Pollution Engineering*, **2005**, *37*(2), 26-28.
15. McMahon C. A., "Investigation of the Quality of Water Treated by Magnetic Fields," University of Southern Queensland, **2009**.
16. Otsuka I. and Ozeki S., "Does Magnetic Treatment of Water Change its Properties?," *The Journal of Physical Chemistry B*, **2006**, *110*(4), 1509-1512.
17. Sueda M., Katsuki A., Nonomura M., and Tanimoto Y., "Effects of High Magnetic Field on Water Surface Phenomena," *The Journal of Physical Chemistry C*, **2007**, *111*(39), 14389-14393.
18. Tomska A. and Wolny L., "Enhancement of Biological Wastewater Treatment by Magnetic Field Exposure," *Desalination*, **2008**, *222*(1), 368-373.
19. Joshi K. and Kamat P., "Effect of Magnetic Field on the Physical Properties of Water," *J. Ind. Chem. Soc.*, **1966**, *43*, 620-622.
20. Parsons S. A., Wang B., Judd S. J., and Stephenson T., "Magnetic Treatment of Calcium Carbonate Scale-effect of pH Control," *Water Research*, **1997**, *31*(2), 339-342.
21. Iwasaka M. and Ueno S., "Structure of Water Molecules under 14 T Magnetic Field," *Journal of Applied Physics*, **1998**, *83*(11), 6459-6461.
22. Nakagawa J. and Shoda M., "Magnetic Field Enhancement of Water Vaporization," *Journal of Applied Physics*, **1999**, *86*(5), 2923-2925.
23. Cho Y. I. and Lee S. H., "Reduction in the

Surface Tension of Water due to Physical Water Treatment for Fouling Control in Heat Exchangers," *International Communications in Heat and Mass Transfer*, **2005**, *32*(1), 1-9.

24. Smith C., Coetzee P., and Meyer J., "The Effectiveness of a Magnetic Physical Water Treatment Device on Scaling in Domestic Hot-water Storage Tanks," *Water SA*, **2004**, *29*(3), 231-236.
25. Busch K. W. and Busch M. A., "Laboratory Studies on Magnetic Water Treatment and their Relationship to a Possible Mechanism for Scale Reductio," *Desalination*, **1997**, *109*(2), 131-148.
26. Tai C. Y., Chang M. C., Shieh R. J., and Chen T. G., "Magnetic Effects on Crystal Growth Rate of Calcite in a Constant-composition Environment," *Journal of Crystal Growth*, **2008**, *310*(15), 3690-3697.
27. Farshad F., Linsley J., Kuznetsov O., and Vargas S., "The Effects of Magnetic Treatment on Calcium Sulfate Scale Formation," *SPE Western Regional/AAPG Pacific Section Joint Meeting*, **2002**.
28. Coey J. and Cass S., "Magnetic water treatment. Journal of Magnetism and Magnetic Materials," **2000**, *209*(1), 71-74.
29. Kobe S., "TEM Examination of the Influence of Magnetic Field on the Crystallization form of Calcium Carbonate: A Magnetic Water-treatment Device," *Acta Chimica Slovenica*, **200,** *48*(1), 77-86.
30. Fathi, A., Tlili M., Claude G., and Maurin G., "Effect of a Magnetic Water Treatment on Homogeneous and Heterogeneous Precipitation of Calcium Carbonate," *Water Research*, **2006**, *40*(10), 1941-1950.
31. Chibowski E., Hołysz L., Szcześ A, and Chibowski M., "Precipitation of Calcium Carbonate from Magnetically Treated Sodium Carbonate Solution," *Colloids and Surfaces A: Physicochemical and Engineering Aspects*, **2003**, *225*(1), 63-73.
32. Tombacz E., Ma C., Busch K. W., and Busch M. A., "Effect of a Weak Magnetic Field on Hematite Sol in Stationary and Flowing Systems," *Colloid & Polymer Science*, **1991**, *269*(3), 278-289.
33. Erickson R. W. and Maksimovic D., "Fundamentals of Power Electronic," Springer, **2001**.
34. Formation Damage Laboratory, Available from: http://petroleum.aut.ac.ir/autcms/labs.
35. Barrett R. A. and Parsons S. A., "The Influence of Magnetic Fields on Calcium Carbonate Precipitation," *Water Research*, **1998**, *32*(3), 609-612.
36. Knez S. and Pohar C., "The Magnetic Field Influence on the Polymorph Composition of $CaCO_3$ Precipitated from Carbonized Aqueous Solutions," *Journal of Colloid and Interface Science*, **2005**, *281*(2), 377-388.

6

Reduction of Reservoir Fluid Equilibrium Calculation for Peng-Robinson EOS with Zero Interaction Coefficients

Mahdi Assareh*

Thermodynamics Research Laboratory, School of Chemical Engineering, Iran University of Science and Technology, Tehran, Iran

ABSTRACT

For some of the EOS models the dimension of equilibrium problem can be reduced. Stability and difficulties in implementation are among the problems of flash calculation. In this work, a new reduction technique is presented to prepare a reduced number of equilibrium equations. Afterwards, a number of appropriate solution variables are selected for the prepared equation system to solve the equations in an efficient numerical scheme. All the derivatives and solution procedures for the new reduced flash calculation framework were prepared based on Peng-Robinson equation of state. One reservoir oil sample and one gas condensate sample were selected from published literature to evaluate the proposed method for the calculations of reservoir fluids equilibrium. The equilibrium calculations with the proposed reduction technique were compared to full flash calculations. The reduced formulation implementation is simple and straightforward as it is derived from full flash fugacity equality criteria. The presented technique not only reduces the number of equations, and hence simplifies flash problem, but also presents a comparable convergence behavior and offers the same solution system for different reservoir fluid types. The results, demonstrates the proposed method performance and the accuracy for modeling with complex equilibrium calculations like compositional reservoir simulation when there are many components available in the mixture fluid description.

Keywords: Reduction, Flash calculation, Reservoir Fluids, Peng-Robinson, Cubic Equation of States

INTRODUCTION

Stability analysis and equilibrium calculations are repeated several times for many reservoir simulation cells for the field studies with compositional reservoir simulation using equations of state (EOS). It is shown that that equilibrium calculation is an important part of the CPU time, depending discretization, non-linearity, process complexity and degree of implicitness [1].

For EOS models, the dimensionality of equilibrium problem can be reduced. Originally reduction techniques were introduced for the mixtures in which all the binary interaction coefficients are zero. Michelsen (1986) found that the dimensionality of the flash calculation by cubic EOS with zero interaction coefficients is three [2]. He explained that zero binary interaction coefficients permit very significant simplifications in the calculation of phase equilibrium for multicomponent mixtures, in particular when the

***Corresponding author**
Mahdi Assareh
Email: assarehm@iust.ac.ir

number of components is high. He proposed to solve three equations with three unknowns to find the equilibrium ratios. This means that the dimensionality of the two phase flash problem is independent of the number of components for cubic EOS. He mentioned that each component having non-zero interaction coefficients with the other components introduces two new equations in case that the added binary interaction coefficient (BIC) is between two components other than previous BIC's component. Two BIC's with one common component add 3 more equations. Michelsen selected three errors (equations) to solve for final equilibrium composition for each phase. In Michelsen method, an initial estimation of the reduced flash variables is used to start calculations. Assuming zero binary interaction coefficients, n_v (the number of moles in vapor phase), a_v and b_v are selected as variables ($\zeta=[n_v, a_v, b_v]^T$):

$$a_v=\sum_{i=1}^{nc}\sum_{j=1}^{nc} y_i y_j (a_j a_j)^{1/2}=(\sum_{i=1}^{nc} y_j (a_j)^{1/2})^2=(\alpha_v)^2 \quad (1)$$

$$b_v=\sum_1^{nc} y_j b_j \quad (2)$$

The liquid phase parameters can be calculated based on vapor phase parameters and feed parameters:

$$\alpha_l=(\alpha_f - n_v\alpha_v)/(1-n_v) \quad (3)$$

$$b_l=(b_f - n_v b_v)/(1-n_v) \quad (4)$$

Michelsen selected three errors (following equations) to solve for final equilibrium composition of each phase ($g=[g_1, g_2, g_3]^T$):

$$g_1=\sum_1^{nc}(u_j^l-u_j^v) \quad (5)$$

$$g_2=\sum_1^{nc}u_j^v\alpha_{iv}-\alpha_v \quad (6)$$

$$g_3=\sum_1^{nc}u_j^v b_{iv}-b_v \quad (7)$$

In Michelsen method, an initial estimation of reduced flash variables is used to start calculations. Based on these variables and with the help of Equations 3 and 4, the fugacity coefficients are calculated to update liquid and vapor composition, namely u_j^v and u_j^l. The new compositions are used to solve the reduced equations. The reduced variables are updated after solution and this cycle is repeated until convergence is obtained. However, for gas condensate systems, it is recommended that the corresponding variables be selected for the liquid phase. This recommendation comes from different sets of equations and different variables selection. It should be emphasized that all of these equations are in fact different versions of the equality of chemical potential equations; therefore, if it is possible to have an analytical solution for the flash calculation problem, the solution must then be the same. But since the flash calculation problems are solved numerically, the selection of the equations and solution variables and numerical scheme for it can affect the final calculation results. Based on these variables, the fugacity coefficients are calculated to update liquid and vapor composition.

Jensen and Fredenslund (1987) extended the Michelsen method for the situations where one non-zero interaction coefficient exists in the mixture by solving a system of only five equations [3]. They assumed one BIC for non-hydrocarbon–hydrocarbon interactions. As previously discussed, the non-zero BIC brings two equations in addition to 3 existing equations. This procedure is advantageous for CO_2 and nitrogen flooding in the reservoirs. To solve the equations, they selected five variables.

The variables in the flash problem are vapor mole fraction, n_v, and first component (non-hydrocarbon component) mole fraction as well as:

$$al_v=\sum_2^{nc} y_j a_j^{1/2} \quad (8)$$

$$dl_v=\sum_2^{nc} y_j a_j^{1/2} Kij \quad (9)$$

$$gl_v=\sum_2^{nc} y_j b_j \quad (10)$$

With these variables, the attraction and repulsion terms for gas phase can be calculated as:

$$a_v = x_1^2 a_1 + 2x_1 a_1^{1/2} (al_v - dl_v) + al_v^2 \quad (11)$$

$$b_v = x_1 b_1 + gl_v \quad (12)$$

In this relations, x_1 and y_1 are the mole fraction of the first component (non-hydrocarbon component). The following equations are used to calculate corresponding parameters for liquid phase:

$$al_l = (al_f - n_v al_v)/(1-n_v) \quad (13)$$

$$dl_l = (dl_f - n_v dl_v)/(1-n_v) \quad (14)$$

$$gl_l = (gl_f - n_v gl_v)/(1-n_v) \quad (15)$$

$$x_1 = (z_1 - y_1 n_v)/(1-n_v) \quad (16)$$

where, al_v, dl_v, gl_v, y_1, and n_v are taken as the solution variables $(\zeta=[n_v, al_v, dl_v, gl_v, y_1]^T)$. An initial guess is used to start reduced flash procedure. The corresponding liquid phase parameters are computed according to Equations 13-16. Afterwards, the liquid and vapor compositions are calculated with fugacity equations using above variables. Jensen and Fredenslund used the following set of equations to be solved for the five variables $(g=[g_1, g_2, g_3, g_4, g_5]^T)$:

$$g_1 = \Sigma_2^{nc} u_j a_j^{1/2} - al_v \quad (17)$$

$$g_2 = \Sigma_2^{nc} u_j a_j^{1/2} K_{1i} - dl_v \quad (18)$$

$$g_3 = \Sigma_2^{nc} u_j b_j - gl_v \quad (19)$$

$$g_4 = u_1 - y_1 \quad (20)$$

$$g_5 = \Sigma_{j=1}^{N} u_j - y_j \quad (21)$$

The updated five variables after equation solution are then used to calculate the liquid and gas composition and the loop is repeated until the convergence. Therefore, again one of the drawbacks of Jessen and Fredenslund method is the different sets of equations relative to full equilibrium equations. This causes different solution behaviors for different thermodynamic conditions and reservoir fluid types relative to full flash approach as it is described in Michelsen technique.

Hendriks (1988) presented a general mathematical theorem in phase equilibrium equations to show how for a well-defined class of thermodynamic models (e.g. cubic EOS) the dimensionality of various phase equilibrium problems can be reduced [4].

Li and Johns developed an effective method for reduced flash calculation in cases that non-zero interaction coefficients are existing in the fluid description [5]. They presented a new method to increase the speed of flash calculations for any numbers of nonzero BIP's. The approach requires the solution of up to six reduced parameters regardless of the number of components and is based on decomposing the BIP's into two parameters using a simple quadratic expression. The new approach is exact in that the equilibrium-phase compositions for the same BIP's are identical to those with the conventional flash calculation; no eigenvalue analysis is required. However, the calculated equilibrium ratios are estimations of full flash equilibrium ratios. The drawback of such approach is that the equilibrium ratios are the estimation of full flash equilibrium ratio. Different applications of the reduction methods have been investigated by Nichita et al. to prove their exactness for a wide range of equilibrium calculation problems such as stability analysis, saturation pressure, extrapolation of the results of previously performed flashes with different application from synthetic hydrocarbon to real petroleum reservoir mixtures, and multiphase flash calculations [6].

Hoteit and Firoozabadi (2006) presented an algorithm for the stability analysis in the reduction method. In their technique, the Newton method is used in the solution of nonlinear equations except in some isolated iterations when one single successive substitution iteration may be required to avoid nonphysical conditions [7].

Nichita and Graciaa (2011) developed a general workflow for flash calculation with reduction techniques for cubic EOS in cases that there are more than one non-zero interaction coefficients [8]. Also, comparisons of the proposed numerical schemes for reduction techniques are available in the literature [9].

They described that, in a classical reduction method for phase equilibrium calculations, there are M reduction parameters, $Q_p=(Q_{p1},Q_{p2},\ldots,Q_{pM})^T$:

$$Q_p=\sum_{i=1}^{nc} q_{\alpha i} x_{ip},\ \alpha=1, M,\ p=L \text{ or } V \quad (22)$$

The M+1 error equations are given by:

$$e_\alpha=\sum_{i=1}^{nc} q_{\alpha i} x_{ip}-Q_{p\alpha},\ \alpha=1, M,\ p=L \text{ or } V \quad (23)$$

$$e_{M+1}=\sum_{i=1}^{nc} (y_i-x_i)=0 \quad (24)$$

choice between these two is made depending on which phase is the least abundant.

Gaganis et al. (2012) presented a general framework for developing solutions to Rachford-Rice equation utilizing function classes, the members of which exhibit a behavior similar to that of the equation to be solved. Fitting such functions, instead of a simple straight line at each iteration of the solution process, as is the case of the Newton–Raphson approach, leads to better subsequent estimates and thus to faster convergence [10]. In this approach, the calculated equilibrium ratios with the reduced technique are the estimation of full flash equilibrium ratios.

Gaganis and Varotsis (2013) replaced the conventional spectral decomposition [8] using basis vectors by new ones so that the approximation error of the energy parameter is minimized. The new reduced variables set leads to improved flash calculations accuracy, thereby allowing them to be performed at a given accuracy level using fewer reduced variables compared to the conventional approach [11].

Gorucu and Johns (2014) presented new reduced parameters using the two-parameter binary interaction parameter formula originally proposed by Li and Johns (2006) [12]. The new reduced parameters were employed to solve two-phase flash calculations for five different fluid descriptions. Later on, an applied phase-related equilibrium (APPLE) solver using only the Peng–Robinson equation of state was developed based on rigorous classical thermodynamics [13] with a theoretically and thermodynamically consistent solver with the stringent equilibrium criterion.

Gaganis and Varotsis (2014) presented an integrated approach to provide direct answers to both the phase stability and phase split problems during compositional modeling using a regression approach. The non-linear transformation of the equilibrium equations is combined to explicit expressions, which are generated to provide directly the solution of the phase-split problem within the reduced variables. Regression models are developed so as to estimate pressure and temperature to the prevailing values of the reduced variables set with a direct function relating feed composition [14].

Petitfrere and Nichita (2015) provided an improvement of two-phase reduced flash calculations [15] to multiphase equilibrium calculation with any number of phases. They proposed their algorithms with two approaches, a direct extension of the reduction for two-phase flashes and a Gibbs free energy constrained minimization. They examined their proposed algorithms for several multiphase systems containing hydrocarbon components, carbon dioxide, and hydrogen sulfide. Numerical experiments showed that the reduction methods for multiphase flash calculations become faster than the conventional methods when the number of components increases, when the number of equilibrium phases increases, and when the number of BIC decreases. They claimed that for samples with many components and few BIC's, their method can improve the equilibrium calculation performance [16].

Petitfrere and Nichita (2015) evaluated stability testing and flash calculations separately. It is shown that reduction methods are more efficient than conventional techniques, so they are suitable for compositional simulation. This is beneficial specially because, in most of the reservoir simulators, no explicit phase split is required and equilibrium calculations are parts of a larger problem (coupled with flow equations), while phase stability analysis must be performed in most grid blocks [17].

The non-linear nature of phase equilibrium calculations requires an iterative solution procedure. Stability and difficulties in implementation are among the problems with flash calculation reduction [18]. In this work, a new reduction technique is presented. In this respect, a reduced number of equilibrium equations was prepared. Afterwards, appropriate solution variables are selected for the prepared equations system to solve the equations in an efficient numerical scheme. All the derivatives and solution procedures for the new reduced flash calculation framework were prepared based on Peng-Robinson equation of state. The proposed approach has two advantages over conventional reduction technique. The first is the unique set of equations and variables for different types of the reservoir fluids. Secondly, having the same equations and the same variables ensures that the convergence behavior of the reduced flash numerical scheme is the same as full flash scheme, and there is no need to change the equation systems and solution variables. Besides, the numerical scheme thresholds can be provided directly from the full flash approach. The other advantage of the reduction technique is that the formulations are just extracted from the full flash algorithms; therefore, there is no considerable difficulties regarding implementation.

Reservoir Fluid Samples

Three main hydrocarbon groups, namely paraffins, naphthenes, and aromatics, are usually available in any petroleum reservoir fluid mixtures. Paraffins are divided into two groups of normal-paraffins and iso-paraffins. Naphthenes or cycloalkanes are rings or cyclic saturated hydrocarbons. Cyclo-pentane, cyclohexane, and their derivatives are normally found in crude oils. The content of naphthenes in petroleum may vary up to 60% [19]. Aromatics are an important series of hydrocarbons found in almost every petroleum mixture from any area of the world. Usually a sample of petroleum fluid is characterized as discrete components, petroleum cuts, and a plus fraction. Each fraction contains paraffinic, naphthenic, and aromatic components. Usually the fractions molecular weight, specific gravity, and normal boiling point are measured and reported. Specific gravity (S) and molecular weight (MW) are useful parameters for the characterization of petroleum fluids to determine PNA composition and also to estimate other properties such as critical constants, density at various temperatures, viscosity etc. In this work, the new reduction technique on equilibrium calculations is examined for two samples of reservoir fluid. Sample 1 (S1) is an oil reservoir fluid characterized with 24 components. Samples S1 data were taken from published data [20]. The sample characterization data are presented in Table 1. The data for sample 2 is obtained from North Sea gas condensate by Leibovici et al. (1993) including 27 components [6]. The characterization data are reported in Table 2 for this sample

Table 1: S1, Fluid Characterization

Components	ZI (percent)	MW	P_c (bar)	T_c (K)	AF	V_c (m^3/kg.mol)
N_2	0.1800	28.0130	33.9440	126.2000	0.0400	0.0900
CO_2	0.8200	44.0100	73.8660	304.7000	0.2250	0.0940
C_1	22.9200	16.0430	46.0420	190.6000	0.0130	0.0980
C_2	7.2100	30.0700	48.8390	305.4300	0.0986	0.1480
C_3	7.3700	44.0970	42.4550	369.8000	0.1524	0.2000
IC_4	1.5800	58.1240	36.4770	408.1000	0.1848	0.2630
nC_4	5.2300	58.1240	37.9660	425.2000	0.2010	0.2550
IC_5	2.2500	72.1510	33.8930	460.4000	0.2270	0.3080
nC_5	3.6000	72.1510	33.7010	469.6000	0.2510	0.3110
C_6	4.8400	86.0000	31.7970	500.4300	0.2675	0.3551
C_7	4.7200	92.0000	31.8080	523.8600	0.2913	0.3721
C_8	4.5200	102.0000	30.4830	552.2700	0.3254	0.4068
C_9	4.0700	117.0000	28.1190	586.8300	0.3752	0.4623
C_{10}	3.5000	134.0000	25.5580	619.6700	0.4315	0.5271
C_{11}	2.9100	151.0000	22.7690	642.7500	0.4898	0.5977
C_{12}	2.4200	165.0000	21.0210	662.6900	0.5370	0.6539
C_{13}	1.9900	178.0000	19.6780	681.2900	0.5801	0.7050
C_{14}	1.7400	194.0000	18.2980	704.1200	0.6323	0.7666
C_{15}	2.0200	208.0000	17.4260	726.2000	0.6757	0.8174
C_{16}	1.3900	225.0000	15.9750	743.9100	0.7322	0.8871
C_{17}	1.2100	239.0000	14.9500	758.4800	0.7780	0.9439
C_{18}	0.9000	249.0000	14.3960	770.6900	0.8092	0.9820
C_{19}	1.1700	270.0000	13.2320	793.3500	0.8754	1.0643
C_{20+}	11.4400	560.0000	4.9241	1021.2000	1.5499	2.2181

Table 2: S2 fluid characterization

Components	ZI (percent)	T_c (K)	P_c (Bar)	AF	MW
N_2	0.3243	126.2000	33.9400	0.0400	28.0100
CO_2	1.9549	304.2000	73.7700	0.2250	44.0100
C_1	76.1530	190.6000	46.0000	0.0115	16.0400
C_2	7.7367	305.4000	48.8400	0.0908	30.0700
C_3	3.6177	369.8000	42.4500	0.1454	44.1000
iC_4	0.5667	408.1000	36.4800	0.1760	58.1200
nC_4	1.3258	425.2000	38.0000	0.1928	58.1200
iC_5	0.4480	460.2600	33.8300	0.2271	72.1500
nC_5	0.6100	469.6000	33.7400	0.2273	72.1500
CC_5	0.0580	511.6000	45.0900	0.1923	70.1400
PC_6	0.6001	503.7900	30.0700	0.2860	86.1800
CC_6	0.4148	547.4100	39.9000	0.2215	84.1600
AC_6	0.1735	562.1000	48.9400	0.2100	78.1100
PC_7	0.4239	536.4400	27.6000	0.3364	100.2100
CC_7	0.5056	566.2700	34.6900	0.2451	98.1900
AC_7	0.3063	591.7000	41.1400	0.2566	92.1400
PC_8	0.3291	565.0500	25.0200	0.3816	114.2300
CC_8	0.2540	594.0500	29.7400	0.2391	112.2100
AC_8	0.2564	619.4600	35.8400	0.3228	106.1600
PC_9	0.2630	590.6400	23.2900	0.4230	128.2500
CC_9	0.1640	621.2100	28.3900	0.2998	125.9700
AC_9	0.1217	644.0600	32.0800	0.3725	120.1600
PC_{10}	0.2411	613.7200	21.4600	0.4646	142.2800
CC_{10}	0.0427	621.5800	26.2500	0.4058	140.2000
AC_{10}	0.1049	670.8300	29.7200	0.3642	133.8000
CN_{-1}	2.6703	711.0400	18.7500	0.8000	240.0000
CN_{-2}	0.3338	848.0800	16.3300	1.3000	450.2600

Reduction Technique

For a two phase flash calculation problem, the following system of nonlinear equations is solved:

$$g_i = \ln K_i + \ln\varphi_i^v(y,P,T) - \ln\varphi_i^l(x,P,T) = 0 \quad i = 1,\ldots,n_f \quad (25)$$

where, φ_i denotes the fugacity coefficient of component *i* calculated using EOS. To calculate the fugacity coefficients for each phase *p*, one should recall PR EOS. If we define the following constants:

$$p_1 = -2\sqrt{2},\; p_2 = 1-\sqrt{2},\; p_3 = 1+\sqrt{2} \quad (26)$$

Compressibility factor is calculated based on PR EOS as for each phase *p*:

$$Z_p^3 - (1-B_p)Z_p^2 + (A_p - 2B_p - 3B_p^2)Z_p - (B_pA_p - B_p^2 - B_p^3) = 0 \quad (27)$$

In this cubic equation of state, capital EOS parameters, A_p and B_p) are calculated according to small EOS parameters, a_p and b_p:

$$A_p = \frac{a_p}{(RT)^2},\; B_p = \frac{b_p}{(RT)^2} \quad (28)$$

The small EOS parameters are resulted from *VdW* mixing rules, if the binary interaction coefficients are zero

$$a_p = \sum_{i=1}^{nf}\sum_{j=1}^{nf} x_{pi}x_{pj}\sqrt{a_ia_j},\; b_p = \sum_{i=1}^{nf} x_{pi}b_i \quad (29)$$

If binary interaction coefficients are not zero, the outcome of the reduced flash problem can be used as an initial estimate of the full flash problem. This makes advantages when we are dealing with near critical fluid with lots of components in the mixture. Moreover, usually the petroleum reservoir fluids contain non-well-defined fractions and plus fractions. The critical and thermos-physical properties of such fractions are estimated through correlations and readjusted through tuning procedures. That is why there are cases in which the petroleum engineers prefer to tune these estimated properties instead of binary interaction coefficients to match saturation pressure and PVT experiments. The fugacity coefficients for phase p for each component are given by:

$$Ln\varphi_i^p = \frac{b_i}{b^p}(Z^p - 1) - Ln(Z^p - B^p) - \frac{A^p}{B^pP_1}\left(\frac{2s^p}{a^p} - \frac{b_i}{b^p}\right).Ln\left(\frac{Z^p + B^pp_2}{Z^p + B^pp_3}\right) \quad (30)$$

in which,

$$s^p = \sum_{j=1}^{nc} x^p{}_j a_j^{0.5} \quad (31)$$

The fugacity coefficients for each phase can be decomposed into three parts:

$$Ln\varphi_i^p = A_1p + A_2^pb_i / bmix + A_3^pa_i^{0.5} \quad (32)$$

The coefficients, required for each component, are:

$$A_1^p = -Ln(Z^p - B^p) \quad (33)$$

$$A_2^p = bmix.(\frac{(Z^p - 1)}{b^p} + \frac{A^p}{B^pb^pp_1}Ln\left(\frac{Z^p + B^pp_2}{Z^p + B^pp_3}\right) \quad (34)$$

$$A_3^p = -\frac{A^p}{B^pp_1}\left(\frac{2s^p}{a^p}\right)Ln\left(\frac{Z^p + B^pp_2}{Z^p + B^pp_3}\right) \quad (35)$$

In this relation, *bmix* has a fixed value and calculated based on feed mole fractions:

$$bmix = \sum_{i=1}^{NC} z_ib_i \quad (36)$$

(This constant factor is added so that the magnitude of A_1^p, A_2^p, and A_3^p changes to the same order required for future calculation. By defining the below equations, one can easily calculate the equilibrium ratios (Equation 40):

$$\zeta_1 = A_1^L - A_1^V \quad (37)$$

$$\zeta_3 = A_3^L - A_3^V \quad (38)$$

$$\zeta_2 = A_2^L - A_2^V \quad (39)$$

$$\ln K_i = \zeta_1 + \zeta_2.\frac{b_i}{bmix} + \zeta_3.a_i^{0.5} \quad (40)$$

This equation provides the framework for the reduction technique. The liquid and vapor compositions are changing in each non-linear iteration of flash problem. Therefore, in each iteration, ζ_1, ζ_2, and ζ_3 are updated, and consequently the equilibrium ratios are updated; at the end of flash calculation, there is no more change in equilibrium ratios and in these parameters. Generally, in conventional flash calculation algorithms, the K-values in each iteration are updated with the following relation:

$$\delta Lnk = \begin{bmatrix} \delta \ln K_1 \\ \delta \ln K_2 \\ \vdots \\ \delta \ln K_{nf} \end{bmatrix} = \begin{bmatrix} \frac{\partial g_1}{\partial \ln K_1} & \frac{\partial g_1}{\partial \ln K_2} & \cdots & \frac{\partial g_1}{\partial \ln K_{nf}} \\ \frac{\partial g_2}{\partial \ln K_1} & \frac{\partial g_2}{\partial \ln K_2} & \cdots & \frac{\partial g_2}{\partial \ln K_{nf}} \\ \vdots & \vdots & \ddots & \vdots \\ \frac{\partial g_{nf}}{\partial \ln K_1} & \frac{\partial g_{nf}}{\partial \ln K_2} & \cdots & \frac{\partial g_{nf}}{\partial \ln K_{nf}} \end{bmatrix}^{-1} \cdot \begin{bmatrix} g_1 \\ g_2 \\ \vdots \\ g_{nf} \end{bmatrix} \quad (41)$$

where, $\frac{\partial g_i}{\partial \ln K_j}$ can be calculated as:

$$\frac{\partial g_i}{\partial \ln K_j} = \delta_{ij} + \frac{\partial \ln \phi_i (y, P, T)}{\partial y_j} K_j \left[\frac{(1-n_v)}{(1-n_v)+n_v K_j} \right] + \frac{\partial \ln \phi_i (x, P, T)}{\partial x_j} K_j \left[\frac{n_v}{(1-n_v)+n_v K_j} \right] \quad (42)$$

in which, δ_{ij} is defined as:

$$\delta_{ij} = \begin{cases} 0 & i \neq j \\ 1 & i = j \end{cases} \quad (43)$$

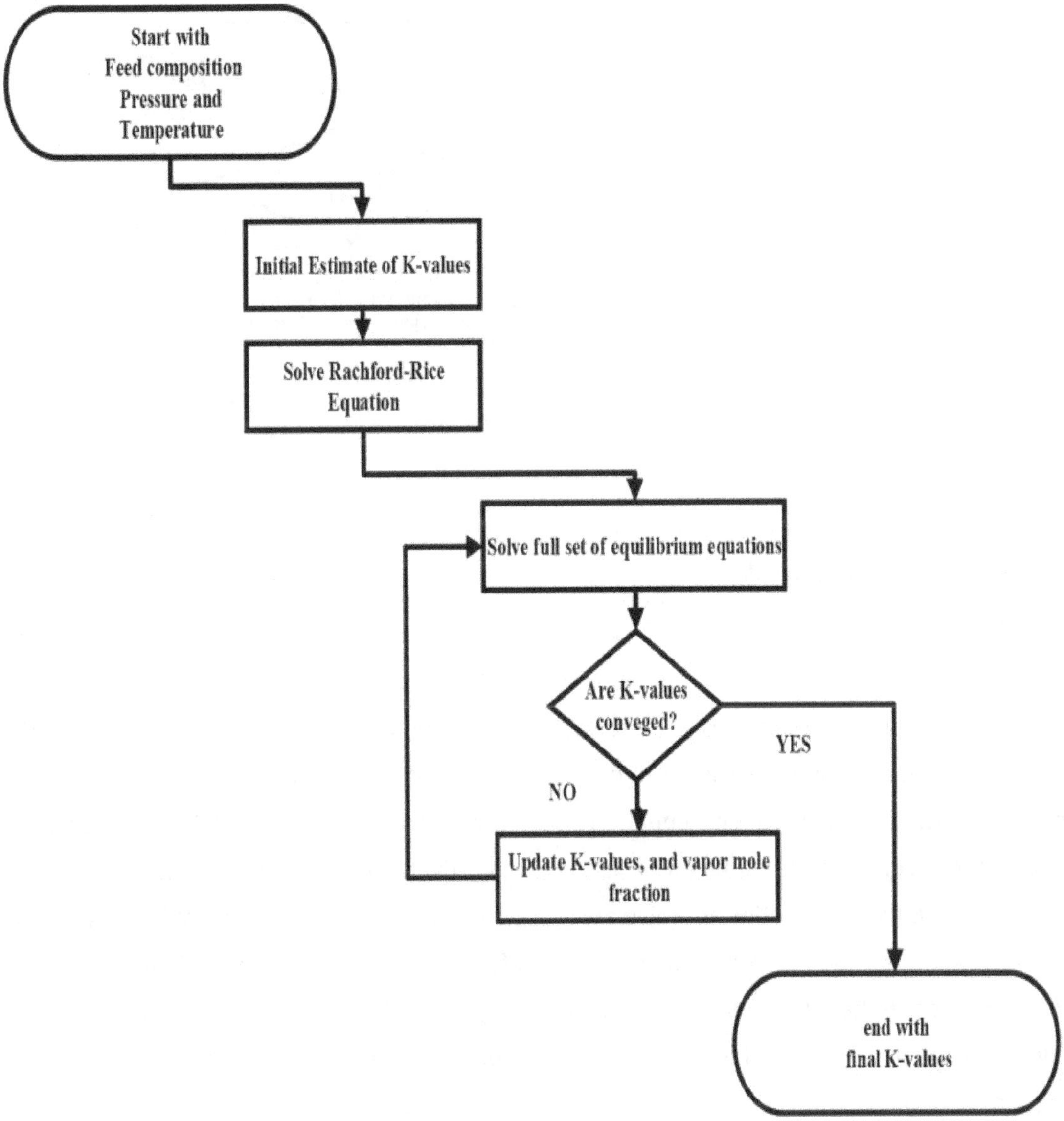

Figure 1: Flowchart of the full flash as a basis for reduced flash.

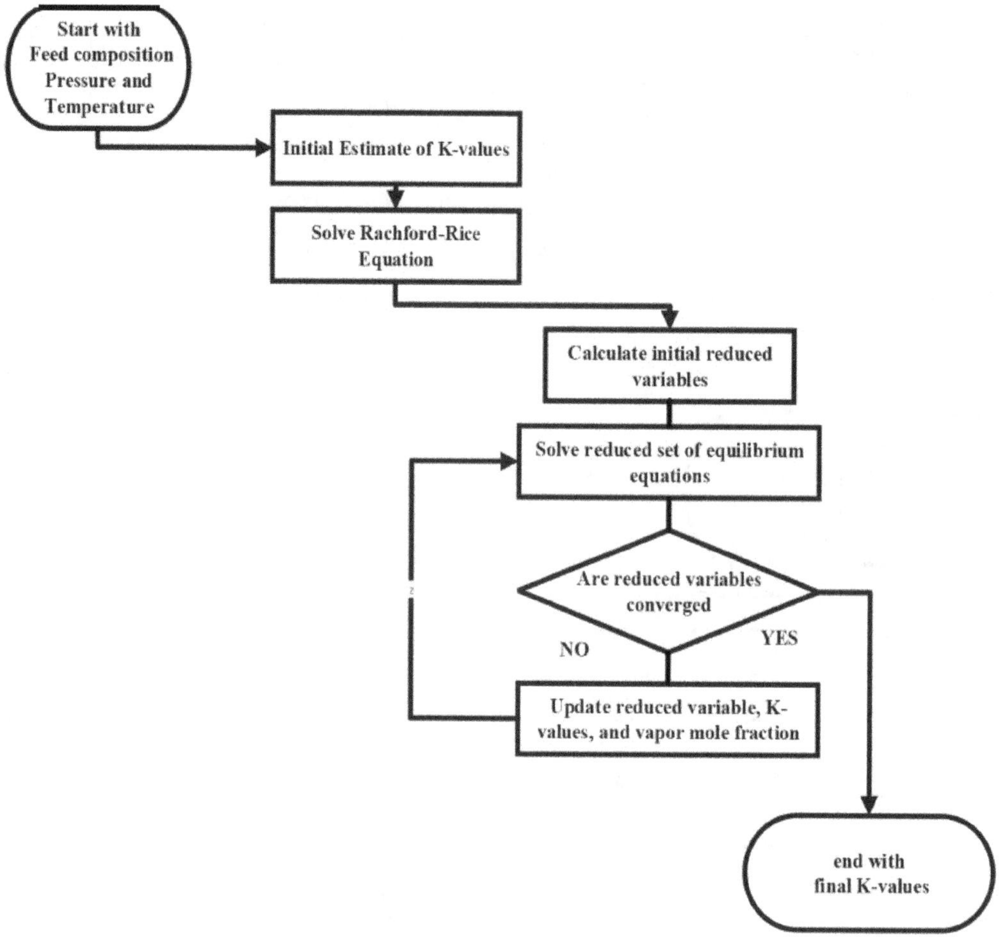

Figure 2: Flowchart of the reduced flash.

For which, the required derivatives are derived in the appendix. It should be noted that the equilibrium equations and Rachford-Rice equation can be solved sequentially. This means that with a set of updated equilibrium ratios, the Rachford-Rice equation is solved, and the resulting mole numbers of vapor of the next set of equilibrium ratios are calculated. In this way, $n_v = 1-n_l$ can be regarded independent of equilibrium ratio in the internal loop for K-value updating. The work flow for the full flash calculation is presented in Figure 1.

The Rachford-Rice equation, a nonlinear equation, is solved for the molar fraction of each phases using Newton-Raphson method:

$$f(n_l) = \sum_{i=0}^{NC}(y_i - x_i) = \sum_{i=0}^{NC} z_i(k_i - 1) / (n_l + (1 - n_l)k_i) = 0, \quad (44)$$

This equation is solved in an external loop. The solution to flash calculation problem is started with pressure, temperature, and feed composition. The initial estimate of K-Values is provided from Wilson correlation. The Rachford-Rice equation is solved iteratively until the mole number of vapor phase does not change anymore. With this in hand, the full set of equilibrium equations, as formulated in Equation 25, is solved based on a Newton approach for a set of non-linear solution using Equation 41. In the updating step, the difference in K-values is simply added to the K-values of the previous step. It should be mentioned that to update the mole number of vapor phase, there is no need to completely solve the Rachford-Rice in each iteration. One can use the following relation (with no need to iterate):

$$nv^{new} = nv^{old} - f(nv) / f'(nv) \quad (45)$$

where, *f(nv)* and *f ′(nv)* are calculated using Equations 44. Now if one considers Equation 40, all the equilibrium constants are functions of ζ_1, ζ_2, and ζ_3. ζ_1, ζ_2, and ζ_3 are the same for all of the mixture components. Therefore, one can set up the following set of variables:

$$\zeta_r^T = [\zeta_1, \zeta_2, \zeta_3]_{nr} \quad (46)$$

In fact, if one could find the above vector, the K-values can be exactly calculated. Consequently, if there is no non-zero interaction coefficient, the problem dimensionality is 3. The proper reduced equilibrium equations to be solved for the reduced flash calculation are defined as:

$$g_r^T = \left[g_{IK(1)}, g_{IK(2)}, g_{IK(3)} \right]_{nr} \quad (47)$$

where, *IK* returns the serial number of related equilibrium equation according to Equation 25. For components with zero interaction coefficients, the first three equations can be selected from three optional components. The solution of the reduced system is:

$$\delta\zeta_{nr} = [\delta\zeta_1, \delta\zeta_2, \delta\zeta_3]_{nr}^T = -J_{nr*nr}^{-1} \cdot g_{nr} \quad (48)$$

To solve the reduced equation with the reduced number of variables, one needs to have the below relation for Jacobian calculation (J_{nr*nr} is extracted from J_{nf*nr}):

$$J_{nf*nr} = \left[\frac{\partial g}{\partial \zeta} \right]_{nf.nr} = \left[\frac{\partial g}{\partial Lnk} \right]_{nf.nf} \cdot \left[\frac{\partial Lnk}{\partial \zeta} \right]_{nf.nr} \quad (49)$$

where, $[\partial g / \partial Lnk]_{nf.nf}$ is calculated from Equation 41. The second part of the calculation of these derivative, $[\partial Lnk / \partial \zeta]_{nf.nr}$, is in fact constant during flash iteration and is defined as the following relation:

$$\frac{\partial g_i}{\partial \zeta_1} = 1 \quad (50)$$

$$\frac{\partial g_i}{\partial \zeta_2} = \frac{b_i}{bmix} \quad (51)$$

$$\frac{\partial g_i}{\partial \zeta_3} = a_i^{0.5} \quad (52)$$

The initial estimate of K-Values and the mole number of vapor phase are calculated based on Wilson and Rachford-Rice equation correspondingly. These values are then used as the basis for the initial estimation of the reduced variables. Afterwards, the reduced equation system in Equation 48 is solved to calculate the new reduced variables. These reduced set of variables are used to update the new full set of equilibrium ratio. The schematic diagram is shown in the Figure 2.

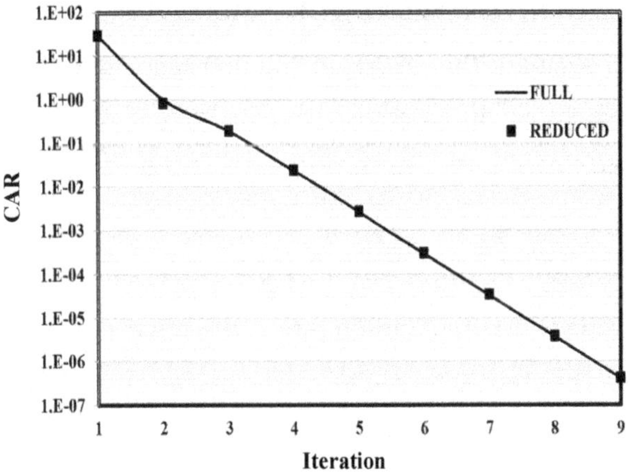

Figure 3: Convergence behavior of full flash calculation and the reduced technique for cumulative absolute residual (CAR); S1; P=50 bar; T= 550 K.

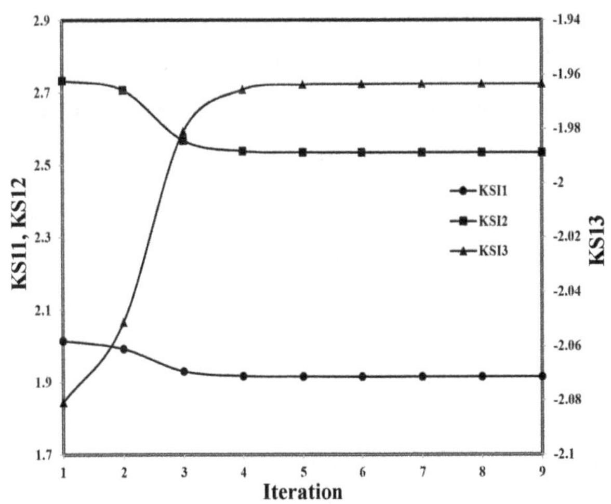

Figure 4: Variable changes during Newton-Raphson iteration; S1; P=50 bar; T= 550 K.

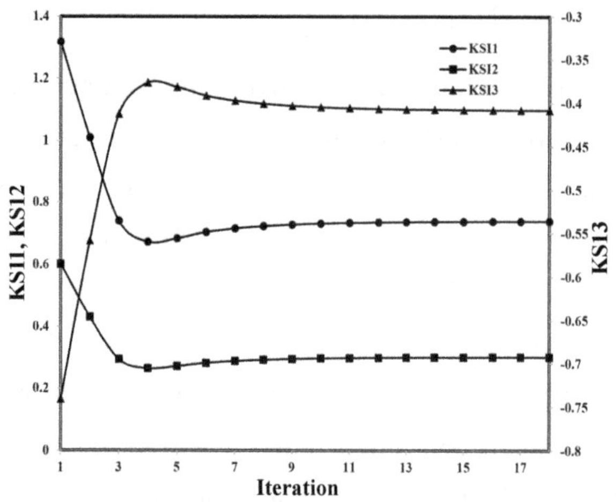

Figure 5: Variable changes during Newton-Raphson iteration; S1; P=45 bar; T=900 K.

RESULTS AND DISCUSSION

To evaluate the proposed solution technique, two real reservoir samples, namely S1 and S2, were used. The initial equilibrium ratios were calculated with Wilson correlation.

$$K_i = (Pc_i / P).exp[5.37\ (1+\omega_i)(1-Tc_i / T)] \quad (53)$$

Using these equilibrium ratios, the initial estimate of the reduced variable is calculated and the reduction flash calculation starts the mentioned numerical scheme. The convergence criteria for both flash types, i.e. the full flash and reduced calculations, are calculated based on:

$$\sum_{1}^{nf} \left| \ln K_i + \ln \varphi_i^{v}(y,P,T) - \ln \varphi_i^{l}(x,P,T) \right| < 10^{-6} \quad (54)$$

which is the cumulative absolute residual (CAR) and ensures the equality for the components in oil and gas phase due to the equilibrium conditions. The oil sample equilibrium reduction was tested in several pressures and temperatures inside a two phase region. At a pressure of 50 bar and a temperature of 550 K, the reduction results are presented. Figure 3 shows the close convergence behavior of the reduced and full flash approach. Both algorithms started from the same cumulative absolute residual due to the same initial estimates from Wilson equation.

The reduced variable changes during the Newton iterations are plotted in Figure 4. It shows that nearly after four or five nonlinear iterations, the reduced variables become very close to their final values. The same results but at a temperature of 900 K, where we have a little amount of oil, are plotted in Figure 5. The equilibrium ratios of the reduced approach are essentially the same as those shown in Figure 6.

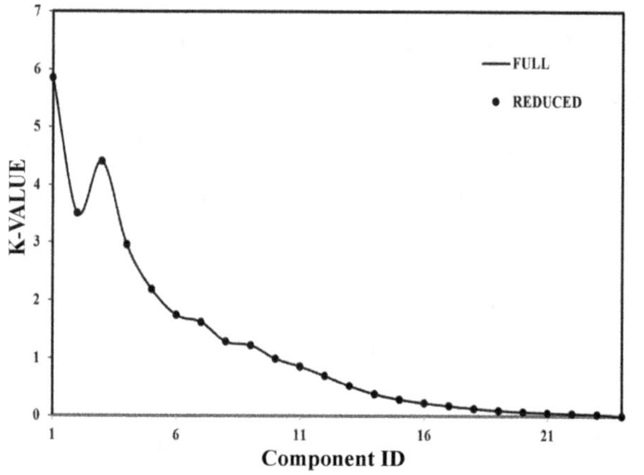

Figure 6: Equilibrium ratio comparison for the reduced and full flash calculation; S1; P= 50 bar; T= 550 K.

For the gas condensate sample, the equilibrium calculations at several pressures and temperatures inside the two phase region were performed. The convergence results are presented at 200 bar and 350 K. The decrease in the cumulative absolute residual during iterations in the full and reduced flash is plotted in Figure 7. The similar convergence behavior for both approaches is deduced from this figure. As it is expected for the gas condensate sample, the program needs a higher number of iterations to satisfy the equality of the fugacity coefficients in equilibrium. The changes in reduced variable during the Newton iterations in the conditions of 200 bar/350 K and 100 bar/300 K, where we have two completely different mole numbers of vapor phase, are plotted in Figure 8 and Figure 9 correspondingly. Here again, in both figures, the reduced variables become very close to their final values in less than five iterations.

The calculated reduced equilibrium ratios are plotted versus the full flash equilibrium ratios in Figure 10

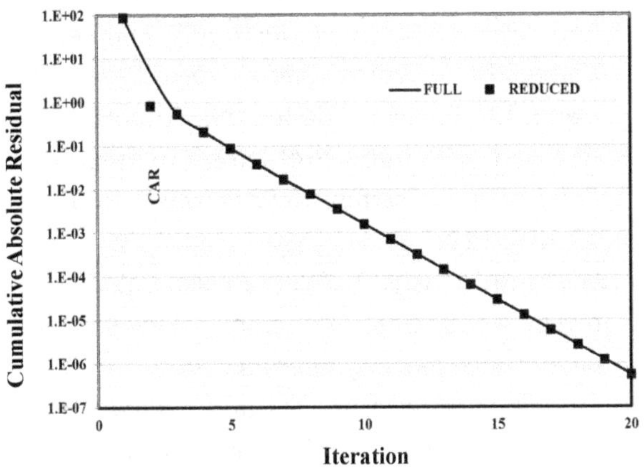

Figure 7: Convergence behavior of full flash calculation and reduced technique for cumulative absolute residual (CAR); S2; P= 200 bar and T= 350 K.

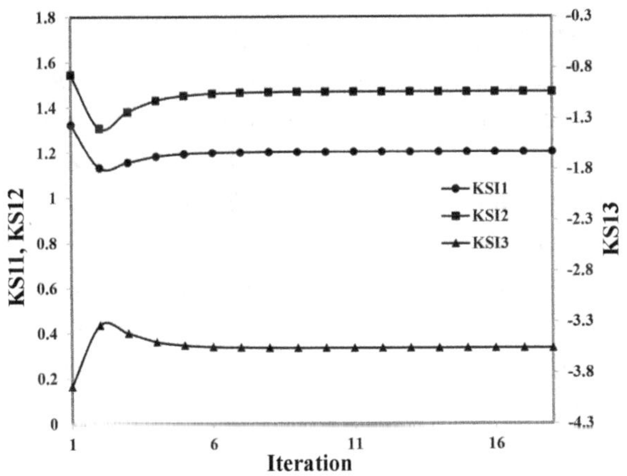

Figure 8: Variable changes during Newton-Raphson iteration; S2; P= 200 bar; T= 350 K.

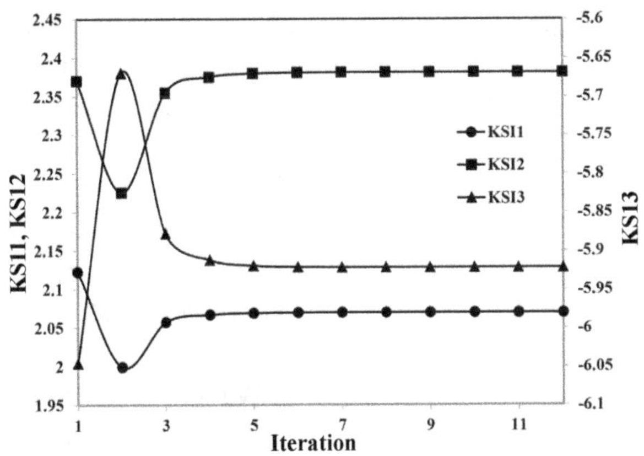

Figure 9: Variable changes during Newton-Raphson iteration; S2; P= 100 bar; T=300 K.

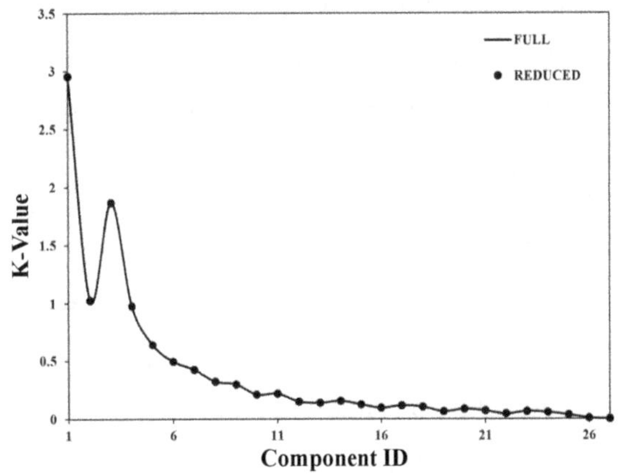

Figure 10: Equilibrium ratio comparison for the reduced and full flash calculation; S2; P=200 bar; T=350 K.

CONCLUSIONS

A new flash reduction technique is presented and verified against the full flash procedure. The proposed approach offers a unique set of equations and variables to reduce the dimension of the equilibrium calculation problem. In this method, the selection of the solution variables and reduced equations extend the stability and convergence of the proposed technique to support different reservoir fluid types and thermodynamic conditions. To demonstrate these capabilities, two real samples of reservoir fluid were selected from reliable published data. The convergence behaviors of the new reduction technique for both samples are close to the full flash calculations according to the performed numerical experiments. The calculated equilibrium ratios and phases mole fractions for the reduced set of equations are very close to the results calculated for the conventional approach. The approach implementation procedure is simple and straightforward. This means that with full set of equilibrium equations and their corresponding derivatives, the reduced equation and their derivatives with respect to solution variables can be obtained, and there is no considerable difficulty regarding implementation. The method development was based on the compressibility and fugacity coefficients of Peng-Robinson equation of state; however, the approach is general and can be applied to any two-parameter cubic equation of state.

NOMENCLATURE

AAD%	Average absolute deviation percent
A, B	EOS capital attraction repulsion parameters
a, b	EOS small attraction repulsion parameters
AD%	Absolute deviation percent
AF	Acentric factor
CAR	Cumulative absolute residual
EOS	Equation of state
f	Fugacity
G	Residual equation
K	Equilibrium ratio
J	Jacobian Matrix
MW	Molecular weight
P	Pressure
P_c	Critical pressure
T	Temperature
T_c	Critical temperature
V_c	Critical Volume
Z	Compressibility factor
nv	Phase mole fraction
ζ	Solution variable
ω	Acentric factor
x_i^p	Component mole fraction for phase p
x_i, y_i	Phases mole fraction
μ_i	Chemical potential
ϕ_i	Fugacity coefficient

REFERENCES

1. Wang P. and Stenby E. H., "Non-iterative Flash Calculation Algorithm in Compositional Reservoir Simulation," *Fluid Phase Equilibria*, **1994**, *95*, 93-108.
2. Michelsen M. L., "Simplified Flash Calculations for Cubic Equations of State," *Industrial & Engineering Chemistry Process Design and Development*, **1986**, *25*, 184-188.
3. Jensen B. H. and Fredenslund A., "A Simplified Flash Procedure for Multicomponent Mixtures Containing Hydrocarbons and One Non-hydrocarbon using Two-parameter Cubic Equations of State," *Industrial & Engineering Chemistry Research*, **1987**, *26*, 2129-2134.
4. Hendriks E. M., "Reduction Theorem for Phase Equilibrium Problems," *Industrial & Engineering Chemistry Research*, **1988**, *27*, 1728-1732.
5. Li Y. and Johns R. T., "Rapid Flash Calculations for Compositional Simulation," *SPE Reservoir Evaluation and Engineering*, **2006**.
6. Nichita D. V., Broseta D., and Leibovici C. F., "Reservoir Fluid Applications of a Pseudo-component Delumping New Analytical Procedure," *Journal of Petroleum Science and Engineering*, **2007**, *59*, 59-72.
7. Hoteit H. and Firoozabadi A., "Simple Phase Stability-testing Algorithm in the Reduction Method," *AIChE Journal*, **2006**, *52*, 2909-2920.
8. Nichita D. V. and Gracia, A., "A New Reduction Method for Phase Equilibrium Calculations," *Fluid Phase Equilibria*, **2011**, *302*, 226-233.
9. Belkadi A., Yan W., Moggia E., Michelsen M. L., Stenby E. H., Aavatsmark I., et al., "Speeding Up Compositional Reservoir Simulation through an Efficient Implementation of Phase Equilibrium Calculation," SPE-163598-MS, Woodlands, Texas, USA, **2013**.
10. Gaganis V., Marinakis D., and Varotsis N., "A General Framework of Model Functions for Rapid and Robust Solution of Rachford–Rice type of Equations," *Fluid Phase Equilibria*, **2012**, *322–323*, 9-18.
11. Gaganis V. and Varotsis N., "An Improved BIP Matrix Decomposition Method for Reduced Flash Calculations," *Fluid Phase Equilibria*, **2013**, *340*, 63-76.
12. Gorucu S. E. and Johns R. T., "New Reduced Parameters for Flash Calculations Based on Two-parameter BIP Formula," *Journal of Petroleum Science and Engineering*, **2014**, *116*, 50-58.
13. Qiu L., Wang, Y., Jiao, Q., Wang, H., and Reitz, R. D., "Development of a Thermodynamically Consistent, Robust and Efficient Phase Equilibrium Solver and its Validations," *Fuel*,

2014, *115*, 1-16.

14. Gaganis V. and Varotsis, N., "An Integrated Approach for Rapid Phase Behavior Calculations in Compositional Modeling," *Journal of Petroleum Science and Engineering*, **2014**, *118*, 74-87.
15. Nichita D. V. and Petitfrere, M., "Phase Stability Analysis using a Reduction Method," *Fluid Phase Equilibria*, **2013**, *358*, 27-39.
16. Petitfrere M. and Nichita, D. V., "Multiphase equilibrium calculations using a reduction method," *Fluid Phase Equilibria*, **2015**, *401*, 110-126.
17. Petitfrere M. and Nichita D. V., "A Comparison of Conventional and Reduction Approaches for Phase Equilibrium Calculations," *Fluid Phase Equilibria*, **2015**, *386*, 30-46.
18. Haugen K. B. and Beckner B. L., "Highly Optimized Phase Equilibrium Calculations," SPE-163583-MS, *Society of Petroleum Engineers*, Woodlands, Texas, USA **2013**.
19. Riazi M. R., "Characterization and Properties of Petroleum Fractions," *ASTM*, **2005**.
20. Jaubert J.N. I. A. and Laurent S., "A Crude Oil Data Bank Containing more than 5000 PVT and Gas Injection Data," *Journal of Petroleum Science and Engineering*, **2002**, *34*, 65-107.

APPENDIX

To calculate the required EOS derivatives for the full and reduced flash, the compressibility factor derivatives can be written as:

$$\frac{\partial Z^P}{\partial x^P{}_i} = \frac{\partial Z^P}{\partial A^P}\frac{\partial A^P}{\partial x^P{}_i} + \frac{\partial Z^P}{\partial B^P}\frac{\partial B^P}{\partial x^P{}_i} \quad (1)$$

The compressibility factor derivatives can be formulated as:

$$\frac{\partial Z^p}{\partial A^p} = -\frac{Z^p - B^p}{3Z^p.Z^p - 2Z^p\left(1-B^p\right) + A^p - 2B^p - 3B^p.B^p} \quad (2)$$

$$\frac{\partial Z^p}{\partial B^p} = -\frac{Z^p.Z^p - 2Z^p - 6Z^pB^p - A^p + 2B^p + 3B^p.B^p}{3Z^p.Z^p - 2Z^p\left(1-B^p\right) + A^p - 2B^p - 3B^p.B^p} \quad (3)$$

The derivative of compressibility factor for each phase with respect to capital EOS parameters for PR must be calculated implicitly:

$$\frac{\partial B^P}{\partial x^P{}_i} = \frac{\partial b^P / \partial x^P{}_i}{RT} \quad (4)$$

$$\frac{\partial A^P}{\partial x^P{}_i} = \frac{\partial a^P / \partial x^P{}_i}{(RT)^2} \quad (5)$$

The capital EOS parameters derivatives with respect to phase composition are defined by:

$$\frac{\partial a^P}{\partial x^P{}_i} = 2a_i^{0.5}\sum_j x_j a_j^{0.5} \quad (6)$$

$$\frac{\partial b^P}{\partial x^P{}_i} = b_i \quad (7)$$

Using the above derivatives, the fugacity coefficient derivatives with respect to composition can be calculated as:

$$\frac{\partial Ln\varphi_i^P}{\partial x_j^L} = \frac{\partial A^P{}_1}{\partial x_j^P} + \frac{\partial A^P{}_2}{\partial x_j^P}b_i + \frac{\partial A^P{}_i}{\partial x_j^P}a_i^{0.5} \quad (8)$$

In this relation each term is defined as following:

$$\frac{\partial A^p{}_1}{\partial x_i^p} = -\frac{\left(\partial Z^p / \partial x_i^p - \partial B^p / \partial x_i^p\right)}{\left(Z^p - B^p\right)} \quad (9)$$

$$\begin{aligned}\frac{\partial A^p{}_2}{\partial x_j^p} &= \frac{\left(b^p\partial Z^p / \partial x_j^p - Z^p\partial b^p / \partial x_j^p\right)}{b^p.b^p} + \\ &+ \frac{B^pb^p\partial A^p / \partial x_j^p - A^p\left(b^p\partial B^p / \partial x_j^p + B^p\partial b^p / \partial x_j^p\right)}{B^p.B^p.b^p.b^pp_1} Ln\left(\frac{Z^p + B^pp_2}{Z^p + B^pp_3}\right) \\ &+ \frac{A^p}{B^pp_1}\left(\frac{\partial Z^p / \partial x_j^p + p_2\partial B^p / \partial x_j^p}{Z^p + B^pp_2} - \frac{\partial Z^p / \partial x_j^p + p_3\partial B^p / \partial x_j^p}{Z^p + B^pp_3}\right)\end{aligned} \quad (10)$$

where,

$$LnZEDB^p = Ln\left(\frac{Z^p + B^pp_2}{Z^p + B^pp_3}\right) \quad (11)$$

$$\left[\begin{aligned}\frac{\partial A^p{}_3}{\partial x_j^p} &= -\left(\frac{B^p\partial A^p / \partial x_j^p - A^p\partial B^p / \partial x_j^p}{B^pp_1}\left(\frac{2s^p}{a^p}\right)LnZEDB^p\right) \\ &- \frac{A^p}{B^pp_1}\left(\frac{2a^p\partial s^p / \partial x_j^p - 2s^p\partial a^p / \partial x_j^p}{a^p.a^p}\right)LnZEDB^p \\ &- \frac{A^p}{B^pp_1}\left(\frac{2s^p}{a^p}\right)\left(\frac{\partial Z^p / \partial x_j^p + p_2\partial B^p / \partial x_j^p}{Z^p + B^pp_2} - \frac{\partial Z^p / \partial x_i^p + p_3\partial B^p / \partial x_j^p}{Z^p + B^pp_3}\right)\end{aligned}\right. \quad (12)$$

7

To Express Required CT-Scan Resolution for Porosity and Saturation Calculations in Terms of Average Grain Sizes

Ahmed Zoeir*[1] and Jafar Qajar[2]

[1] Faculty of Petroleum and Natural Gas Engineering, Sahand University of Technology, Tabriz, Iran

[2] School of Chemical and Petroleum Engineering, Shiraz University, Shiraz, Iran

ABSTRACT

Despite advancements in specifying 3D internal microstructure of reservoir rocks, identifying some sensitive phenomenons are still problematic particularly due to image resolution limitation. Discretization study on such CT-scan data always has encountered with such conflicts that the original data do not fully describe the real porous media. As an alternative attractive approach, one can reconstruct porous media to generate pore space representations. The reconstructed structures are then used for simulations using some sort of discretization. In this paper, It is examined the effect of discretization on porosity and saturation calculations in porous media models. Some 3D Boolean models of random overlapping spheres of fixed and variable diameters in three dimensions are used. The generated models are then discretized over 3D grids with different number of voxels. The porosity can be calculated and saturation of the discretized models are then compared with the analytical solutions. The results show that when meshgrid sizes are 8% of smallest grains, porosity is calculated with 95% precision. In addition to that, meshgrid sizes of 5% and

3% of average grain diameter are adequate to calculate non-wetting and wetting phase saturations with at least 95% precision. This helps in choosing the optimum voxel size required in imaging for efficiently use of available computational facility.

Keywords: Boolean Model, Discretization, Micro-CT, Mesh-Grid, Model Porous Media, Porosity, Saturation

INTRODUCTION

The characterization of the microstructure of natural and synthetic porous materials has been of great importance for scientists and engineers over a wide range of disciplines and hence, it has been the subject of numerous studies in various fields. In petroleum engineering, for example, an understanding of the interactions among rock microstructure, mineralogy, and fluids in pore space is crucial to us for better interpretation and prediction of reservoir rock properties and flow processes in underground formations. Development in structural study of porous materials has a long history. Over the past two decades, direct measurements of 3D structures at resolutions down to a few microns using X-ray micro-

***Corresponding author**

Ahmed Zoeir

Email: ah_zoeir@sut.ac.ir

computed tomography (micro- CT) is being used increasingly. CT is considered as a non-destructive imaging technique that uses X–ray technology (XRT) and mathematical reconstruction algorithm to view cross sectional slices of internal structure of rock samples. CT- scan generates slices through the object for different positions during stepwise rotation whereby either the source and detector or the object is moved. From sets of these projections, cross sectional images and then 3D images are reconstructed by applying Fourier transform algorithm. Although the micro-CT technique has evolutionised, the amount of information that we can obtain from a rock sample and several challenges behind the implementation of the method for characterizing the rock microstructure which still exist, limit the usability of the acquired data. In particular, a major limitation of the imaging techniques is finite resolution of the images. First, the finite image resolution makes ambiguity in identifying different materials/phases within the sample. Second, the high-resolution imaging gives rise to very large datasets and hence requires intensive computations, large computer memory, and considerable processing, time, as stated in 2006 [1]. Third, the higher the resolution the smaller the sample and the smaller the sample the fewer the structural properties can be captured in a representative way as declared by, Vogel, Weller, and Schluter in 2010 [2]. This leads to the challenge of detecting the microstructure of the material at finer resolution and then to connect it to the macro-scale where continuum mechanics applies, Liu and Regenauer in 2011 [3]. Due to such difficulties, flow simulations in complex sedimentary rocks based on images are problematic and it is of great interest to use alternative methods to the direct imaging or to propose the appropriate use of the imaging techniques that can optimize computational resources usage.

As an alternative to direct visualization and characterization, numerical simulations based on simplified representations (models) of rock microstructure have long been employed to obtain an understanding of static and dynamic behavior of rocks at the pore scale, as studied by Adler in 1992 [4], and Sahimi in 2011 [5].

In this regard, a large number of models have been developed to simulate the structure of porous materials. Among them, the Boolean model of randomly distributed grains has successfully employed to represent irregular spatial structures, by Bilodeau, Meyer, and Schmitt in 2007 [6], Torquato in 2005 [7] including reservoir rocks in particular, Arns, Knackstedt, and Mecke in 2009 [8]. Either in the case of micro-CT images or artificial image models of porous media, spatial discretization is required to compute physical properties or to solve differential equations that govern fluid flow. The rock microstructure obtained by digital images has been already naturally discretized as voxels. The artificial image models have also be discretized into a mesh-grid domain. In general, using smaller voxel size in image-based calculations as well as smaller grid- block size in simulation permits more detailed description of porous media. In spite of this, using small voxels/mesh-grids leads to large numbers of voxels/mesh-grids in a fixed size domain and hence large requirements of computational resources. In this regard, it is of great interest to determine the optimum voxel size to which structural details can be sufficiently resolved. In practice, a certain resolution is enough to calculate different reservoir rock properties like porosity, contact area, and wetting and non-wetting phase saturation. On the other hand, an optimum mesh-grid size in simulation leads to properly use the computational resources. A number of studies have addressed the discretization effects and the associated errors for estimation of the properties of materials. Latief, Irayani, and Fauzi in 2012 [9] used a Skyscan 1173 μ-CT scanner at the Basic Science Advanced Laboratory of the Bandung Institute of Technology to produce images of a sandstone sample at three voxel sizes of 30, 15, and 7.5 μm. They calculated a number of macro- and micro-scale properties including porosity, permeability, and specific surface area to examine the effect of scanning resolution on the physical properties. They showed that the voxel size has a direct influence on the calculated properties. In particular, porosity and tortuosity increase with an increase in voxel size whereas specific surface area and permeability decreases. Ferréol and Rothman in 1995 [10] studied the effect of grid resolution on image-based calculation of permeability using the lattice-Boltzmann method and qualitatively concluded that finer space discretization leads to more accurate estimation of porous media properties. In another work, Pan, Hilpert, and Miller in 2001 [11] studied the discretization effects of permeability calculation during simulation of single phase flow in simulated sphere packing with varying porosity and pore-size distribution. They applied several different sphere pack models that vary in sphere-size distribution to study primitive media properties. They used pore-network calibration onto the size distribution of pore bodies and throats. They also varied voxel sizes to reach discretization independent properties, but their study on discretization effects was only limited to permeability. Arns et al in 2001 [12] considered major sources of errors, namely finite size effects, discretization errors, and statistical fluctuations in calculation of transport properties of rocks from digitized images. They showed that using different voxel sizes led to different conductivity values. They argued that the discretization errors are due to a number of factors including inaccurate description of curved grain boundaries and closing of narrow pores. They finally concluded that considering discretization errors may lead to scale out the potential error due to channels not being resolved and hence an accurate continuum value for transport properties was obtained. Silin and Patzek in 2009 [13] considered sandstone images as discretized porous media and predicted fluid flow properties using some spherical models for voxels of 5 and 10 percent of grains diameter. They showed good agreements between computed parameters and direct experimental measurements.

The aim of this study is to investigate the effects of discretization on porosity and saturation calculations in order to address a proper voxel/ mesh-grid size for practice.

The paper is organized as follows: First, It will be generated some 3D Boolean models of random overlapping spheres of fixed and variable diameters in three dimensions. Then the generated model is discretized over a 3D grid of different number of voxels. After that, It is compared the calculated porosity and saturation of the discretized model with analytical solutions and evaluate the associated errors. Finally, It is applied the proposed procedure for the micro-CT images of two natural sedimentary core samples. The novelty of this study is that we focused on saturation calculation of wetting and non- wetting phases, say water and oil, and expressed calculated adequate resolution in terms of average grain sizes. Knowing oil and water distributions in each section and every pore channel is vital for calculation of capillary pressure curves and also relative permeability curves. Since any preventable error in calculation of saturation distributions that comes from the selection of inadequate resolution will impact on the calculation of relative permeability and capillary pressure curves.

EXPERIMENTAL PROCEDURES

The Boolean Model

We used some 3D Boolean models of random overlapping spheres as porous medium prototypes. The Boolean model is a random closed set, a topic belong in the realm of stochastic geometry which deals with mathematical models for random geometric structure, (Schneider and Weil in 2008) [14]. In essence, the Boolean model is constructed by a collection of points, called germs, distributed according to a Poisson point process of intensity λ and a system of objects, called grains, with the specified shape, size, and orientation, Bilodeau, Meyer, and Schmitt in 2005 [15]. The Poisson intensity indicates the mean number of objects per unit volume. The Boolean model includes two types of randomness, i.e., a spatial Poisson distribution of germs characterized by an intensity value, and randomness of the grains specified by the probability distribution of the grains, as Sobczyk, and Kirkner declared in 2012 [16]. In addition, in the Boolean model of overlapping spheres, the grains are spheres with a random radius, and the sphere centers are located on points of a Poisson point process of intensity λ, (Molohanov in 1990) [17]. In this case, the probability distribution of the grains is the distribution of their radius. The density of sphere centers is related to the sphere volume fraction.

Computational Methods

In this paper, It is considered a collection of randomly overlapping spheres of fixed and variable diameter in three dimensions. In the spherical model, the position and diameter of grains are taken into consideration while the orientation is irrelevant. In order to generate the model of overlapping spheres, the sphere centers are distributed randomly according to the Poisson distribution. Three random Poisson-distributed numbers represent the position of center of each spherical grain in a 3D cubic lattice of arbitrary size in Cartesian coordinates. In the model of overlapping spheres of fixed diameter, the diameter of all grains is fixed to a desired value. Here as case studies, two and three spheres mutually overlap each other are considered and illustrated in Figure 1 (2D slices in the top row). It is also possible to adjust ellipses instead of spheres to generate realizations which model rocks in which grains have no more spherical shape perhaps due to inadequate displacement time before sedimentation.

One can use specific functions to distribute grains according to desired algorithms. In order to do that, position and diameter of generated grains within Cartesian coordinates obey given functions instead of random function. In the model of overlapping spheres of variable diameter, four random numbers specify the generated grains. The first three random numbers denote the center of each grain in the Cartesian coordinate system according to the Poisson-point process. To have an appropriate diameter range, a maximum and a minimum values are chosen. The fourth random number is multiplied by the difference of the largest and the smallest grain diameters and then is summed with the smallest possible diameter to give a range of random values for grain diameter. The slices in the bottom row of Figure 1 shows the examples of 2D images of the 3D overlapping spheres of variable diameter.

To develop the prescribed model, first It is considered a particular density of sphere centers within a 3D lattice. Then the first sphere is generated, and the associated volume fraction and surface area are exactly calculated by analytical formulae. Procedure continues until the specified number of grains reaches, and then grain generation terminates. If the generated grains overlap each other, shared volume and surface area are calculated and subtracted from grain volume and surface area. The result is an artificial rock model with known exact porosity and surface area. In the next step, the Boolean model is discretized over a 3D grid of nxnxn voxels as illustrated in Figure 2.

The 3D representations of the generated Boolean models of spheres with variable diameters are shown in Figure 3.

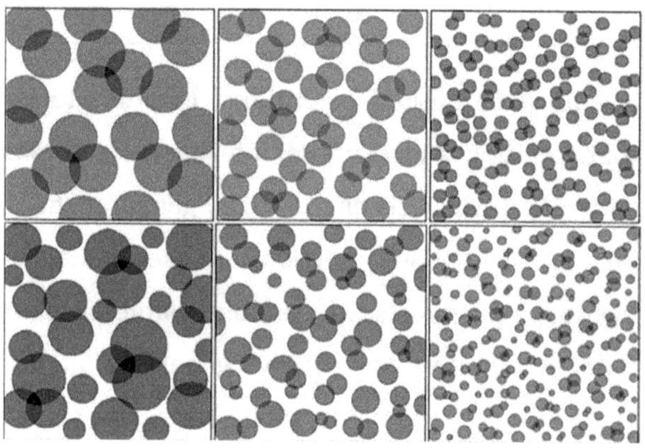

Figure 1: Slices through the 3D Boolean model of spheres in a lattice of 10003. Top row: fixed diameter sphere model, (left) 200, (middle) 100, and (right) 50 units, Bottom row: variable diameter sphere model.

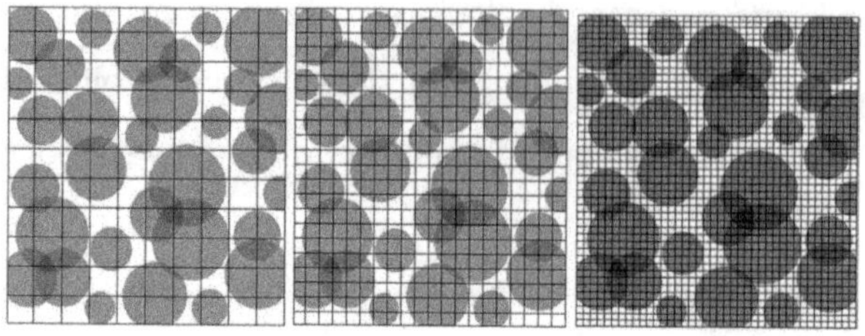

Figure 2: Examples of 2D slices representing discretization of the Boolean model of overlapping spheres in a lattice of 10003, Mesh-Grid sizes are 100, 50, and 25 from left to right respectively.

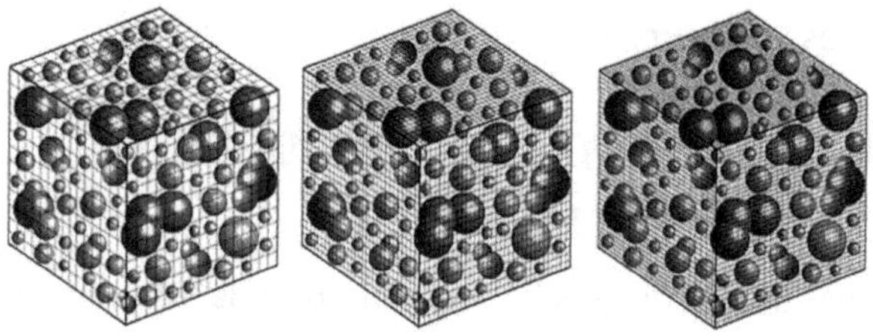

Figure 3: 3D representations of the discretized Boolean model with the Mesh-Grid sizes of 100, 50, and 25 from left to right respectively.

If all eight corners of a voxel fully fall into the pore space of the discretized model, this voxel contributes in pore volume calculation, while if some of the corners of a voxel place in the pore region and some of them place in the solid region, this voxel lies in pore-grain interface region and is excluded from porosity computation. The remaining voxels are counted as grain region.

Fluid saturation is one of the important physical properties which its accurate prediction is essential in various reservoir engineering calculations. The initial (static) saturation distribution as well as the dynamic saturation redistribution during production and flow Processes in petroleum reservoirs highly depends on the wetting state of the reservoir rock. In this paper, we also aim at investigating the effect of discretization on wetting and non-wetting phase saturation calculation precision. We assume that the wetting phase surrounds solid grains, and non-wetting phase occupies the remaining pore space of porous media. As a consequence, the wetting and non-wetting regions are considered as two phases along with the solid grain as a third phase. One can consider two main ways to generate n-phase systems. The first method is to place particles of different phases in the lattice. The second way employs a simple geometrical overgrowth algorithm based on parallel surfaces,

Arns in 2002 [18]. In this approach, particles are covered by a parallel layer with the thickness determined by saturations desired.

The second method is adopted to generate wetting phase in the developed models. The parallel layer represents water and oil in water-wet and oil- wet rocks respectively. Thickness of the parallel layer along each curved surface and in all corners is calculated according to two phase distribution formulas that determine local phase saturation in each pore. The total integration of local saturations must be equal to overall saturation of wetting and non-wetting phases.

It should be noted that the remaining pore-phase is considered to be filled with the non-wetting phase. Since the variable-diameter grain models are more realistic, we developed saturated models for grains with variable diameters. If all eight corners of a voxel fully fall into the wetting or non-wetting phase of the discretized model, this voxel contributes in the wetting-phase or non-wetting phase saturation calculation respectively. Since the wetting phase may overlap other phases, and the corners of a voxel may place in the wetting phase, solid, or non-wetting regions. In this case, such a voxel which is in the interface is excluded from saturation calculation. The generated model is then discretized over a 3D grid of n×n×n voxels to study the effects of mesh-grid size on wetting and non-wetting phase saturation calculations. Figure 4 shows a slice of a 3D model with artificial saturation. All the process of model development and the related calculation were coded using MATLAB R2012a.

RESULTS AND DISCUSSION

Effects of Discretization on Porosity Calculation

In order to prepare Boolean models, generate spheres with fixed and also variable diameters (from a Poisson distribution with λ=200) are imaginary dropped into an assumed cubic lattice of side L=1000 units to generate multiple realizations of the model.

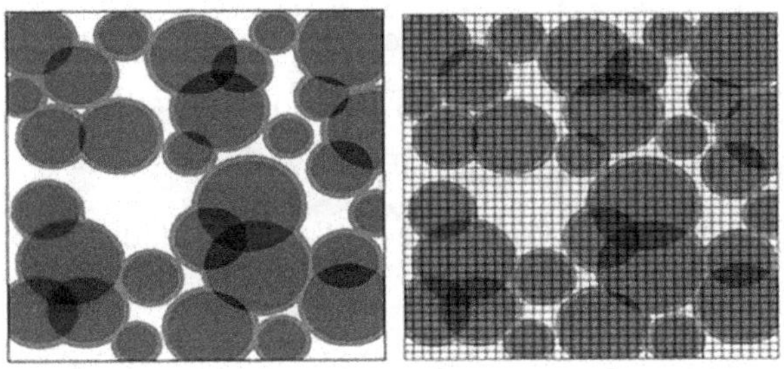

Figure 4: An example of a 2D slice through the 3D Boolean model of overlapping spheres with added artificial saturation in a lattice of 10003 with mesh- grid size of 25.

The chosen lattice size accompanying of spheres overlap leads to porosities lower than 20% to 30% in our work. The models are then discretized over a 3D grid of n×n×n voxels. Since the fixed number of voxels in the lattice gives results with varying accuracy for large and small grains, we consider the voxel size as a portion of the grain diameter. Hence, the number of voxels differs for each specific realization, and this leads to a more general result independent of the grain diameter. Table 1 shows the calculated porosities from several realizations of discretized models that are compared with analytically calculated values in fixed diameter spherical models. For multiple realizations of the fixed diameter grain models, the grain diameter range from from 20 to 200 is considered with arbitrary length unit. The 3D mesh-grid (voxel) sizes range from 1 to 20 percent of the specified grain diameters as illustrated in Table 1. As a result, the number of voxels decreases as the grain diameter increases. For instance, if the voxel size is 10% of the grains with diameters of 50 and 200, the number of voxels will be 2003 and 503 respectively in a lattice of side 1000 arbitrary length unit. Note that the calculated porosity for the discretized model and the actual porosity mathematically calculated are the averages from multiple realizations. As indicated in Table 1, when the voxel size becomes smaller, values of the calculated porosities from discretized models converge to analytical values.

The results also show the similarity in relative error for all cases when the voxel size is presented as fractions of the grains diameter. For example, the relative error is about 5% when voxels are set to 5% of grains diameter. This error is about 15% and 30% for voxels which are set to 10% and 20% of the grain sizes respectively. If we consider the relative error of 5% as a satisfactory threshold for porosity calculation, the fixed diameter grain model limits us to choose voxels with size of at most 5% of grain diameter. It seems from tabulated results that when CT-scan imaging power is used to provide pixel sizes smaller than 3% of grain average diameter, negligible effects appear in calculated results. Then we generated multiple realizations of variable size models with diameters ranged from 20 to 80 units, as shown in Table 2. For each average grain diameter, we allow grains to be generated with diameters of ±10%, ±20%, and ±50% larger and smaller than the average value. For example, for average diameter of 20 units, we allow random generation of grains with 20 ±2, 20 ±4, and 20 ±10 units.

Table 1: Porosity values of the Boolean models of fixed-diameter overlapping spheres computed on various grid sizes.

Voxel size as a fraction of grain diameter	Grains diameter (arbitrary length unit)						
	20	30	50	80	100	150	200
0.2	13.3	11.4	9.7	7.8	6.6	5.6	4.8
0.15	14.9	12.8	10.8	8.6	7.4	6.3	5.4
0.10	16.2	13.8	11.8	9.5	8.1	6.9	5.9
0.05	18.1	15.5	13.2	10.6	9.0	7.7	6.5
0.03	18.4	15.8	13.4	10.8	9.2	7.8	6.6
0.02	18.7	16.0	13.6	10.9	9.3	7.9	6.7
0.01	19.0	16.2	13.8	11.1	9.4	8.0	6.8
Actual average porosity (%)	19.1	16.3	13.9	11.2	9.5	8.1	6.9

Table 2: Porosity values of the Boolean models of variable-diameter overlapping spheres computed on various grid sizes.

Voxel size as a fraction of grain diameter	Grains diameter (arbitrary length unit)								
	20 ±2	20 ±4	20±10	50 ±5	50 ±10	50 ±25	80 ±8	80 ±16	80 ±40
0.1 of the largest grain	13.6	13.5	13.4	10.9	10.8	10.6	9.5	9.2	9.1
0.1 of the smallest grain	17.5	17.4	17.1	14.0	13.9	13.7	12.1	11.9	11.7
0.05 of the largest grain	17.9	17.7	17.5	14.3	14.1	14.0	12.4	12.1	12.0
0.05 of the smallestgrain	18.9	18.7	18.5	15.1	14.9	14.7	13.1	12.8	12.6
Actual average porosity (%)	19.5	19.3	19.1	15.6	15.4	15.2	13.5	3.2	13.0

Similar to the fixed diameter model, the calculated porosity for the discretized model and the actual porosity mathematically calculated are compared together in Table 2.

In variable diameter grain models, it is essential to choose voxel sizes with attention to distributed range limits. If voxel sizes are 10% of the largest grains in each realization, about 30% relative error results in calculated porosity. But if we set voxel sizes as 10% of smallest grain, the relative error becomes only about 10%. Table 2 also shows that when voxel sizes are 5% of the largest grains, the relative error becomes 8% while if we set voxel sizes as 5% of the smallest grains, we will have only 3% error. If relative error of 5% is considered to be a satisfactory threshold, the variable diameter grain models suggest voxels with size of 8% of the smallest grains.

Effects of Discretization on Saturation Calculation

The variable diameter grain model is used to study the effects of discretization on wetting phase and non-wetting phase saturations.

Table 3 shows the calculated average wetting phase saturations of the models of variable-diameter spheres for the case where the grain diameters range from 16 to 25 (i.e., 20 ± 4).

Table 3: Average wetting phase saturation values of the Boolean models of variable-diameter overlapping spheres (grain diameters: 20 ± 4) computed on various grid sizes.

Voxel size as a fraction of the smallest grain diameter (i.e. d = 16)	Calculated wetting phase saturation (%)						
0.15	3.0	6.2	10.6	14.3	21.5	28.3	37.0
0.10	3.5	7.1	12.0	16.2	24.3	32.4	41.5
0.05	4.2	8.5	14.3	19.1	28.5	38.2	47.8
0.03	4.6	9.4	14.6	19.5	29.2	39.0	48.7
0.02	4.8	9.8	14.8	19.8	29.7	39.7	49.3
Actual wetting phase saturation (%)	5	10	15	20	30	40	50

For relative error of 5% as a satisfactory threshold for saturation calculation, and for realizations in which the wetting phase saturation is larger than 15%, model offers voxels with size, at most, 5% of the diameter of the smallest grains. Conversely, for realizations in which saturation is lower than 15%, the voxel size should be 3% or less of the smallest grains diameter to have a relative error less than or equal to 5%. Authors believe this difference is caused due to this fact that for lower saturations the wetting fluid covers grain surface as form of a thin layer and hence, is more sensitive to voxel size, and relatively smaller voxels are required to distinguish wetting phase from grains and non-wetting phase. The same procedure is performed to investigate the effects of resolution and voxel sizes on non-wetting phase saturation calculations, with this difference that if all eight corners of voxels are placed in non-wetting phase, voxel is considered to count non-wetting phase saturation.

As indicated in Table 4, non-wetting phase saturation is less sensitive to voxel size, and grids with sizes, 5% of grains diameter, satisfy 95% accuracy as our chosen threshold. Authors believe that this less sensitivity, which can be explained with respect to this fact that the porous media space that occupy the non-wetting phase has less corners and sharp regions and thus do not require very small voxels for saturation calculation. Presenting required voxel sizes for calculation of wetting and non-wetting phase saturation with acceptable accuracy, in terms of average grain sizes is very valuable when we want to advance work to predict relative permeability and capillary pressure curves.

Case Studies

Several sandstones and many other sedimentary rock types like silts, shale, and even some carbonates can be represented with variable size grain Boolean models. In order to examine the applicability of the results of this paper to real sedimentary rock samples, the micro-CT images of two core samples are used and compared with the results of the artificial models.

Figure 5 shows examples of the 2D slices through the 3D micro-CT images of sandstone (Figure 4a) and carbonate (Figure 4b) samples with voxel sizes of 2.67 and 2.44 μm respectively.

Table 4: Average non-wetting phase saturation values of the Boolean models of variable-diameter overlapping spheres (grain diameters: 50 ± 10) computed on various grid sizes.

Voxel size as a fraction of the smallest grain diameter (i.e. d = 40)	Calculated non-wetting phase saturation (%)						
0.15	14.9	22.5	29.8	37.2	44.1	51.5	57.6
0.10	17.0	25.4	34.0	42.3	50.8	59.2	67.4
0.05	19.1	28.7	38.3	47.8	57.1	66.6	76.2
0.03	19.4	29.2	38.9	48.5	58.0	67.8	77.7
0.02	19.6	29.5	39.3	49.0	58.9	68.6	78.5
Actual non-wetting phase saturation (%)	20	30	40	50	60	70	80

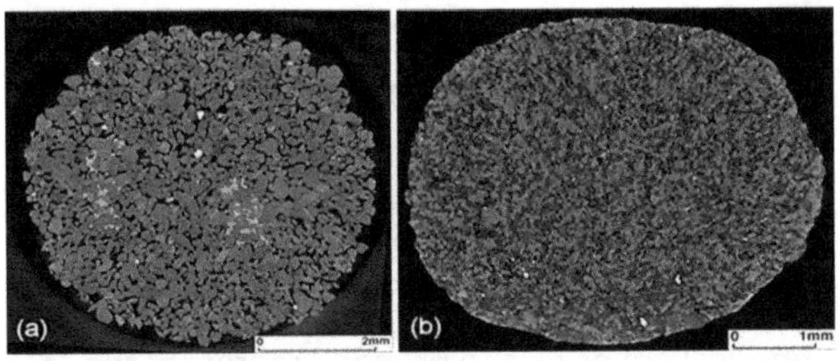

Figure 5: Examples of 2D slices through the 3D micro-CT tomograms of (a) sandstone and (b) carbonate core samples.

Visual inspections of the images reveal that the sandstone sample is relatively a clean core with some clay regions near the core center. Its grains can be considered to be 0.05 to 0.3 millimeter in size. It is apparent that the sandstone sample consisted of distinct grains with approximately well-rounded shape and a variety of sizes, adequate to be represented with spherical Boolean models. The carbonate sample seems to contain three main minerals in which grains range from 0.02 to 0.1 millimeters. Despite the carbonate sample exhibits more complex grains structure, the sample may be assumed to have a potential to be compared with a Boolean model. From visual inspection of the tomograms, the voxel sizes of both images are apparently smaller than 2% of the smallest grain diameters of each core samples.

The procedure adopted in this study to examine the effect of discretization on the tomographic images is as follows. First, the original grey-scale images are exported to MATLAB. Next, we simply binarize the images and calculate the porosity based on the original resolution. Then, neighbor voxels within pore or grain regions will be merged to give upgrided voxels with side lengths of 5 % and 10 % of the smallest and largest grains. In other words, a mesh-grid is superimposed on the image in which the grid size is chosen by visual inspection of the original image and identification of the smallest and largest grains.

Moreover, It is simply taken an average voxel values to denote the values of the upgrided voxels. The new mesh-grids are used to calculate the porosity of case study samples by the same procedure as the one used for the original image. Results are presented in Table 5. In addition, a comparison graph between calculated porosities from the Boolean models of spherical grains and from the case studies is shown in Figure 6. The horizontal axis of the graph refers to the generated artificial mesh-grids as a fraction of smallest and largest grain diameters. It is evident form Figure 6 that the predicted porosities from both carbonate and sandstone case studies has a good agreement with artificial model results. Figure 6 clearly shows that the discretization errors introduced by superimposing the artificial mesh-grids over the models or actual rock images are almost the same. This is because the rock samples can be properly represented by the spherical grain models (SGMs).

Table 5: Calculated porosities for case study samples.

Rock sample	Sandstone	Carbonate
Smallest and largest grains diameter (μm)	50-300	20-100
Using upgrided pixels with side length of 10% of largest grain	12.8	4.6
Using upgrided pixels with side length of 10% of smallest grain	16.5	5.9
Using upgrided pixels with side length of 5% of largest grain	16.8	6.0
Using upgrided pixels with side length of 5% of smallest grain	17.8	6.3
Actual average porosity (calculated based on the original images)	18.3	6.5

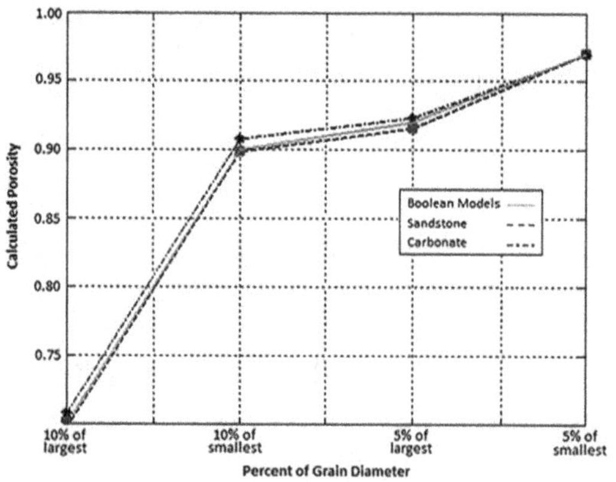

Figure 6: A comparison of the predicted porosities from the Boolean models, carbonate and sandstone case studies.

CONCLUSIONS

Despite the simplicity of the Boolean model of spherical grains, it can be applicable to represent a variety of rock samples. An appropriate model of a sandstone or carbonate rock can be used to predict their properties and interrelationships. Several key findings can be summarized as follow;

(1) In simulated fixed-size grain models, the relative error for calculated porosity was about 15% when mesh-grid sizes were 10% of the grains diameter. This is while only 5% relative error was achieved when applying mesh-grids with the sizes of 5% of the grains diameter. Conversely in variable size grain models, when mesh-grids were 10% of the upper limit of grains diameter range, the relative error was about 30%. The relative error decreased to 10% when mesh-grid sizes were 10% of the lower limit grains diameter.

(2) To achieve the relative error of 5% error in porosity calculation when using the variable size grain models, one should pay a close attention to the lower limit of grain size distribution and adjust the mesh-grid size to about 8% of the smallest diameter grains.

(3) In variable size grain models, if the grain size distribution covers a wide range, it is recommended to adjust the mesh-grid size to <5% of the smallest grains.

(4) In the case of clay- or cement-rich rock samples, the relative error in porosity calculation increases since the inter granular pore-filling materials are irregular in shape and the Boolean model of spherical grains fails to properly represent the rock structure.

(5) In the case of high wetting phase saturation, i.e. larger than 15%, the mesh-grid size of 5% of the smallest grains would be sufficient to accurately calculate the wetting phase saturation using the Boolean model. Conversely, if wetting phase saturation is low (e.g., for the case of connate water saturation), mesh-grid size should be adjusted to 2 to 3% of the smallest grains diameter to decrease the relative error of saturation calculation.

(6) The calculation of non-wetting phase saturation using the Boolean model is less sensitive to the chosen mesh-grid size and mesh-grids with sizes of about 5% of the grains diameter would be sufficient.

The results of this paper show that the Boolean model of spherical grains can quantitatively characterize the porosity and saturation distributions in clastic and carbonate rocks using the pre-defined mesh-grid sizes based on the rough knowledge about the grain size distribution of the rock samples. In addition, the present study can be considered as a guidance for choosing a proper voxel/ mesh-grid size in advance in image acquisition or numerical simulation of the rock samples in order to avoid gratuitously consuming computational and memory resources.

ACKNOWLEDGEMENTS

The authors acknowledge the support of Reservoir Rock Technologies Pty Ltd., Perth, Australia for providing micro-CT images of sandstone and carbonate cores.

NOMENCLATURES

Micro-CT	: Micro-Computed Tomography
XRT	: X–Ray Technology
SGMs	: Spherical Grain Models

REFERENCES

1. Lawrence A., Bouwer J. C., Perkins G., and Ellisman M. H., "Transform-based Backprojection for Volume Reconstruction of Large Format Electron Microscope Tilt Series," *Journal of Structural Biology*, **2006**, *154*(2), 144-167.
2. Vogel H. J., Weller U., and Schluter S., "Quantification of Soil Structure Based on Minkowski Functions," *Computers and Geosciences*, **2010**, *36*(10), 1236-1245.
3. Liu J. and Regenauer-Lieb K., "Application of Percolation Theory to Microtomography of Structured Media: Percolation Threshold, Critical Exponents, and Upscaling," *Physical Review E-Statistical, Nonlinear, and Soft Matter Physics*, **2011**, *83*(1), 016-106.
4. Adler P., "Porous Media: Geometry and Transports," (1st ed.) Stoneham MA: Butterworth-Heinemann, **1992**, 1-196.
5. Sahimi M., "Flow and Transport in Porous Media and Fractured Rock (2nd ed.)," Boschstr Germany: Wiley-VCH Verlag GmbH and Co. KGaA., **2011**, 1-217.
6. Bilodeau M., Meyer F., and Schmitt M., "Structure and Randomness: Contributions in Honor of Georges Matheron in the Fields of Geostatistics, Random Sets and Mathematical Morphology", (1st ed.) New York: Springer, **2007**, 1-354.
7. Torquato S., "Random Heterogeneous Materials: Microstructure and Macroscopic Propertie," (1st ed.) New York: Springer, **2005**, 1-237.
8. Arns C. H., Knackstedt M. A., and Mecke K. R., "Boolean Reconstructions of Complex Materials: Integral Geometric Approach," *Physical Review E*, **2009**, *80*(5), 051-303.
9. Latief F. D. E., Irayani Z., and Fauzi U., "Resolution Dependency of Sandstone's Physical Properties," *Bruker MicroCT Meeting*, Brussels, Belgium, **2012**, 1-8.
10. Ferréol B. and Rothman D. H., "Lattice-Boltzmann Simulations of Flow through Fontainebleau Sandstone," *Transport in Porous Media*, **1995**, *20*(1), 3-20.
11. Pan C., Hilpert M., and Miller C. T., "Pore-Scale Modeling of Saturated Permeabilities in Random Sphere Packings," *Physical Review E*, 066702, **2001**.
12. Arns C. H., Knackstedt M. A., Val Pinczewski M., and Lindquist W. B., "Accurate Estimation of Transport Properties from Microtomographic Images," *Geophysical Research Letters*, **2001**, *28*(17), 3361-3364.
13. Silin D. and Patzek T., "Predicting Relative-Permeability Curves Directly From Rock Images," SPE 124974, *SPE Annual Technical Conference and Exhibition New Orleans*, Louisiana, USA, **2009**, 1-8.
14. Schneider R. and Weil W., "Stochastic and Integral Geometry," (1st ed.) Berlin Germany: Springer-Verlag, **2008**, 1-265.
15. Bilodeau M., Meyer F., and Schmitt M., "Space, Structure, and Randomnes," (1st ed.) New York: Springer, **2005**, 1-346.
16. Sobczyk K. and Kirkner D. J., "Modeling of Microstructures," (1st ed.) Cambridge: Birkhäuser Boston, **2012**, 1-238.
17. Molohanov I. S., "Estimation of the Size Distribution of Spherical Grains in the Boolean Model," *Biometrical Journal*, **1990**, *32*(7), 877-886.
18. Arns C. H., "The Influence of Morphology on Physical Properties of Reservoir Rocks," PhD Thesis, University of New South Wales, Sydney, Australia, **2002**, 1-71.

8

A Nonlinear Creep-damage Constitutive Model of Mudstone Based on the Fractional Calculus Theory

Zhang Jun and Zeng Jia*

School of Petroleum Engineering, Northeast Petroleum University, Heilongjiang, China

ABSTRACT

During the flood development in an oil field, the creep characteristic of mudstone is one of the important factors causing casing damage. In this study, based on the theory of fractional order differential and taking into account the creep damage evolution rules, a fractional nonlinear creep-damage model is proposed to reflect the instantaneous deformation in loading processes and the accelerated creep stage of mudstone. After assuming that the relationships between strain and time can be expressed by the exponential functions, the creep-damage constitutive model to describe the relationship between stress and strain stress of mudstone is established under the condition of accelerated strain rate loading. Furthermore, uniaxial creep tests and conventional triaxial compression tests were carried out to validate the proposed model. It is found that the fractional nonlinear creep-damage model can effectively describe the characteristics of the three stages of mudstone creep; moreover, this model can reduce the number of components and the parameters. Both the fractional nonlinear creep model and the creep-damage constitutive model have a high fitting relationship with the test results. Also, the initial elastic modulus and confining pressure are in a good linear relationship. Finally, the parametric sensitivity analysis of the theoretical model is carried out. The correctness and applicability of models were confirmed from three respects of the derivation process, test results, and the theoretical analysis.

Keywords: Fractional Calculus, Stress-Strain Relationship, Damage Mechanics, Mudstone Accelerated Creep

INTRODUCTION

Creep behavior is one of the most important mechanical properties of rock. During the flood development in oil field, mudstone creep is an important factor causing casing damage. After the water absorption and softening in mudstone, the mechanical properties and the stress state of which changed, the diagenetic cementation force gradually disappeared, and became plastic. Under the action of far field stress, the mudstone begins to creep and produces displacement and deformation around the borehole that makes the field stress acting on the cement sheath and casing. Therefore, the casing transfigured and damaged [1,2]. An accurate description of the deformation constitutive relation of mudstone is the main problem to solve the casing damage. In recent years, many scholars have made a lot of

***Corresponding author**

Zeng Jia

Email: herogail@163.com

researches on the rock creep process which mainly consider the first and second creep stage. But in fact, in the process of rock creep, the third stage of creep characteristics has a more important significance [3]. At present, most studies tend to use integer order models to describe the creep behavior and creep constitutive relation of rock and soft soil; these models are intuitive and easy to understand and have clear physical concepts [4-6]. However, a large number of experimental results show that the integer order differential creep model cannot well be consistent with the experimental data. For example, the standard linear creep model can describe the instantaneous and delayed elasticity of solids. But the whole process of the creep compliance and relaxation modulus cannot fit well with the experimental data, especially in the initial stage of creep or relaxation [7,8]. Furthermore, these models cannot describe the whole process curve of rock creep very well, especially for the description of the accelerated creep stage [9-11]. The fractional calculus theory is an extension of the integer order calculus to any order and can be used to study the differential and integral of any order. Therefore, the fractional order calculus theory has obvious advantage in establishing the rock creep constitutive model and describing the rock creep characteristics [12-14]. In recent years, some scholars have used the modeling approach of the element combination model for reference and used the fractional order viscous bodies instead of the classical viscous materials to describe the creep behavior of rock and soil materials. It provides a new idea for the study of the creep model of rock and soil [15,16]. Based on the fractional calculus theory, Yin et al. proposed a kind of software element to simulate the rock and soil material between ideal solid and fluid. The combination model consists of this element and the classical linear mechanical element can describe the nonlinear behavior of the rock decay creep and steady creep [17]. Yin et al. established an improved triaxial creep model of CCG on the basis of a Nishihara model and another visco-elasto-plastic model, parameters of which were fitted on test data [18]. Zhou et al. proposed a new creep constitutive model on the basis of time-based fractional derivative by replacing a Newtonian dashpot in the classical Nishihara model with the fractional derivative Abel dashpot of a variable viscosity coefficient [19,20]. Yao et al. constructed a fractional order derivative Burgers creep model to simulate the steady stage of creep by replacing the serial mode of Burgers model with a fractional order derivative function model; they also used a fractional order derivative acceleration Burger creep model to simulate the acceleration stage of creep by connecting the acceleration element with the fractional order derivative Burgers model [21]. Chen et al. proposed a new four-element nonlinear visco-elasto-plastic rheological model based on an expression with a fractional calculus form [22]. Wu et al. presented a fractional nonlinear dashpot element on the basis of a non-Newton fluid viscous damping element which can reflect the nonlinear acceleration characteristic of the creep curve. Afterwards, a new nonlinear creep constitutive model was obtained by introducing the element [23,24]. Li and Yue built a fractional order derivative nonlinear rheological model which consists of five components and obtained the analytical solution of the model [25]. Kang et al. proposed a fractional non-linear model to describe the creep behavior of coal taking into account the visco-elasto–plastic characteristics and the damage effect [26]. The models in the above research are almost established by replacing the classical element with the fractional order element, which can well fit rock creep test curve of decay creep stage and steady creep stage; however, they cannot reflect the accelerated creep characteristics well.

In this paper, a nonlinear creep-damage model of mudstone with differential expression was proposed combining the damage mechanics theory and the fractional calculus theory. The model has some advantages such as a simple structure and a few numbers of parameters, and it can reflect the whole process of rock creep. After assuming that the relationships between strain and time can be expressed by the exponential functions, the creep-damage constitutive model is established under the condition of accelerated strain rate loading. Finally, the rationality and applicability of the models were verified by the mudstone creep test and the conventional triaxial compression test results.

EXPERIMENTAL PROCEDURES

Establishment of Nonlinear Model

Fractional Calculus Basic Element

Fractional calculus is to extend the order of calculus to the field of fractions and even negative numbers. The software element with fractional calculus is considered to be an ideal element model between solid and fluid, which can well reflect the visco-elastic characteristics of rock and soil materials. Based on the Riemann-Liouville fractional calculus operator theory, the β order integral of the function f(t) is defined as:

$$\frac{\mathrm{d}^{-\beta} f(t)}{\mathrm{d}t^{-\beta}} = D_t^{\beta} f(\mathrm{t}) = \frac{\mathrm{d}^n}{\mathrm{d}t^n}\left[D_t^{-(\mathrm{n}-\beta)} f(\mathrm{t})\right] \quad (1)$$

And the β order differential is defined as:

$$\frac{\mathrm{d}^{\beta} f(t)}{\mathrm{d}t^{\beta}} = D_t^{\beta} f(\mathrm{t}) = \frac{\mathrm{d}^n}{\mathrm{d}t^n}\left[D_t^{-(\mathrm{n}-\beta)} f(\mathrm{t})\right] \quad (2)$$

where, $\beta > 0$, and n-1 $< \beta \le n$;

$\Gamma(\beta)$ is the Gamma function defined as:

$$\Gamma(\beta) = \int_0^{\infty} e^{-t} t^{\beta-1} dt \ \left(\mathrm{Re}(\beta) > 0\right) \quad (3)$$

The Laplace transformation formula of fractional calculus is:

$$L\left[D_t^{-\beta} f(\mathrm{t}), \mathrm{p}\right] = p^{-\beta} \bar{f}(p)$$
$$L\left[D_t^{\beta} f(\mathrm{t}), \mathrm{p}\right] = p^{\beta} \bar{f}(p) \quad (4)$$

(The function f(t) can be integrated when t is close to 0, $0 \le \beta \le 1$)

where, $\bar{f}(p)$ is the Laplace transformation of f(t). According to the classical solid mechanics and the fluid mechanics theory, the constitutive relation of the ideal solid should satisfy Hooke law $\sigma(t)$ - $\varepsilon(t)$, and the ideal fluid should satisfy Newton's law of viscosity $\sigma(t) - d^1\varepsilon(t)/dt^1$. The fractional calculus can be used to study the characteristics of the differential and integral operators of any order. If we transform $\sigma(t) - \varepsilon(t)$ into $\sigma(t) - d^0\varepsilon(t)/dt^0$, the fractional order differential form of the stress-strain relationship of rock mass between the ideal solid and ideal fluid can be expressed as:

$$\sigma(t) = \xi \frac{\mathrm{d}^{\beta}\varepsilon(t)}{\mathrm{d}t^{\beta}} \quad (5)$$

where, ξ is the visco-elastic coefficient which is similar to the elastic modulus of Hooke law. The dimension of ξ is "stress-strain β".

When $0 < \beta < 1$, Equation 5 describes the state of the matter between ideal solid and ideal fluid; When $\beta > 1$, Equation 5 describes the state of the accelerating rheology. In this paper, we mainly study the accelerating rheological state of the material. According to this expression method, we define the element which $\beta > 1$ as a software element (Figure 1).

Figure 1: Fractional calculus viscous element.

When $\sigma(t)=const$, the stress is a fixed value. Based on the Riemann-Liouville fractional operator theory, Equation 5 can be transformed into:

$$\varepsilon(t)=\frac{\sigma}{\xi}\frac{t^{\beta}}{\Gamma(1+\beta)} \quad (6)$$

Under the condition of constant stress, the creep curves of fractional calculus viscous element are shown in Figure 2. When $\beta=0$ with $\beta=1$, the curves describe the creep characteristics of ideal solid and ideal fluid respectively. When $0<\beta<1$ the strain increases slowly, but the strain rate decreases gradually, showing the characteristic of slow speed change. When $\beta>1$, both strain and strain rate increase significantly with time. Moreover, the creep characteristic is gradually strengthened with an increase in β. By considering the deformation characteristics of three stages of rock creep, it is vivid when $\beta=0$, the fractional order differential element describes the instantaneous elastic deformation. In addition, the element describes the visco-elastic deformation (deceleration creep stage and steady creep stage) when $0<\beta<1$ and visco-plastic deformation (accelerated creep stage) when $\beta>1$.

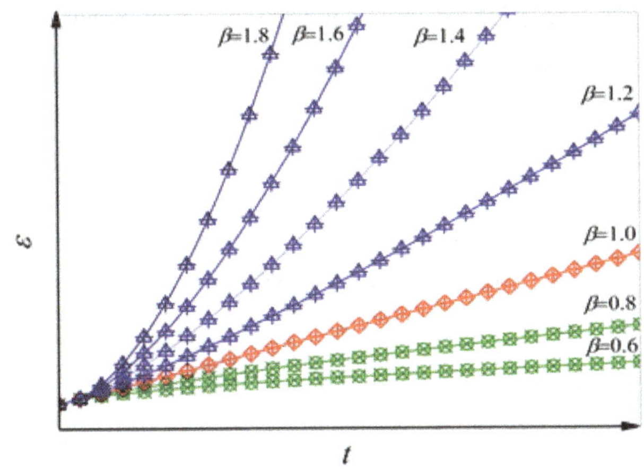

Figure 2: The creep curves of fractional calculus element.

Based on the derivation and the change rule of fractional order model, combining with the typical rock creep deformation characteristics, we can use the element to simulate the creep curve of rock. When *ε(t)=constant*, by using the fractional calculus, the relaxation equation is derived as:

$$\sigma(t)=\xi\,\varepsilon_0\frac{t^{-\beta}}{\Gamma(1-\beta)} \quad (7)$$

where, ε_0 is the initial strain under the initial stress. For the different materials, we can change the creep curve or the relaxation curve by adjusting element parameters β and ξ so as to accurately fit the experimental results of the material.

Damage in the Process of Rock Creep

The critical value of rock creep damage stress is the long-term strength of rock. When the external stress is greater than the long-term strength of rock itself, rock will produce creep damage with an increase in time. According to the basic theory of damage mechanics, under the action of external stress, the micro cracks in the rock mass expand, encrypt, and increase the ability of rock mass to resist damage, and deformation decreases gradually. Therefore, the damage degree of rock mass is related to the magnitude of the external stress and the time. The damage variable in this paper is defined by the method of elastic modulus:

$$D(\sigma,t)=1-\frac{E(\sigma,t)}{E_0} \quad (8)$$

where, E_0 is initial elastic modulus of rock mass; $D(\sigma, t)$ is damage variable of rock mass at time t; $E(\sigma, t)$ represents elastic modulus of rock mass at time t and is mainly related to the external stress level at the moment. The rock has lost its carrying capacity after creep and failure. Based on the study on the creep damage of mudstone by Liu et al., $E(\sigma, t)$ is defined using the following relation:

$$E(\sigma,t)=E_0\exp\left[-\langle\sigma-\sigma_\infty\rangle t/b\right] \tag{9}$$

where, σ_∞ is the long term strength of rock material, and it can be determined by an experiment; b is a rock material constant; $\langle\sigma-\sigma_\infty\rangle$ is a step functions defined by:

$$\langle\sigma-\sigma_\infty\rangle=\begin{cases}0 & (\sigma\le\sigma_\infty)\\ \sigma-\sigma_\infty & (\sigma>\sigma_\infty)\end{cases} \tag{10}$$

After inserting Equation 9 into Equation 8, the following relation is obtained:

$$D(\sigma,t)=1-\exp\left[-\langle\sigma-\sigma_\infty\rangle t/b\right] \tag{11}$$

According to Equation 11, when the external load stress is greater than the long-term strength of rock mass $(\sigma>\sigma_\infty)$ and $t\to 0$, the damage variable (*D*) is equal to 1, which means the complete damage of rock. As can be seen, with the increase of time (*t*) and stress (σ), the elastic modulus of the material decreases gradually, but the damage variable increases gradually. Based on the definition of damage variable *D* in Equation 11, the effective stress and nominal stress should meet the following relationships:

$$\tilde{\sigma}=\frac{\sigma}{1-D(\sigma,t)} \tag{12}$$

where, $\tilde{\sigma}$ is the effective stress, and σ is the nominal stress. Inserting Equation 7 into Equation 8:

$$\tilde{\sigma}=\sigma\exp\left[\langle\sigma-\sigma_\infty\rangle t/b\right] \tag{13}$$

The Fractional Order Creep-damage Model of Mudstone

Due to the phenomenon of elastic deformation of rock mass in the early loading stage, the elastic element is used to describe the instantaneous elastic deformation. The accelerated creep characteristic of rock mass is gradually changing with time and external load stress, and there is a critical value of damage in this process. Therefore, the accelerated deformation process of rock mass is described by using the fractional order damage element. The model structure diagram is shown in Figure 3.

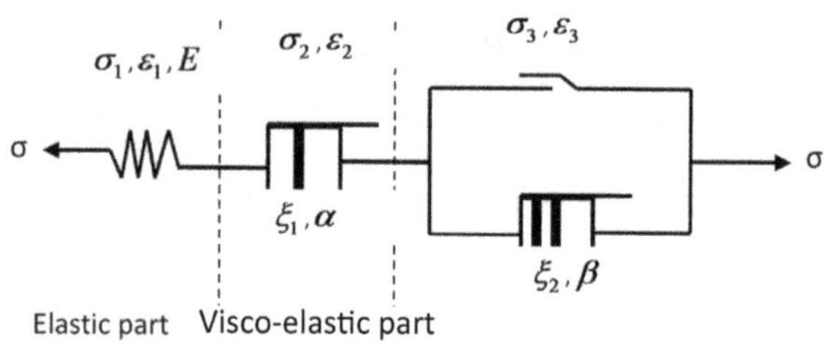

Figure 3: A fractional order damage creep model.

When the external load stress is greater than the long-term strength of rock mass, mudstone is in the stage of accelerated creep. According to the model diagram, we can obtain:

$$\begin{aligned}&\sigma=\sigma_1=\sigma_2=\sigma_3\\&\varepsilon=\varepsilon_1+\varepsilon_2+\varepsilon_3\end{aligned} \tag{14}$$

$$\begin{cases}\sigma_1=E\varepsilon_1\\ \sigma_2=\xi_1\dfrac{d^\alpha\varepsilon_2}{dt^\alpha}\\ \sigma_3=\sigma_\infty+\xi_2\dfrac{d^\beta\varepsilon_3}{dt^\beta}\end{cases} \tag{15}$$

When the external stress is less than the long-term strength of mudstone, the mudstone is in the stages of creep of I and II, and viscoelastic deformation occurs. In this situation, $0<\alpha<1$. After rewriting Equation 15 by Laplace transform and Laplace inverse transform and inserting it into Equation 14, the nonlinear rheological constitutive equation can be obtained:

$$\varepsilon=\frac{\sigma}{E_0}+D_t^{-\alpha}\left(\frac{\sigma}{\xi_1}\right)+D_t^{-\beta}\left(\frac{\sigma-\sigma_\infty}{\xi_2}\right) \tag{16}$$

Under the condition that the external load stress is greater than the long-term strength of the rock mass, the rock mass will enter the stage of creep damage. Equation 16 can be rewritten as:

$$\varepsilon=\frac{\sigma}{E_0}+D_t^{-\alpha}\left(\frac{\sigma}{\xi_1}\right)+D_t^{-\beta}\left(\frac{\tilde{\sigma}-\sigma_\infty}{\xi_2}\right) \tag{17}$$

By considering the initial conditions of t=0 and $\sigma = \sigma_0$, the nonlinear creep constitutive equation can be obtained based on the Riemann-Liouville fractional calculus theory.

$$\varepsilon = \frac{\sigma_0}{E_0} + \frac{\sigma_0}{\xi_1}\frac{t^\alpha}{\Gamma(1+\alpha)} + \frac{\tilde{\sigma} - \sigma_\infty}{\xi_2}\frac{t^\beta}{\Gamma(1+\beta)} \quad (18)$$

By inserting Equation 13 into Equation 18, the creep damage constitutive model based on the fractional calculus theory was obtained:

$$\varepsilon = \frac{\sigma_0}{E_0} + \frac{\sigma_0}{\xi_1}\frac{t^\alpha}{\Gamma(1+\alpha)} + \frac{\sigma_0 \exp\left[\langle \sigma - \sigma_\infty \rangle t/b\right] - \sigma_\infty}{\xi_2}\frac{t^\beta}{\Gamma(1+\beta)} \quad (19)$$

The Constitutive Model of Mudstone under Accelerated Strain Rate Loading

According to the actual situation, assuming that $\varepsilon = ae^t$ and inserting into Equation 19, the stress-strain relationship of mudstone creep under the condition of accelerated strain rate loading can be obtained:

$$\varepsilon = \frac{\sigma}{E_0} + \frac{\sigma}{\xi_1}\frac{(\ln\varepsilon - \ln a)^\alpha}{\Gamma(1+\alpha)} + \frac{\sigma \exp\left[\langle \sigma - \sigma_\infty \rangle (\ln\varepsilon - \ln a)/b\right] - \sigma_\infty}{\xi_2}\frac{(\ln\varepsilon - \ln a)^\beta}{\Gamma(1+\beta)} \quad (20)$$

where, a is a constant. The creep damage constitutive model of mudstone under the condition of accelerating strain rate loading can be derived by the Equation 20.

$$\sigma = \frac{E_0\xi_1\Gamma(1+\alpha)\xi_2\Gamma(1+\beta)\varepsilon + E_0\xi_1\Gamma(1+\alpha)(\ln\varepsilon - \ln a)^\beta \sigma_\infty}{\xi_1\Gamma(1+\alpha)\xi_2\Gamma(1+\beta) + E_0\xi_2\Gamma(1+\beta)(\ln\varepsilon - \ln a)^\beta + E_0\xi_1\Gamma(1+\alpha)(\ln\varepsilon - \ln a)^\beta \exp\left[\langle \sigma - \sigma_\infty \rangle (\ln\varepsilon - \ln a)/b\right]} \quad (21)$$

When the external load stress is greater than the long-term strength of mudstone ($\sigma > \sigma_\infty$), the mudstone will enter the creep-damage stage and complete damage occurs. At this time, $\exp[-(\sigma-\sigma_\infty) t/b]=0$ and Equation 21 can be rewritten as:

$$\sigma = \frac{E_0\xi_1\Gamma(1+\alpha)\xi_2\Gamma(1+\beta)\varepsilon + E_0\xi_1\Gamma(1+\alpha)(\ln\varepsilon - \ln a)^\beta \sigma_\infty}{\xi_1\Gamma(1+\alpha)\xi_2\Gamma(1+\beta) + E_0\xi_2\Gamma(1+\beta)(\ln\varepsilon - \ln a)^\beta + E_0\xi_1\Gamma(1+\alpha)(\ln\varepsilon - \ln a)^\beta} \quad (22)$$

According to Equation 22, the stress-strain curves of mudstone under the creep condition are shown in Figure 4. With an increase in the strain, the stress increases, but the increase rate decreases gradually. It shows that in the process of mudstone creep, the greater the strain is, the greater the stress becomes. The rock stress-strain curves under different fractional orders (β) can also be obtained (Figure 4). As can be seen from the figure, the stress and strain present the exponential relation, and the increasing rate of stress decreases with an increase in β. The smaller the β is, the more obvious the stress changes.

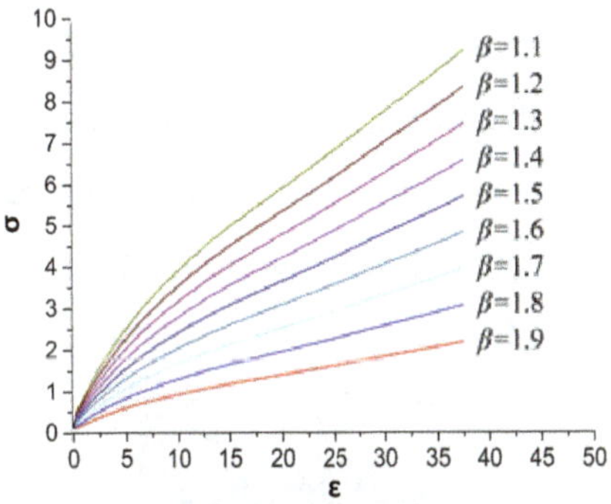

Figure 4: Stress-strain curves from theoretical derivation under different fractional orders (β).

Model Validation and a Parametric Sensitivity Analysis

Experimental Verification

In order to verify the correctness and rationality of the models, the creep test and triaxial compression test are carried out. The sample is the mudstone of casing damaged wells from Daqing Oilfield in China. The test is used the experimental set-up of creep test of rock from Northeast Petroleum University (Figure 5). Test instrument parameters are as follows: confining pressure which ranges from 0 to 30 MPa; axial load which ranges from 0 to 600 kN; indoor temperature and humidity which are controlled by air conditioning. The standard size of mudstone samples is 75 mm × 150 mm (diameter × height).

Figure 5: The experimental set-up of the creep test of rock.

First, the creep test of mudstone is carried out by the step loading. The load imposed on the sample increased gradually. The first step loading stress is 4 MPa and then increases by 2 MPa during each step. The loading time of each step is about 14 days and test is stopped until the specimen fails. The whole process creep curves of mudstone were obtained after about 142 days. From the experimental curves in Figure 6, we can see the obvious nonlinear acceleration characteristics at the last loading step, where the loading stress is about 18 MPa. In this step, based on the experimental data, the time to enter the acceleration creep stage can be roughly judged, t=12.34 day and ε_a=5.03%. Based on the nonlinear creep damage model, the experimental data of 8th loading step were fitted and analyzed by using nonlinear least square method. Other initial parameters are assumed as follows: E_0=2.0 GPa; $\xi_1 = 18Gpa.h^{\alpha}, \xi_2 = 18GPa.h^{\beta}, \alpha = 0.3; \beta = 1.5.$

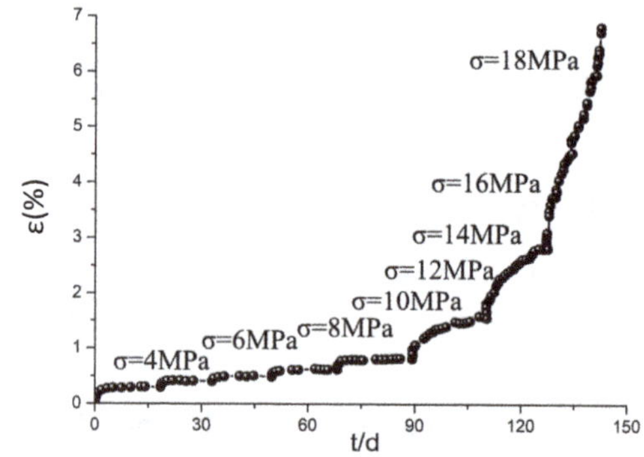

Figure 6: The whole process creep.

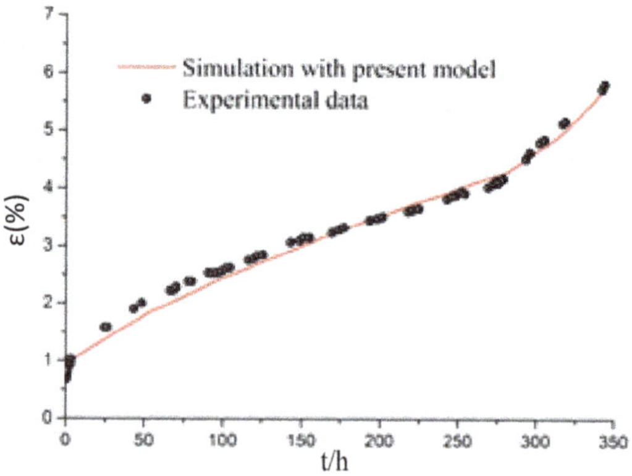

Figure 7: Experimental data and the fitting curves by nonlinear fractional derivative model curves.

The fitting analysis results of the experimental data are shown in Figure 7. It can be seen that the fractional order nonlinear creep-damage model is in good agreement with the experimental data. The fractional order nonlinear creep-damage model can not only reflect the early stages of the creep nonlinear gradual process, but it also describes rock nonlinear accelerative creep properties.

Then, the conventional triaxial compression tests were carried out on the mudstone specimens, and the stress-strain curves at different confining pressures were obtained (Figure 8). Equation 22 is used to fit and analyze the experimental data and found the following features: (1) the experimental results can be well fitted to the constitutive model of mudstone derived in this paper; (2) at different confining pressures, β of each curve is 1.37. It means that β does not change with confining pressure, and it can reflect the "soft and hard" degree of mudstone; (3) at different confining pressures, there is a very good linear relationship between initial elastic modulus and confining pressure of each curve (Figure 9).

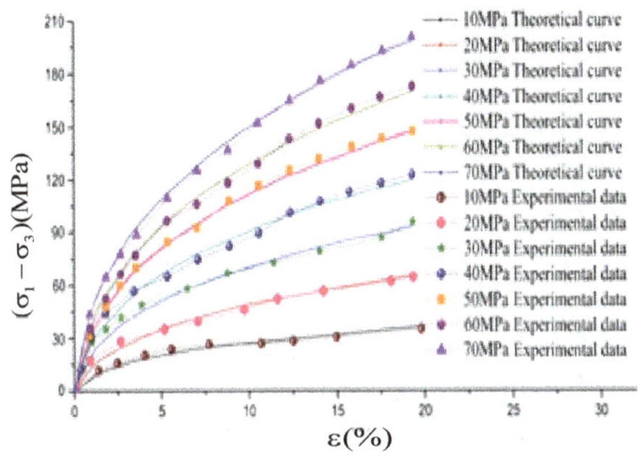

Figure 8: Stress-strain curves of triaxial compression tests.

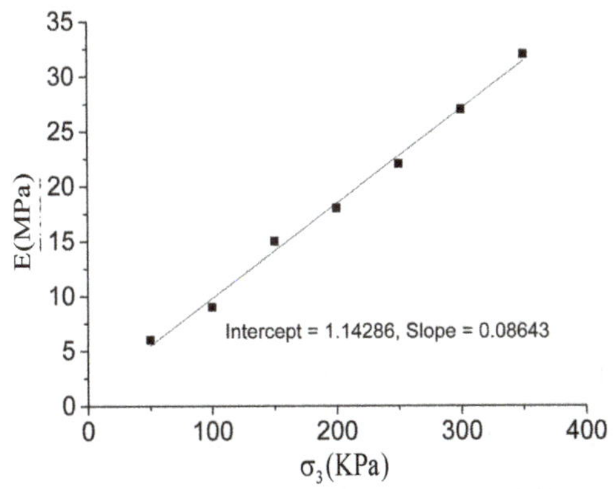

Figure 9: Confining pressure and initial tangential.

The Parametric Sensitivity Analysis

From the fractional nonlinear creep-damage model of mudstone and the discussion of previous sections, it is obvious that the performance of the proposed model depends on several material parameters, among which the fractional order parameters α and $\boldsymbol{\beta}$, the damage factor ω [$\omega=(\sigma-\sigma_{\infty})/b$], the fractional viscosity coefficients ξ_1 and ξ_2 are especially important. The sensitivity of creep strain to these parameters is shown in Figures 10-12, in which one parameter takes three different values to show its effect on the strain-time curve of mudstone creep, while other parameters use fixed values.

The effect of fractional order α on the variation of strain with creep time is shown in Figure 10(a). It can be seen that when α increases, indicating the transition of fractional element in the visco-elastic body from an elastic element to a viscous element, the strain rate corresponding to each creep stage rises accordingly. However, the increase of fractional order $\boldsymbol{\beta}$ only reduces the strain rate corresponding to the accelerating creep stage, while the transient and steady creep stages are scarcely affected, as shown in Figure 10(b). This is because the fractional element in the visco-plastic body takes effect only when the mudstone damage dominates the process of creep deformation.

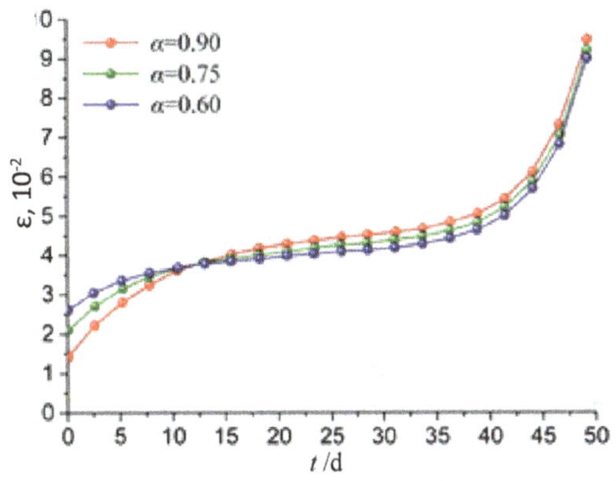

(a) The effect of α on the mudstone creep curves.

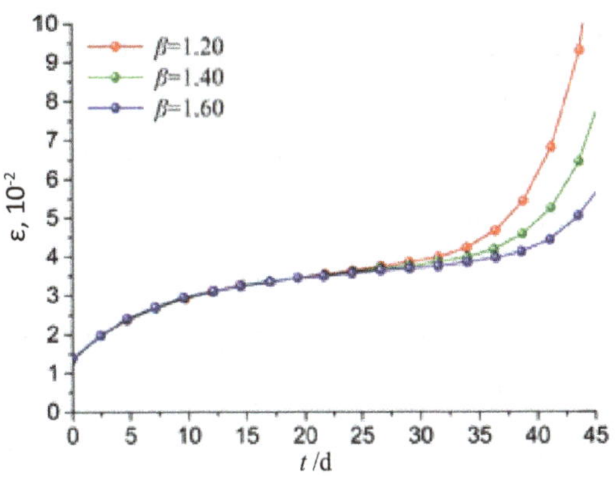

(b) The effect of the β on the mudstone creep curves.

Figure 10: Sensitivity of the creep strain to fractional order parameters.

The effect of damage factor ω [$\omega=(\sigma-\sigma_\infty)/b$] on the creep curves is illustrated in Figure 11. Since the factor ω controls the damage evolution with creep time, the greater the factor ω is, the more rapidly the mudstone damage evolves. Consequently, it can be seen from Figure 11 that the curve corresponding to the larger value of ω branches earlier from the transient creep stage into the accelerating creep stage. It means that the occurrence of accelerating creep is much earlier with an increase in ω. Moreover, the duration of each creep stage is also reduced with an increase in ω.

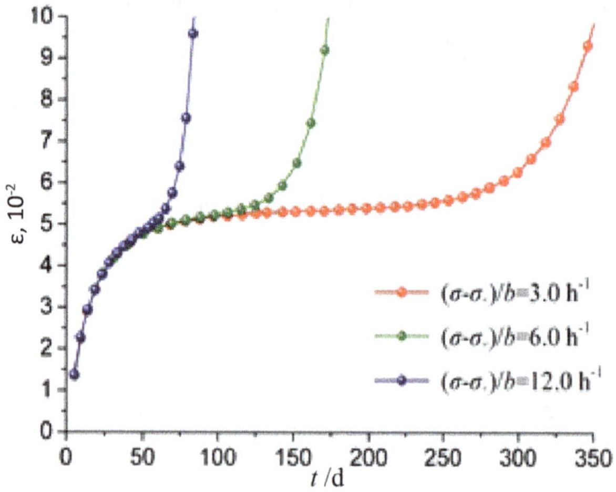

Figure11: The effect of damage factor ω [$\omega=(\sigma-\sigma_\infty)/b$] on the mudstone creep curves.

Finally, the sensitivity of creep strain to fractional viscosity coefficients ξ_1 and ξ_2 is demonstrated in Figure 12. It is apparent that the creep strain increases with the decrease of viscosity. The parameter ξ_1 has a global effect on the creep evolution process, while the effect of ξ_2 is limited to the II and III creep stages, which is similar to the case of fractional orders α and β. The effect of ξ_1 on the accelerating creep is smaller compared to its effect on transient and steady creep. Furthermore, a smaller value of ξ_2 leads to a shorter period of transient and steady creep.

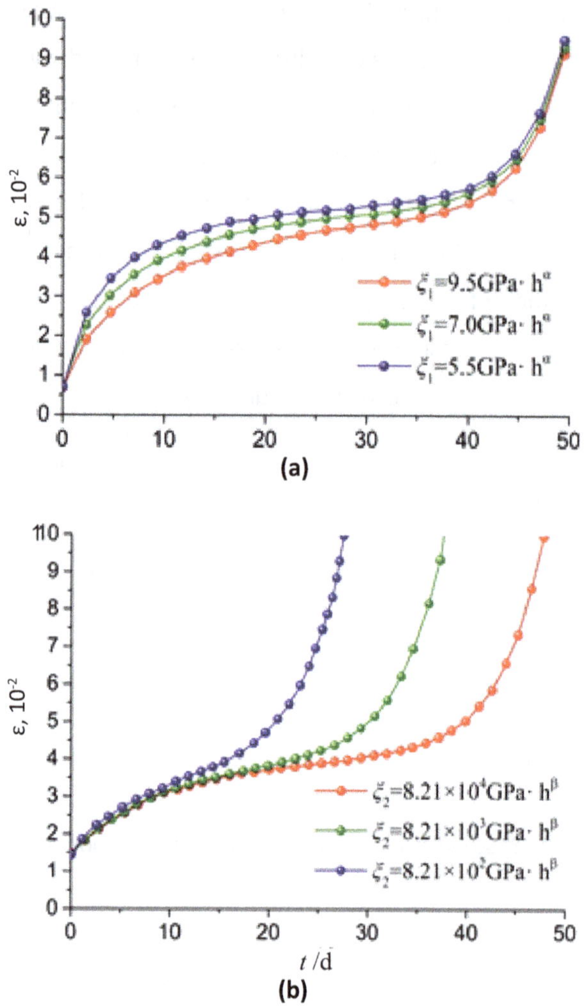

Figure 12: Sensitivity of the creep strain to fractional viscosity coefficients.

CONCLUSIONS

(1) Based on the fractional calculus theory, two software elements with a fractional order differential and a spring element were connected in series. Also, by introducing the creep damage evolution law, a fractional nonlinear creep-damage model of mudstone was proposed which can well describe the instantaneous deformation and the accelerated creep characteristics in loading process.

(2) Based on the nonlinear creep-damage model, assuming that the relationships between strain and time can be expressed by the exponential functions, the creep-damage constitutive model to describe

the relationship between stress and strain stress of mudstone is established under the condition of accelerated strain rate loading. The results show that under the accelerated creep condition, the stress and strain present the exponential relation, and the increasing rate of stress decreases with the increase of β. In addition, the smaller the β is, the more obvious the stress changes.

(3) Through the mudstone creep test, the whole creep curve was obtained. Then, the nonlinear creep-damage model is fitted with the data of creep test. The results show that the model based on fractional differential element can not only reflect the initial creep stage effectively, but it can also reflect the nonlinear acceleration creep characteristics. The model and the experimental data are in good agreement, so this good agreement can reduce the number of components and the parameters.

(4) The creep-damage constitutive model of mudstone is validated by the data obtained from the conventional triaxial compression tests. The results show that the theoretical model is consistent with the experimental data. Moreover, it is found that β does not change with confining pressure, and it can reflect the "soft and hard" degree of mudstone. There is a very good linear relationship between initial elastic modulus and confining pressure.

(5) The fractional calculus theory in describing the mechanical properties of rock and soft soil has other advantages such as the description of the volume change. Due to the limitation of this research, we will next carry out further researches.

REFERENCES

1. Yan X. Z., Yang H. L., and Yang X. J., "The Reason Analysis of Mudstone Creep on Casing Damage," *Drilling & Production Technology*, **2003**, *26*(3), 65-69.
2. Huang X. L., Yang C. H., Liu J. J., and et al., "Experimental Study on Mudstone Creep Behavior under Different Water Contents and Its Effect on Casing Damage," *Rock Mechanics and Engineering*, **2008**, *2*(2), 3477-3487.
3. Song Y. J. and Lie S. Y., "Mechanical Model of Rock Nonlinear Creep Damage Based on Fractional Calculus," *Chinese Journal of Underground Space and Engineering*, **2013**, *9*(1), 91-95.
4. Yang C. H., Daemen J. K., and Yin J. H., "Experimental Investigation of Creep Behavior of Salt Rock," *International Journal of Rock Mechanics and Mining Sciences*, **1999**, *36*(2), 233-242.
5. Wang G. J., "A New Constitutive Creep-Damage Model for Rock and its Characteristic," *Int. J. Rock Mech. Min. Sci*, **2004**, *41*(1), 61-67.
6. Shao J. F., Zhu Q. Z., and Su K., "Modeling of Creep in Rock Materials in Terms of Material Degradation," *Computers and Geotechnics*, **2003**, *30*(7), 549-555.
7. Shao J. F., Chau K. T., and Feng X. T., "Modeling of Anisotropic Damage and Creep Deformation in Brittle Rock," *International Journal of Rock Mechanics and Mining Sciences*, **2006**, *43*(4), 582-592.
8. Li W. Q., Li X. D., Han B., and et al., "Recognition of Creep Model of Layer Composite Rock Mass and Its Application," *Journal of Central South University of Technology*, **2007**, *17*(1), 329-331.
9. Xu W. Y., Yang S. Q., Xie S. Y., and et al., "Investigation on Triaxial Rheological Mechanical Properties of Green Schist Specimen(II) Model Analysis," *Rock and Soil Mechanics*, **2005**, *26*(5), 693-698.
10. Zhou H. W., Wang C. P., Han B. B., and et al., "A Creep Constitutive Model for Salt Rock Based

on Fractional Derivatives," *Int. J. Rock Mech. Mining Sci.*, **2011**, *48*(1), 116-121.

11. Li Y. L., Yu H. C., and Liu H. D., "Study of Creep Constitutive Model of Silty Mudstone under Triaxial Compression," *Rock and Soil Mechanics*, **2011**, *33*(7), 2035-2040.
12. Sun H. Z. and Zhang W., "Analysis of Soft Soil with Viscoelastic Fractional Derivative Kelvin Model," *Rock and Soil Mechanics*, **2007**, *28*(9), 1983-1986.
13. He L. J., Kong L. W., and Wu W. J., "A Description of Creep Model for Soft Soil with Fractional Derivative," *Rock and Soil Mechanics*, **2011**, *32*(2), 239-249.
14. Xu G. W., He C., Hu X. Y., and Wang S. M., "A Modified Nishihara Model Based on Fractional Calculus Theory and its Parameter Intelligent Identification," *Rock and Soil Mechanics*, **2015**, *43*, 132-147.
15. Wang J. B., Liu X. R., Rong J. Q., and et al., "Creep Properties of Salt Rock and Its Nonlinear Constitutive Model," *Journal of China Coal Society*, **2014**, *39*(3), 445-451.
16. Ding J. Y., Zhou H. W., Li C., and et al., "The Fractional Derivative Approach of Creep Constitutive Model of Salt Rock Based on WEIBULL Distribution," *Journal of Solid Mechanics*, **2013**, *34*(5), 675-679.
17. Yin D. S., Ren J. J., and He C. L., "Stress-stain Relation of Soft Soil Based on Fractional Calculus Operators Theory," *Chinese Journal of Rock Mechanics and Engineering*, **2009**, *28*(1), 2973-2979.
18. Yin G. Z., Wang D. K., Huang G., Zhang D. M., and et al., "A Triaxial Creep Model for Coal Containing Gas and its Stability Analysis," **2009**, *15*(3), 248-251.
19. Zhou H. W., Wang C. C., Duan Z. Q., and et al., "Time-based Fractional Derivative Approach to Creep Constitutive Model of Salt Rock," *Scientia Sinica Physica, Mechanica & Astronomica*, **2012**, *42*(3), 310-318.
20. Zhou H. W., Wang C. P., Mishnaevsky J. L., and et al., "A Fractional Derivative Approach to Full Creep Regions in Salt Rock," *Mechanics of Time-Dependent Materials*, **2013**, *17*(3), 413-425.
21. Yao Z. M., Xu Y., Wang H. L., and Shao W., "Fractional Order Derivative Constitutive of Artificial Frozen Soil," *Journal of Sichuan University of Science and Technology* (Natural Science).
22. Chen L., Chen S. G., Zhang H., and Yang J. S., "A Nonlinear Visco-elasto-Plastic Creep Model Based on Fractional Calculus," *Journal of Sichuan University of Science and Technology (Natural Science)*, **2013**, *45*(7), 7-14.
23. Wu F., Liu J. F., Bian Y., and Zhou Z. W., "Fractional Derivative Creep Model of Salt Rock," *Journal of Sichuan University (Engineering Science Edition)*, **2014**, *46*(5), 22-28.
24. Wu F., Xie H. P., Liu J. F., and et al., "Experimental Study of Fractional Viscoelastic-plastic Creep Model," *Chinese Journal of Rock Mechanics and Engineering*, **2014**, *33*(5), 964-970.
25. Li R. Z. and Le J. Z., "Nonlinear Rheological Constitute of Soft Soil Based on Fractional Order Derivative Theory," *Journal of Basic Science and Engineering*, **2014**, *22*(5), 856-865.
26. Kang J. H. , Zhou F. B., Liu C., and Liu Y. K., "A Fractional Non-linear Creep Model for Coal Considering Damage Effect and Experimental Validation," *International Journal of Non-Linear Mechanics*, **2015**, *76*, 20-28.

9

Hydraulic Fracturing Length Optimization and its Impact on the Reservoir Performance

Mohammad Hossein Zare[1]*, Abbas Helalizade[2], and Gholam Abbas Safiyan[1]

[1] National Iranian South Oilfield Company, Ahwaz, Iran

[2] Department of Petroleum Engineering, Ahwaz Faculty of Petroleum Engineering Petroleum University of Technology (PUT), Ahwaz, Iran

ABSTRACT

Hydraulic fracturing is one of the most common methods of well stimulation for reservoirs with low permeability. Hydraulic fracturing increases the flow capacity, alters flow geometry, bypasses damage and improves recovery factor. The pressure of most of Iranian oil reservoirs is declined and consequently the production is reduced. It is necessary to improve the production by using new stimulation techniques, like hydraulic fracturing. In general, hydraulic fracturing treatments are used to increase the production rate, furthermore increasing recovery factor. In such cases, the fracture length is an appropriate optimization design variable against an economic criterion, e.g., the Net Present Value (NPV). This involves the balancing of incremental future revenue against the cost of operation. The production response in economic terms shows the effect of this design parameter. In this paper, a hydraulic fracturing operation has been designed by the simulator FracCADE 5.1 then its impact on production and ultimate recovery has been investigated by ECLIPSE. According to NPV, the hydraulic fracturing schedule was designed to achieve an optimum fracture half-length.

The results show that hydraulic fracturing increases oil recovery factor and production rate significantly. According to the NPV diagram, the best fracture half-length for AZ-X well is 1100 feet and for MNS-Y well is 900 feet.

Keywords: Hydraulic Fracturing, Optimum Fracture Half Length, Net Present Value, Improve Oil Recovery, Economic Criterion.

INTRODUCTION

Iran has been producing oil for more than 100 years. Most of Iran's fields are mature and it is estimated that 200,000-250,000 bbl/d of crude production is lost annually through natural pressure declines of the fields. Moreover, with available technology, it is only possible to extract 20% to 25% of the original oil in place. Therefore, looking for new technology to sustain and increase the production is unavoidable. Hydraulic fracturing technology is presented as an effective method that can make a difference [1].

***Corresponding author**

Mohammad hossein Zare

Email: zare2@nisoc.ir

Hydraulic fracturing is now seen as an essential piece of well and reservoir administration and a pillar of production engineering rather than a choice of last resort for depleted or mysteriously underperforming wells. Fracturing has continuously expanded until it has become the completion choice for all types of wells. Possibly, as indicated by Iranian oil reservoir condition, hydraulic fracturing is the appropriate technique to upgrade oil recovery, increase production rate. It can bring back old hydrocarbon fields to life.

The essence of hydraulic fracturing is production improvement, i.e. accelerating production in a depleted drainage area. The question, of course, is whether this acceleration of the production, compared to an unfractured well and evaluated through their respective present values of the revenue streams (one for a fractured well and another for an unfractured well), can be adjusted against the expenses of the treatment. In addition to the increase in production, an assessment of the economics of a hydraulic fracturing treatment must consider many factors, including treatment costs, additional reserves that may be produced before the well reaches its economic limit and reservoir risks associated to mechanical problems that could cause the treatment to be unsuccessful. While there are other criteria for assessing economic attractiveness or lack thereof, for the purposes of this work, we will use the Net Present Value as the criterion for the optimization and the evaluation of the desirability of the specific hydraulic fracture treatment. The NPV criterion has been already selected by several researchers to monetize the incremental production obtained after a fracturing treatment. No matter how sophisticated is the model used to predict fracture propagation and performance, the economic optimization requires a trial and error process to determine the optimum treatment design from a set of the calculated physical designs. According to the studies which done on two reservoirs in Iran, besides selecting the best layer for making fracture, the most optimized case designed and effect of hydraulic fracturing on production studied.

The first commercial hydraulic fracturing treatment aimed to enhance production was conducted in the Hugoton Gas Field in 1947 on Kelpper Well 1. As a fracturing fluid, a gasoline-based napalm-gel was used. However, these unpropped treatments did not increase production leading to the belief that hydraulic fracturing did not represent any improvement in the performance of the wells [2]. In 1949, 11 out of 23 wells where hydraulic fracturing treatments had been applied reported a significant increase in productivity in fields located in Wyoming, Colorado, Oklahoma and Texas [3]. In 1957, mathematical relations between fracturing efficiency, injection rate, pumping time and fracture width were developed to predict fracture extent. This explained why some fluids were more efficient than others and why some pumping rates yielded better results. From this time forward, fracturing treatments changed from an experimental basis to technically-based perspective [4]. Hydraulic fracturing is well-known as a technique to improve productivity in low permeability reservoirs and overcome damage in moderate and high-permeability reservoirs. Hydraulic fracturing provides improved fines control in unconsolidated formations; in addition the pressure drop due to production will be distributed over the surface area of the created surface instead of the surface area of the wellbore or gravel pack radius [5].

EXPERIMENTAL PROCEDURES

Criteria for Selecting a Candidate Well

Choosing an excellent candidate for stimulation results in success whereas choosing a poor candidate leads to economic failure. The mechanical properties can be used in selecting oil and gas wells for fracturing treatments. If these properties are taken into considerations, they will facilitate the task of identifying the proper candidate for this expensive job [6].

Reservoir Permeability

There are several factors which govern reservoir permeability. These factors are lithology, the design of treatment, and available equipment. As these factors differ for different fields so the reservoir permeability also differs. There is not an exact range of reservoir permeability that is appropriate for hydraulic fracturing. Based on experience, there are some classifications which provide a general idea about reservoir permeability for hydraulic fracturing. According to one classification which was based on field experience, good candidates for matrix stimulation are oil reservoirs with a permeability of 10 md or more. For hydraulic fracturing, oil reservoir permeability of 1 md or below is considered a good candidate. If the permeability is between 1 md and 10 md then selecting a suitable stimulation technique is difficult and it needs more study. A permeability less than 0.1 md is a good candidate for gas reservoirs. Other classifications show that a permeability less than 5 md is a good candidate for oil reservoirs whereas less than 0.5 md is appropriate for a gas reservoir [7].

Production History of the Well

It is a good idea to consider the production history of the candidate well and to compare it with the productivity of some of the offset wells. It provides a better understanding of the productivity decline of the candidate well [7].

Oil and Gas in Place Volume, Hydrocarbon Saturation, and Reservoir Pressure

An important factor in the selection of a candidate well for hydraulic fracturing is the volume of hydrocarbons in place. The purpose of hydraulic fracturing is to increase the productivity but, if there is not enough oil and gas then it is not economically viable. For good results, it should have a considerable amount of oil and gas in place and reasonable reservoir pressure. There is no exact value for oil and gas in place, reservoir pressure, and hydrocarbon saturation, but it should be such that doing hydraulic fracturing is economically profitable. Some of the typical values are given in Table 1 [8].

Table1: Typical value of some parameters for hydraulic fracture [8].

Parameter	Oil reservoir	Gas reservoir
Hydrocarbon saturation	> 40%	> 50%
Water cut	< 30%	< 200bbl/ MMscf
Reservoir pressure	< 70% depleted	Twice abandonment press
Gross reservoir height	> 10 m	> 10 m

Two types of characteristic should be considered for a suitable layer for hydraulic fracturing operation. After selecting the suitable well which are Target layers and Barrier layers. Target layer should has characteristics such as high porosity, low in-situ stresses, low water saturation, low and proper uniaxial compressive strength (UCS), and large difference between the minimum and maximum horizontal stresses [9].

Candidate Reservoir General View

Most of Iran›s oil production comes from aging super-giant fields that have been producing for several decades, hence increasingly; they require techniques to sustain production. Although these fields still hold billions of barrels of oil, the production of the remaining reserves needs stimulation methods for improved recovery. Some of the old wells in most part of major fields such as Ahwaz have been cut back and shut-in due to rapid pressure drops and rising water cuts. Since most of the reservoirs in Iran are predominantly made of carbonates, chemical treatment is almost the only reservoir stimulation method that has been conducted routinely in recent decades. Hydraulic fracturing is one of the most effective methods of well stimulation and may have usefulness in these reservoirs if it is designed and executed properly.

Production Forecast and Analysis

The hydraulic fracturing design is completed on the basis of the production forecast and Net Present Value (NPV) analyses. In order to perform Production forecast, Eclipse Simulator was used.

Model Description

The base case is a Cartesian grid dimension 30×30×11, in X, Y and Z directions having the following characteristics. Blocks dimension is 500×500×500 ft, black oil in under-saturated reservoirs. The simulation was run for 4000 days. In this case, one vertical producer was used. Table 2 presented the reservoir parameters used in the model. Refined grid blocks were created around the wells and in the fractured area by using the local grid refinement (LGR) feature in Eclipse. The refinement along the I-direction and J-direction was determined to be NX=9, NY=30, NZ=5.

Table 2: Summary of Input parameters for homogeneous reservoir.

Capillary pressure(psi)	0.00
Permeability(md)	3.00
Porosity	0.10
Initial reservoir pressure(psi)	6300.00
Maximum oil rate(STB/Day)	10000.00
Bottom hole flowing pressure(psi)	1000.00
Radius wellbore (ft)	0.55

Local Grid Refinement Method

LGR method was used to create hydraulic fractures in reservoir simulation model. LGR is a technique within Eclipse which represents the splitting of a coarse grid block into a smaller cell in order to achieve a more detailed simulation in sensitive areas.

Refined grid blocks were created around the wells and in the fractured area by using the LGR feature in Eclipse. The degree of refinement along the wells and in the fracture was determined using sensitivity analyses. The wellbore refinement is specified in eclipse as Nx, Ny, and Nz, which are the level of refinement in I, J and K-direction, respectively. The refinement along the I-direction and J-direction was determined to be Nx=5, Ny=5. This means that the center grid blocks of the fracture LGR, representing the hydraulic fracture itself, is surrounded by 2 gradually refining levels towards the center. The fracture width is 0.5 inch, but for the simulation model, we used a row of cells of 2ft wide to represent the hydraulic fracture in the center-blocks. In addition it will not be practical to refine the grid cells down to the actual fracture width because the wellbore radius was 0.7 ft, and the fracture width had to be higher than this dimension, in order to reduce numerical problems during the simulations. Figure 1 illustrates the center-blocks in the fracture LGR, the hydraulic fracture is represented by a line. However, important parameters such as permeability and porosity of the hydraulic fracture are different from the host cells. This is where the output from the fracture design is applied. Assuming a constant fracture conductivity, the fracture permeability corresponding to a desired fracture grid width can be adjusted and specified in the model using Equation 1.

$$K_{fe} = \frac{K_f \ W}{w_e} \tag{1}$$

where K_{fe} is the equivalent fracture permeability, and w_e is the equivalent fracture width.

The value of the input permeability for the hydraulic fracture has been scaled to preserve the actual fracture designed width and the values of the inputs such as equivalent fracture permeability grid cell were obtained from Equation 1.

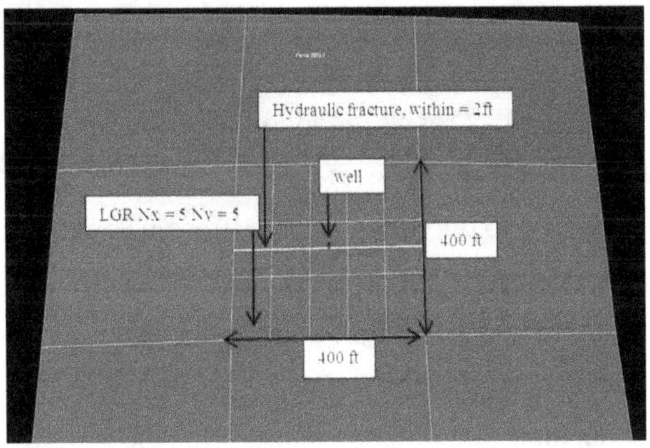

Figure 1: Fracture and wellbore LGR used in the model.

Production Forecast and Net Present Value Analysis for Select Optimum Fracture Half-length

The hydraulic fracturing design is completed on the basis of the production forecast and Net Present Value (NPV) analyses. In order to perform Production forecast, Eclipse Simulator was used. This sensitivity analysis is used to assess the performance of stimulations and guide future design choices.

Project managers, and indeed anyone involved in hydraulic fracturing design must be able to analyze the financial outcome of one's decision. Making a decision requires the measure of performance net present value (NPV), financial analysis methods are tools that enable us to evaluate the total costs and benefits involved in a hydraulic fracturing job in wells. Comparison of the production forecast for the fractured well and the predicted production decline for the unstimulated well allows for calculations of the annual incremental cumulative production for year n for an oil well [10]:

$$\Delta N_{p,n} = N^{f}_{p,n} - N^{nf}_{p,n} \tag{2}$$

$\Delta N_{p,n}$ = the predicted annual incremental cumulative production for year n.

$N^{f}_{p,n}$ = the forecasted annual cumulative production of a fractured well for year n.

$N^{nf}_{p,n}$ = the predicted annual cumulative production of a non-fractured well for year n.

The annual incremental revenue for the unstimulated well is expressed as:

$$\Delta R_n = (\$)\Delta N_{p,n} \tag{3}$$

where (\$) is oil price. The present value of the future revenue is then

$$NPV = \sum_{n=1}^{m} \frac{R_n}{(1+i)^n} - cost \tag{4}$$

where m is the remaining life of the well in years, and i is the discount rate.

The cost should include the expenses for proppant, fracturing fluid, pumping, and a fixed cost for the treatment job. According to the simulation, proppant will be the major job expenses. A financial analysis was then performed using the present-day crude oil price $30/STB. In addition, a future oil price has been predicted. Some costs and operational parameters are shown in Table 3. All cost values are engineering estimates and near to real costs.

Table 3: Cost of fracture treatment used in the economic study.

Components	Cost
fixed Cost ($)	80000
Proppant ($/lb)	1.1
Fluid ($/gal)	0.69
Pumping Charge(4000-5000 psi)	$ 7/HHP
Pumping Charge(5000-6000 psi)	$ 9/HHP
Pumping Charge(6000-7000 psi)	$ 12/HHP

The treatment schedule is designed for different fracture lengths and the fracture features are estimated by the simulator. Oil production is anticipated in both fractured and unfractured well. So the optimal fracture length has been selected by the NPV graph.

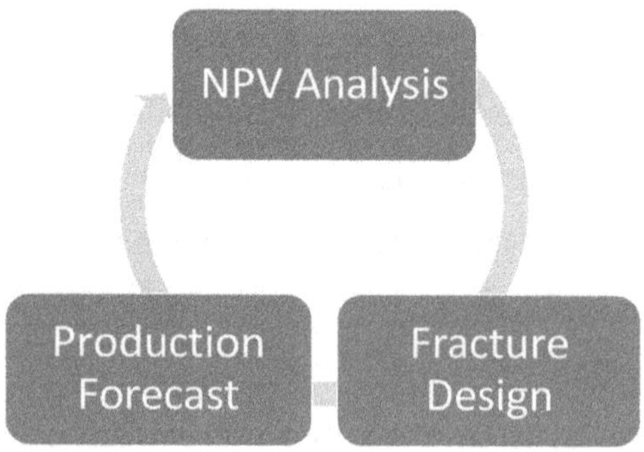

Different scenarios and treatment schedules have been designed to achieve distinctive features of different fracture lengths, then oil production is projected at different scenarios using the Eclipse simulator. With regard to the expenditure and income in various scenarios, NPV is calculated. The best NPVs for all scenarios have been shown in Figure 2 and Figure 3, the best net present value can be obtained from 900 feet fracture half-length for MNS-Y well and for AZ-X well. Also, the NPV increased to 1050 feet fracture half-length and after this, the NPV has not increased significantly. So the optimal fracture half-length for well MNS-Y well is 900 feet, and for AZ-X well is 1050 feet.

Properties of the Optimized Fracture

According to NPV figure for Ahwaz and Mansori reservoir, treatment schedule has been designed to achieve a favorable fracture half-length.

The proppant and fluid have been selected considering the wells and reservoirs conditions. Pump schedule has been designed and hydraulic fracturing operation has been simulated to achieve the optimum fracture length based on Figure 2 and Figure 3. Fracturing fluid plays a vital role in hydraulic fracture treatment because it controls the efficiencies of carrying Proppant and filling in the fracture pad. YF555 fracturing fluid type was chosen for runs. This is one of the fluids included in the model database and is the fluid with the median viscosity. YF555 is a fracturing fluid that has 40 pounds of a polymer per 1000 gallons of water with a Crosslinker added. Proppant must be selected on the basis of in situ stress conditions, According to the ruling of insitu stress in Bangestan Reservoir, Real concerns are compressive quality and the impact of weight on proppant penetrability. Carbolite type proppant has been selected for this simulation. The optimum pump schedule is shown in Tables 4 and 5.

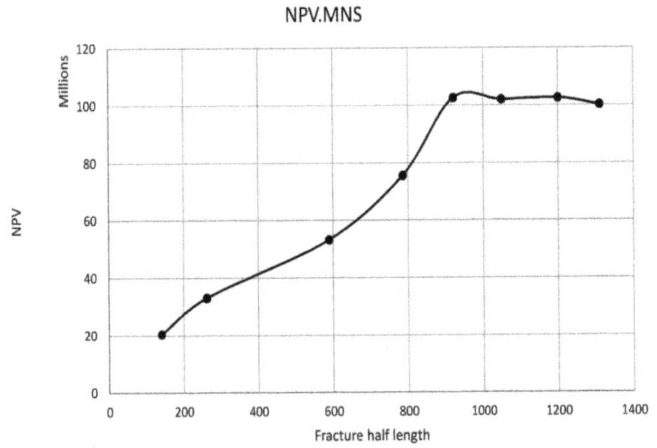

Figure 2: NPV versus fracture length for well MNS-Y.

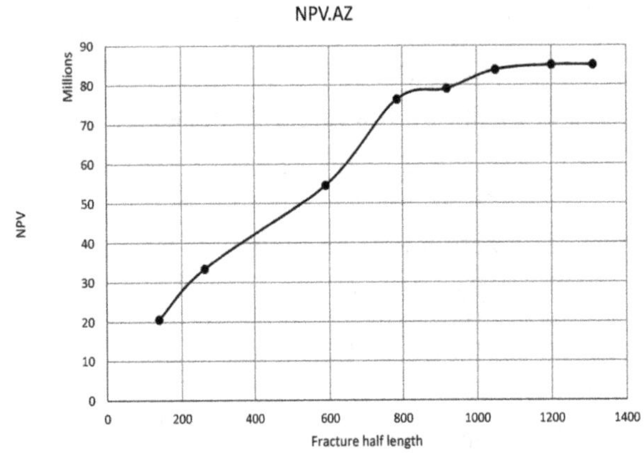

Figure 3: NPV versus fracture length for well AZ-X.

Table 4: Pump schedule for MNS-Y well.

	Stage Name	Pump Rate bbl/ min	Fluid	Fluid Name	Gel Conc. lb/mgel	Fluid Volume gal	Prop	Prop Conc. PPA	Prob Mass lb	Slurry Volume bbl	Pump Time min
1	PAD	30.0	4	Prime FRA	30.0	141700	0	0.0	0	3373.8	112.5
2	1.0 PPA	30.0	5	YF555	30.0	3200	3	0.1	3199	78.8	2.6
3	2.0 PPA	30.0	5	YF555	30.0	4000	3	0.2	7998	101.7	3.4
4	3.0 PPA	30.0	5	YF555	30.0	6000	3	0.3	17996	157.4	5.2
5	4.0 PPA	30.0	5	YF555	30.0	7000	3	0.4	27994	189.2	6.3
6	5.0 PPA	30.0	5	YF555	30.0	8000	3	0.5	39991	222.7	7.4
7	6.0 PPA	30.0	5	YF555	30.0	11000	3	0.6	65985	315.1	10.5
8	7.0 PPA	30.0	5	YF555	30.0	12000	3	0.7	83981	353.4	11.8
9	8.0 PPA	30.0	5	YF555	30.0	13000	3	0.8	103999	393.3	13.1
10	FLUSH	30.0	1	2% KCL W	0.0	20686	0	0.0	0	492.5	16.4

Table 5: Pump schedule for AZ-X well.

	Stage Name	Pump Rate bbl/ min	Fluid	Fluid Name	Gel Conc. lb/mgel	Fluid Volume gal	Prop	Prop Conc. PPA	Prob Mass lb	Slurry Volume bbl	Pump Time min
1	PAD	30.0	10	Prime FRA	30.0	105000	0	0.0	0	2500.0	83.3
2	1 PPA	30.0	2	YF555	30.0	4000	3	0.1	3999	98.6	3.3
3	2 PPA	30.0	2	YF555	30.0	5000	3	0.2	9998	127.6	4.3
4	3 PPA	30.0	2	YF555	30.0	5000	3	0.3	14997	131.8	4.4
5	4 PPA	30.0	2	YF555	30.0	5000	3	0.4	19996	136.1	4.5
6	5 PPA	30.0	2	YF555	30.0	5000	3	0.5	24994	140.3	4.7
7	6 PPA	30.0	2	YF555	30.0	6000	3	0.6	35992	173.5	5.8
8	7 PPA	30.0	2	YF555	30.0	6000	3	0.7	41991	178.6	6.0
9	8 PPA	30.0	2	YF555	30.0	6000	3	0.8	47999	183.7	6.1
10	FLUSH	30.0	1	2% KCL W	0.0	3386	0	0.0	0	80.6	2.7

The hydraulic fracturing properties have been mentioned in Table 6 and Table 7 and proppant concentration profile for two different wells are shown in Figure 4 and Figure 5.

Table 6: Optimum fracture properties for MNS-Y well.

Property	Value
Hydraulic fracture half length (ft)	1127
Propped fracture half length (ft)	918
Hydraulic height at well (ft)	150
Averaged propped width (in)	0.127
Effective FCD	0.9
Effective conductivity (md.ft)	2188
MAX surface pressure (psi)	8590

Table 7: Optimum fracture properties for AZ-X well.

Property	Value
Hydraulic fracture half length (ft)	1201
Propped fracture half length (ft)	1049
Hydraulic height at well (ft)	81
Averaged propped width (in)	0.122
Effective FCD	0.6
Effective conductivity (md.ft)	1698
MAX surface pressure (psi)	8647

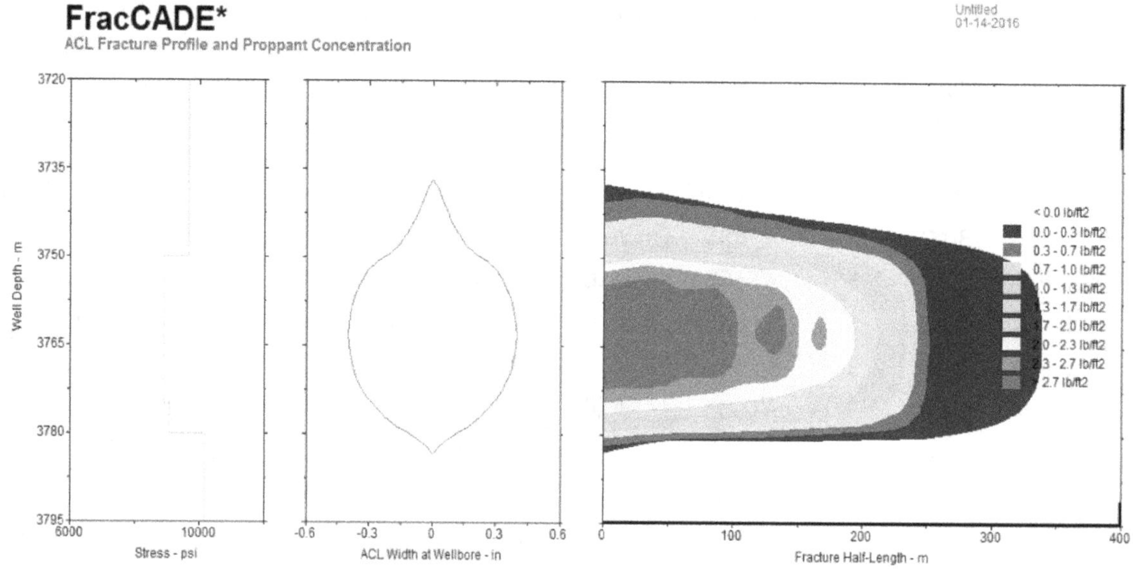

Figure 4: Proppant concentration profile for well MNS-Y.

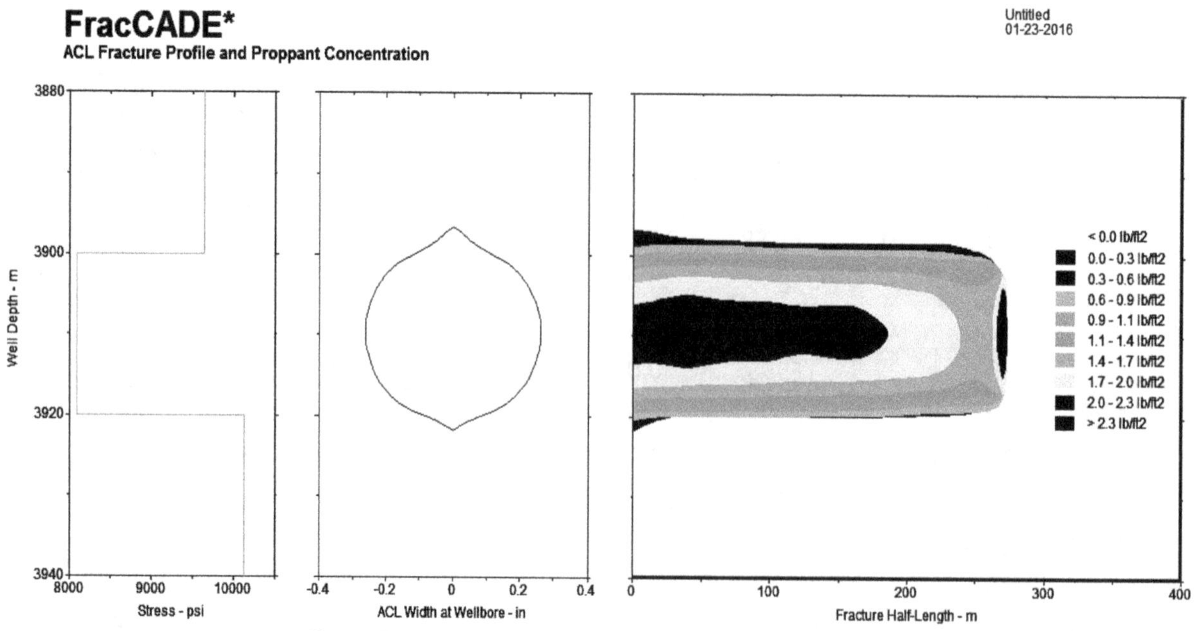

Figure 5: Proppant concentration profile for well AZ-X.

Production Forecast

An unfractured well has been simulated as a base case scenario. Consequently, it's simulation is then compared with the hydraulically fractured well. Both the unfractured well and fractured well had constant formation permeability and constant condition. The main objective of this scenario is to analyze the increase in production of a well fractured with a vertical fracture in relation to unfractured wells. Reservoir with hydraulic fracturing model was made according to the optimum hydraulic fracturing properties.

RESULTS AND DISCUSSIONS

According to FIgures 6 to 9, the results show clearly that hydraulic fracturing improves the oil production rate. The connectivity between wells and reservoir was improved Due to the increase of the permeability near the wellbore. Thus, the fluid flow efficiency improved as well.

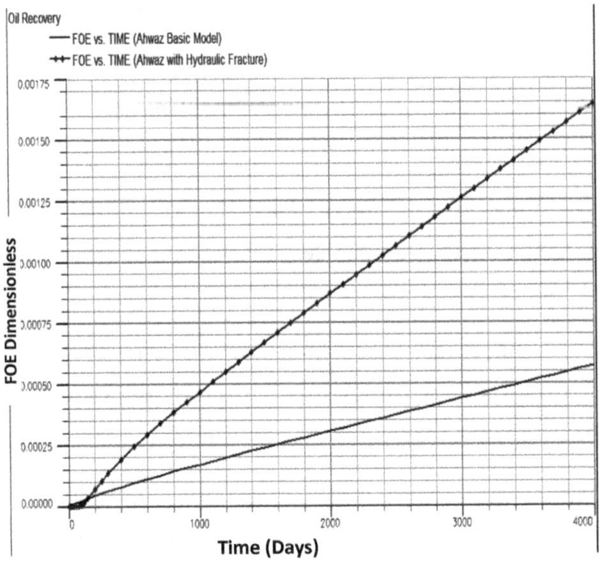

Figure 6: Oil recovery factor for Ahwaz reservoir before and after hydraulic fracturing simulation.

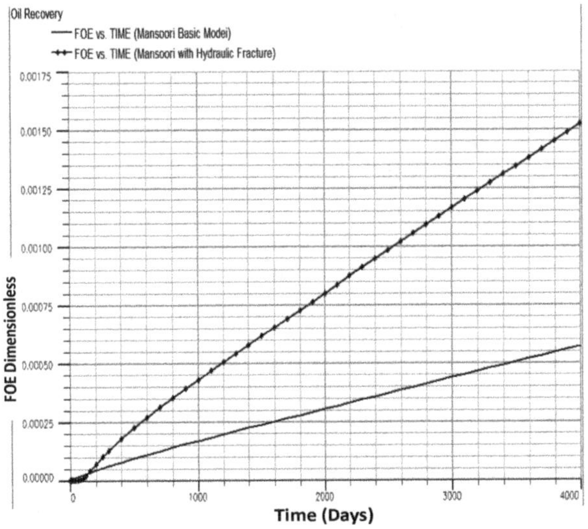

Figure 7: Oil Recovery factor for mansori reservoir before and after hydraulic fracturing simulation.

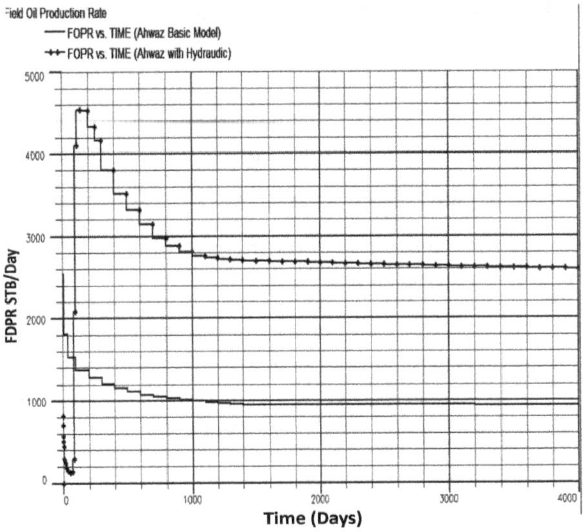

Figure 8: Oil production rate from a vertical well for AZ-X before and after hydraulic fracturing simulation.

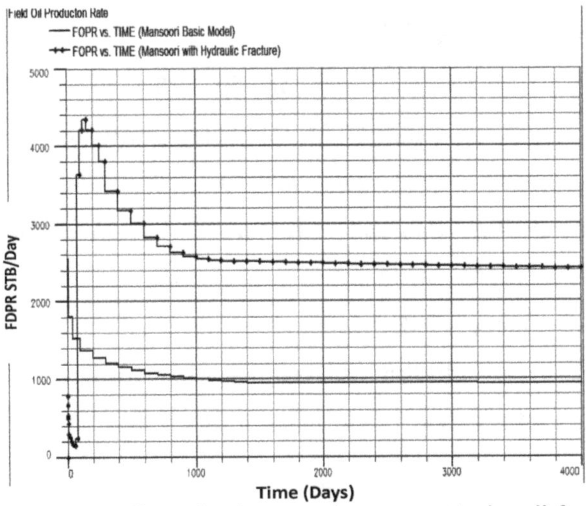

Figure 9: Oil production rate from a vertical well for MNS-Y before and after hydraulic fracturing simulation.

CONCLUSIONS

NPV calculations results confirm the suitability of the procedure presented for the identification of the most profitable fracture half-length to design Pump Schedule to achieve optimum fracture half-length. The case study belongs to two of the Iranian southern oil reservoir selected for the design of hydraulic fracturing operation. The impact of a hydraulic fracturing treatment on the reservoir performance was investigated. The best scenario for hydraulic fracturing according to the NPV diagram was selected to achieve the optimum length. The best fracture half-length for AZ-X well is 1100 feet and for MNS-Y well is 900 feet.

According to reservoir simulations, hydraulic fracturing can significantly improve the recovery factor, as well as oil production rate. Therefore, we can use hydraulic fracturing operations as an effective method for improving oil recovery.

ACKNOWLEDGMENTS

The authors would like to gratefully thank National Iranian Oil Company (NIOC) and National Iranian South Oil Company (NISOC) for their help and support during all stages of this project.

REFERENCES

1. Hashemi A., Shadizadeh S. R., and Zoveidavianpoor M., "A Local Computerized Multi-screening of Vast Amount of Data to Select Hydraulic Fracturing Candidates in Iranian Carbonate Oil Fields," *Int. J. Comput. Appl.*, **2012**, *39*(8), 37-45.
2. Fjar E., Holt R. M., Raaen A. M., Risnes R., and et al., "Petroleum Related Rock Mechanics.," Elsevier, *53*, **2008:**
3. Matthews H. L., Schein G. W., and Malone M. R., "Stimulation of Gas Shales: They're All the Same-Right," in *SPE Hydraulic Fracturing Technology Conference*, Society of Petroleum Engineers, **2007**.
4. Fonseca E. R. and Farinas M. J., "Hydraulic Fracturing Simulation Case Study and Post Frac Analysis in the Haynesville Shale," *in SPE Hydraulic Fracturing Technology Conference*, Society of Petroleum Engineers, **2013**.
5. Abdideh M. and Ahmadifar A., "Prediction of Geomechanical Modeling and Selection of Suitable Layer for Hydraulic Fracturing Operation in Oil Reservoir (South West of Iran)," *European Journal of Environmental and Civil Engineering*, **2013**, *17*(10), 968-981.
6. Ardalan A., "Economic and Financial Analysis for Engineering and Project Management," Taylor & Francis., **1999.**
7. Heydarabadi F. R., Moghadasi J., Safian G. H., and Ashena R., "Criteria for Selecting a Candidate Well for Hydraulic Fracturing," *in Nigeria Annual International Conference and Exhibition*, Society of Petroleum Engineers, **2010.**
8. Gudmundsson A., "Rock Fracturein Geological Processes," Cambridge University Press, **2010**:
9. Darvish H., Nouri-Taleghani M., Shokrollahi A., and Tatar A., "Geo-mechanical Modeling and Selection of Suitable Layer for Hydraulic Fracturing Operation in an Oil Reservoir (South West of Iran)," *Journal of African Earth Sciences*, **2015**, *111*(5), 409-420.
10. Marongiu-Porcu M., Economides M. J., and Holditch S. A., "Economic and Physical Optimization of Hydraulic Fracturing," *Journal of Natural Gas Science and Engineering*, **2013**, *14*, 91-107.

10

Application of M3GM in a Petroleum Reservoir Simulation

Ehsan Taheri [1]*, Amoirodin Sadrnejad[2], and Hasan Ghasemzadeh[2]

[1] Faculty of Engineering, Tarbiat Modares University, Tehran, Iran
[2] Faculty of Civil Engineering, K. N. Toosi University of Technology, Tehran, Iran

ABSTRACT

Reservoir formations exhibit a wide range of heterogeneity from micro to macro scales. A simulation that involves all of these data is highly time consuming or almost impossible; hence, a new method is needed to meet the computational cost. Moreover, the deformations of the reservoir are important not only to protect the uppermost equipment but also to simulate fluid pattern and petroleum production strategy. In this regard, multiscale multiphysic mixed geomechanical model (M^3GM) is recently developed. However, applications of petroleum reservoirs through gas or water injection in the depleted reservoir are in concern. In the present paper, a multiscale finite volume framework and a finite element method are employed to simulate fluid flow and rock deformation respectively. The interactions of solid and fluid phases are instated through the M^3GM framework. Then, its application in the petroleum reservoir through injection process is validated. The numerical results are compared with the fine scale simulations and reasonable agreement with high computational efficiency is obtained.

Keywords: Multiscale, Petroleum Reservoir, Computational Efficiency, Interaction, Injection

INTRODUCTION

Flow transport in petroleum reservoirs widely ranges from micro to macro scales. By considering these scales, interactions with solid phase make a great challenge in a petroleum reservoir simulation. In this regard, various upscaling and multiscaling models were developed in recent two decades [1-8]. However, all models are only in concern of fluid mechanics and geomechanical effects have been neglected [9,10]. It is worth mentioning that geomechanical effects are important not only to predict settlements in order to protect the uppermost facilities but also to simulate flow path and production strategy [10,11]. On the other hand, several geomechanical incidents such as surface subsidence which damages the uppermost installations, wellbore instability, and environmental issues have increased attention in this area [12-14]. In this regard, several models with different trends were developed to incorporate geomechanical effects [15-29]. Nevertheless, in all the models the fluid and solid scales are identical. However, every physic property has major effects on their domains of influence. Therefore, in order to place any event in the right position and save computational costs, it would be efficient to treat multiple physical

***Corresponding author**

Ehsan Taheri
Email: e_taheri@modares.ac.ir

phenomena separately. In this context, recently, multiscale multiphysic mixed geomechanical model (M^3GM) is developed by the authors to simulate different physic through speared scales [9,10]. In the present paper, the fluid mechanic framework with multiscale finite Volume (MSFV) is described. Then, the finite element framework is employed to simulate geomechanical aspects. Moreover, the interaction between the fluid and solid phases is stablished through an appropriate algorithm. After the theoretical basis of the model is justified, its application to petroleum reservoirs, particularly to water injection in depleted petroleum reservoirs, is revealed

MULTISCALE WHY AND WHEN?

Several physical phenomena in the nature such as the mechanical behavior of composite materials and flow in porous media occur on a wide variety of physical scales from millimeter for pore scale to several kilometers with respect to reservoir dimensions [30,31]. The various scales of reservoir are shown in Figure 1. It should be mentioned that a conventional reservoir has computational modules in the order of 10^7-10^8 elements [32]. The number of elements could be increased with respect to the dimensions and complexity pattern of the geological formations. On the other hand, current computers are capable of simulating 10^5-10^6 elements [32]. In order to fill this gap, multiscale simulation is a crucial method [33-35]. Moreover, several physics, namely solid mechanics and fluid transport, are affected with respect to different orders of the influential domain [9,10,30]. In this regard, a framework which could simulate these wide ranges of scale lengths and physics is not available. Therefore, upscaling methods were developed, especially in the petroleum reservoir simulation [36-41]. Nevertheless, in upscaling methods, the reservoir problem have only been solved on a coarse scale, and the fine scale information have been lost [33,34]. On the contrary, in a multiscale framework, the fine scale properties are preserved during the homogenization and are used in resolution in the reference scale [33]. Thus, extensive motivations have been emerged into the multiscale modeling of the reservoir formation during 2003-2015 [9,10,42-54]. Figure 2 shows the methodology of upscale and multiscale methods to simulate reservoir formations [10].

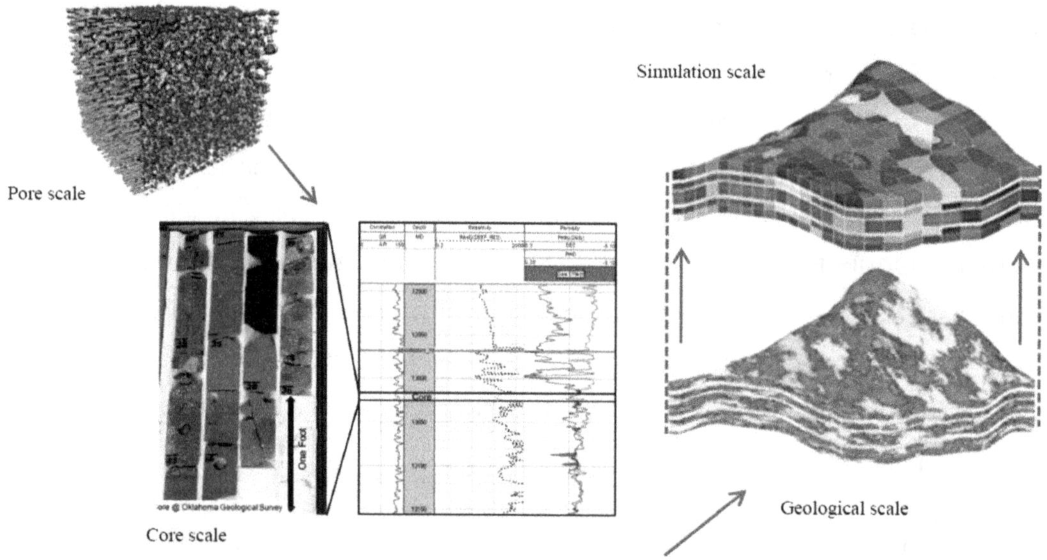

Figure 1: Indicative scales in petroleum reservoirs

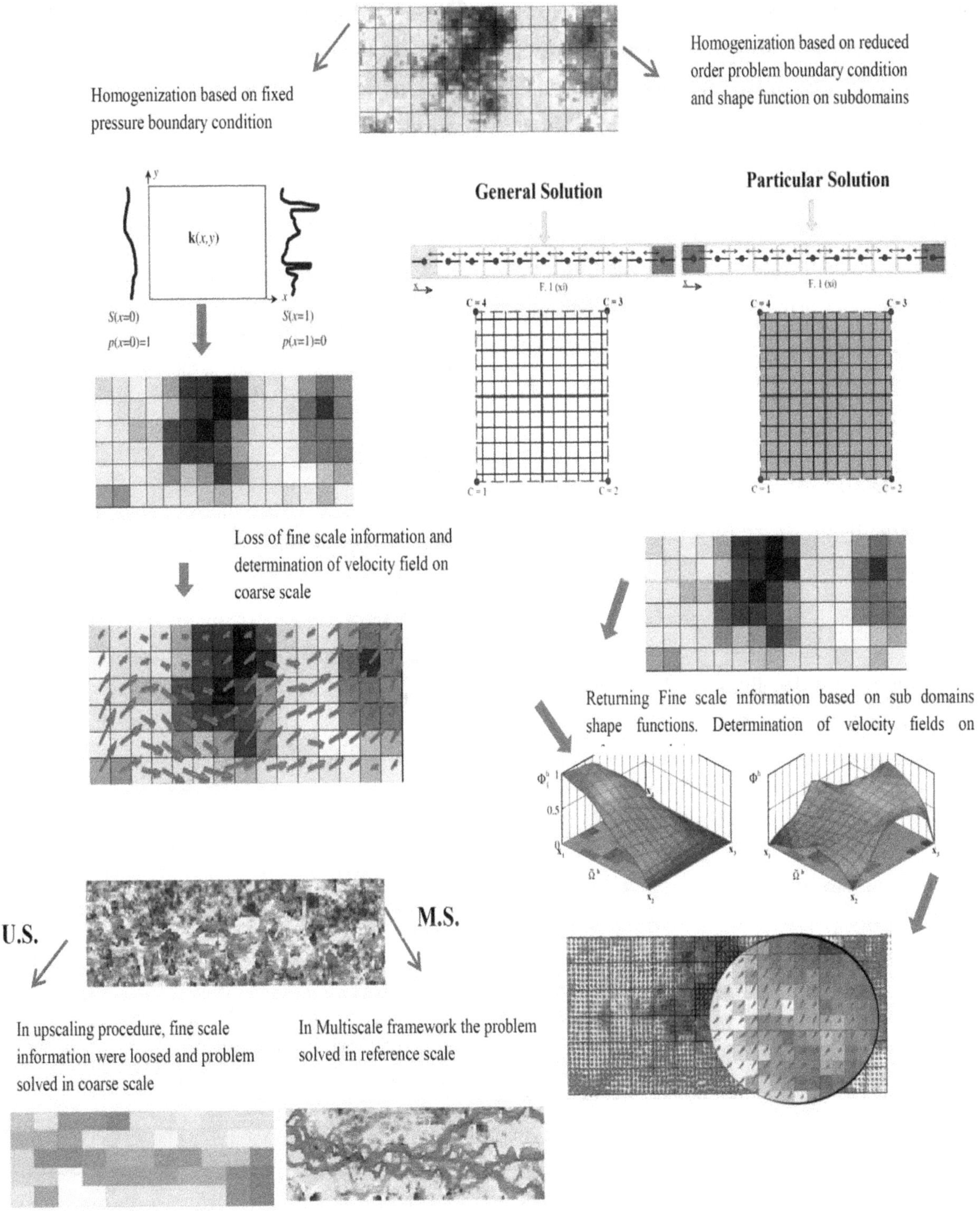

Figure 2: Comparison of upscale and multiscale procedures to simulate reservoir formation.

In the present paper, the MsFVM framework is applied with respect to preserving the fine scale specifications during upscaling and also mass conservative feature of the proposed method. The structure of the utilized Multiscale method relies on separation of pressure and saturation equation the same as streamline method. However, with respect to limitation of streamline method for the depletion derive and deformable porous media [55&56], the adoptive updating of basis function regarding the saturation front is applied in order to solve saturation equation with high efficiency [10].

Governing Equations

In this section, in order to simulate fluid phases and porous media deformation and their interactions, the principle equations of mass and momentum of each phase are explained.

Fluid Phase

It is pointed out in the literature that fluid transport is in Eulerian, but soil transport is in Lagrangian point of view [11]. However, since there are both fluid and solid in the system, the fluid is observed from the deformed porous media not from a fixed mesh.

Mass Balance

The general mass conservation is governed by Equation 1:

$$\frac{dM}{dt} = \frac{d}{dt}\int_{V_t} \rho dV = 0 \tag{1}$$

where, ρ is a phase density, and M stands for total mass. Moreover, with respect to Reynold theorem, one may obtain:

$$0 = \frac{dM}{dt} = \int_{V_t} \frac{\partial \rho}{\partial t} dV + \int_{S_t} \rho \mathbf{v}.\mathbf{n} dS = 0 \tag{2}$$

applying Gauss law will result in:

$$0 = \frac{dM}{dt} = \int_{V_t} \left(\frac{\partial \rho}{\partial t} + \nabla.(\rho \mathbf{v}) \right) dV \tag{3}$$

Thus by defining a relative velocity instead of interstitial velocity, by employing material derivatives, and by doing some mathematical manipulation, one may obtain:

$$\frac{D^s(\phi S_\alpha \rho_\alpha)}{Dt} + \nabla.(\phi S_\alpha \rho_\alpha \mathbf{w}_\alpha) + \phi S_\alpha \rho_\alpha \nabla.\mathbf{v}_s = \dot{m}_\alpha \tag{4}$$

where, ϕ is porosity, and S_α is phase saturation; $\mathbf{w}_\alpha$ represents relative velocity. Also, $\mathbf{w}_\alpha = \mathbf{v}_\alpha - \mathbf{v}_s$; $\mathbf{v}_\alpha$ is velocity, and $\mathbf{v}_s$ is solid skeleton velocity; $\dot{m}_\alpha$ stands for sink or source term. Since the system of concern is made up of two fluid phases, two mass balances, i.e. Equation 4, are taken into account.

Momentum Balance of Fluid Phase

In order to simulate fluid flow and set up interaction between phases, momentum balance should be considered.

$$\varphi_\alpha S_\alpha W_\alpha = \lambda_\alpha.(-\nabla p + \rho_\alpha \mathbf{g}) \tag{5}$$

where, λ_α is phase mobility tensor defined as $\boldsymbol{\lambda}_\alpha = \frac{\mathbf{K}kr_\alpha}{\mu_\alpha}$ and **K** represents the inherent permeability tensor; Kr_α stands for relative permeability and is simulated as a function of the pertinent phase saturation, and μ_α is the phase viscosity.

Solid Phase

With respect to history conservation aspect [11], the solid deformation is considered in Lagrangian point of view.

Mass Balance of Solid Phase

The solid phase mass balance is stated as follows:

$$\frac{\partial}{\partial t}(1-\varphi)\rho_s + \nabla.(1-\varphi)\rho_s \mathbf{v}_s = 0 \tag{6}$$

Neglecting grain compressibility leads to:

$$\frac{\partial(1-\varphi)}{\partial t} + \mathbf{v}_s.\nabla(1-\varphi) + (1-\varphi)\nabla.\mathbf{v}_s = 0 \tag{7}$$

$$\varepsilon_{vol} = \nabla.\mathbf{u}_s \tag{8}$$

Finally, utilizing material derivative and the mechanical definition of deformation rate will lead to:

$$\frac{-1}{1-\varphi}\frac{D(1-\varphi)}{Dt} = \frac{D\varepsilon_{vol}}{Dt} \tag{9}$$

Equation 9 will also be used to state the interaction between solid and fluid phases.

Momentum Balance of Solid Phase

The momentum balance of the solid phase is governed by the following equation:

$$\nabla.\sigma + \rho \mathbf{g} = 0 \tag{10}$$

where, σ is total stress tensor. However, in geomechanical considerations, the effective stress σ' is applied instead.

$$\sigma' = \sigma - \mathbf{I}\, p_t \tag{11}$$

where, P_t is pore fluid pressure.

Multiscale Multiphysic Fluid and Solid Interaction

The interactions of the solid and fluid phases are stated through a Newton-Raphson iteration loop with respect to solid deformation and fluid phase pressure. Furthermore, by integrating Equation 9, one may obtain:

$$\varphi = (1-(1-\varphi_0)\exp[-\varepsilon_{vol}]) \tag{12}$$

As stated in Equation 12, porosity will change with respect to the stress evolution and the related volumetric strain. In general, phase mobility is related to porosity through chain rule derivatives. However, in order to keep the basic functions (which will be defined later) off the iteration loop, the explicitly incorporation of porosity changes is take into account.

Multiscale Multiphysic Geomechanical Model (M³GM)

M³GM is established by coupling the fluid phases transport and solid skeleton deformation as explained below.

Multiscale Finite Element Framework

By incorporating the momentum balance of solid and fluid phases in Equation 4, by the implicit time discretization of the resulted formula, and by integrating over all phases (for detailed mathematical information, interested readers are referred to our previous researches [9,10]) one may obtain:

$$\frac{\varphi^{n+1}}{\Delta t} + \frac{-\varphi^{n}}{\Delta t}\sum_{\alpha=1}^{n_p} B_\alpha^{n+1}\rho_\alpha^n S_\alpha^n - \sum_{\alpha=1}^{n_p} B_\alpha^{n+1}\nabla.(\rho_\alpha^{n+1}\lambda_\alpha(\nabla p^{n+1} - \rho_\alpha^{n+1}\mathbf{g}\nabla z))$$
$$+\varphi^{n+1}\frac{\varepsilon_v^{n+1}-\varepsilon_v^{n}}{\Delta t} = q_t \tag{13}$$

where, n and $n+1$ are two sequential time steps, and B_α is a formation volume factor defined as the inverse of phase density.

The superposition of basis and correction functions (general and specific solutions of homogenous and inhomogeneous parts of Equation 4 with respect to the reduced problem boundary conditions) will lead to:

$$p_f(x) \approx p'(x) = \sum_{h=1}^{N}\left[\sum_{k=1}^{M}\Phi_k^h(x)\bar{\boldsymbol{p}}_k + \Phi^h(x)\right] \tag{14}$$

By integration over all coarse cells, one may obtain:

$$\int_{\bar{\mho}_l}\frac{C_c}{\Delta t}\left(\sum_{h=1}^{N}\left(\sum_{k=1}^{M}\Phi_k^h\bar{\boldsymbol{p}}_k^{v+1} + \Phi^{h^v}\right) - \boldsymbol{p}^{w}\right)d\Omega -$$
$$\int_{\partial\bar{\mho}_l}\left(\lambda_t.\sum_{h=1}^{N}\left(\sum_{k=1}^{M}\Phi_k^h\bar{\boldsymbol{p}}_k^{v+1} + \Phi^{h^v}\right)\right).\tilde{\mathbf{n}}_l d\Gamma$$
$$= \int_{\bar{\mho}_l} RHS^v d\Omega \tag{15}$$

which will form the system of equation:

$$\mathbf{A}_{lk}\mathbf{p}_k^{v+1} = \mathbf{b}_l^v \tag{16}$$

where,

$$\mathbf{A}_{lk} = \sum_{h=1}^{N}\left(\int_{\bar{\mho}}\frac{C_c}{\Delta t}\Phi_k^h d\Omega - \int_{\partial\bar{\mho}_l}(\lambda_t.\nabla\Phi_k^h).\tilde{\mathbf{n}}_l d\Gamma\right) \tag{17}$$

and

$$\mathbf{b}_l^v = \int_{\bar{\mho}}\left(RHS^v + \frac{C_c}{\Delta t}\boldsymbol{p}^{w}\right)d\Omega -$$
$$\sum_{h=1}^{N}\left(\int_{\bar{\mho}}\frac{C_c}{\Delta t}\Phi^{h^v}d\Omega - \int_{\partial\bar{\mho}}\lambda_t.\nabla\Phi^{h^v}.\tilde{\mathbf{n}}_l d\Gamma\right) \tag{18}$$

The above nonlinear system of equations will be sequentially solved until convergence is achieved.

Finite Element Framework

As mentioned earlier, the finite element method is employed for solid deformation framework. Eight node of a serendipity family shape function is applied in this regard (see Figure 3).

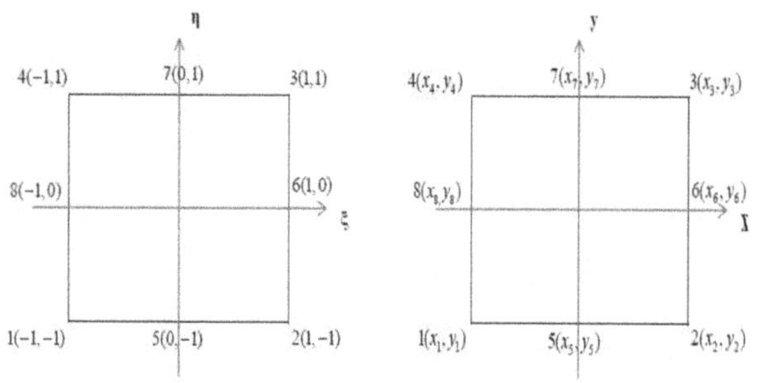

Figure 3: A solid phase element with a serendipity 2D family shape function.

The classical finite element weighted residual discretization of Equation 11 leads to:

$$\int_{\Omega} \mathbf{W}_u^t (\mathbf{L}^T \sigma - \rho \mathbf{g}) d\Omega + \int_{\Gamma_u} \overline{\mathbf{W}}_u^t (\mathbf{L}^T \sigma - \overline{\mathbf{t}}) d\Gamma = 0 \quad (19)$$

By implementing Gauss theorem and stress-strain relation, one may obtain (interested readers are referred to our previous works [9, 10]):

$$\left[\int_{\Omega} \mathbf{B}^T \mathbf{D} \mathbf{B} \; d\Omega\right] \hat{\mathbf{u}} = -\int_{\Gamma^N} \mathbf{N}_u^T \overline{t} d\Gamma + \int_{\Omega} \mathbf{B}^T (p\mathbf{m}) d\Omega \quad (20)$$

which results in a linear deformation system:

$$\mathbf{K}\hat{\mathbf{u}} = \mathbf{F} \quad (21)$$

where:

$$\mathbf{K} = \left[\int_{\Omega} \mathbf{B}^T \mathbf{D} \mathbf{B} \; d\Omega\right] \quad (22)$$

and

$$\mathbf{F} = -\int_{\Gamma^N} \mathbf{N}_u^T \overline{t} d\Gamma + \int_{\Omega} \mathbf{B}^T (p\mathbf{m}) d\Omega = \mathbf{F}_1 + \mathbf{F}_2 \quad (23)$$

The first term in the right hand side of Equation 22 is obtained from the finite element boundary condition. However, the second term is achieved with respect to the integration over the fine pressure attained from the multiscale framework.

$$\mathbf{F}_2 = \sum_{i=1}^{i=nfs} B_i^T p_i . m . A \quad (24)$$

NUMERICAL RESULTS

Water Injection in Homogeneous Rigid Media

The first test is water injection in homogenous reservoirs with the inherent permeability of $2.5 \times 10^{-13} m^2$. The viscosity ratio is 5, and the reservoir is initially filled with oil. In order to verify the model, the results are compared with the simulation of White and Lewis in 1991 [55]. The Brook & Corey's relative permeability function is selected to compare the results. In addition, a quadratic function is also utilized to evaluate the relative permeability constitutive function. As it is clear from Figure 4, although fine scale is employed, the contour is deviated from a quarter circle shape which is expected. In order to modify the results, Geiger and coworkers simulate the case by combining finite element and finite volume methods [56]. The results with 4096 elements are also shown in Figure 4. As it is clear from Figure 4, reasonable agreement is observed with just 100 elements in new multiscale methods. It should be noted that, in petroleum reservoir development, lots of scenarios are performed to allocate the most appropriate situation for injection. In this regard, the computational efficiency of the multiscale models is the crucial matter.

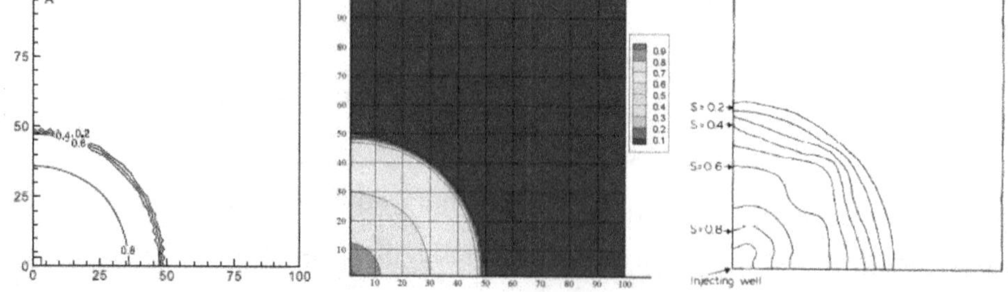

Figure 4: Saturation contours: left: finite element after White & Lewis; middle: after Geiger et al.; right: M³GM.

Moreover, in order to precisely evaluate model accuracy, pressure and saturation contours at three points, namely A (3, 3), B (1, 11), and C (14, 14), are compared with a fine scale. As shown in Figures 5 and 6, the trend and the amount of results are in rational agreement. The pressure at point A (the point closer to injection well) is reduced sharply with respect to a rapid flow front, while at points B and C, the pressure drops smoothly since injection fluid front reaches this area. As shown in Figure 6, the oil saturation curves also follow the same path. It is apparent from Figure 5 that the quadratic constitutive relation of relative permeability with saturation fails to predict the pressure fall correctly. Although the pressure curves have little discrepancies, the saturation curves are fitted closely. It shows that little deviation in the calculation of saturation will lead to more errors in the pressure estimation.

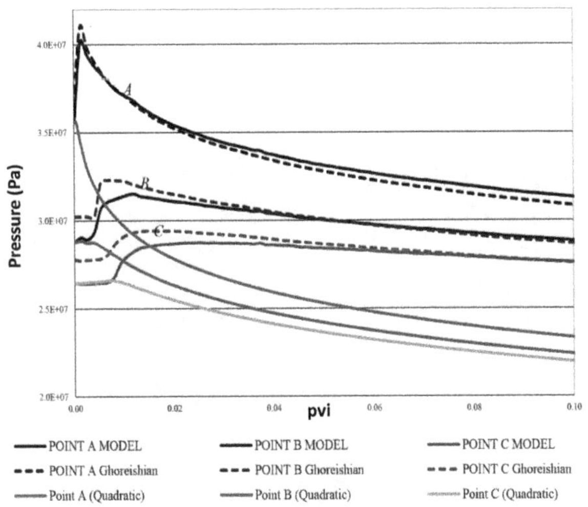

Figure 5: Pressure history curves at pints A, B, and C from fine scale and M³GM simulations.

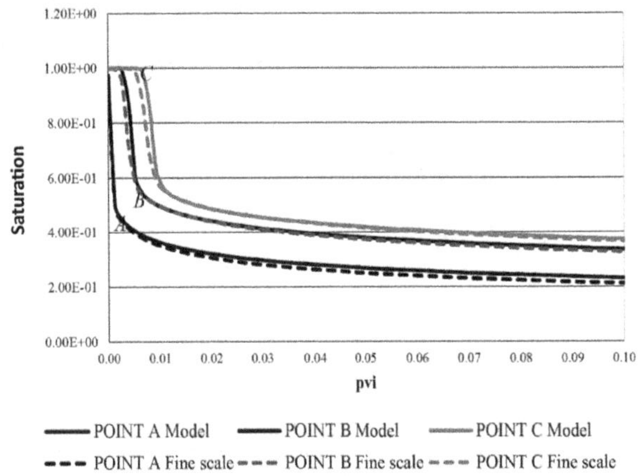

Figure 6: Comparison of oil saturation history curves at points A, B, and C of fine scale and M³GM simulations.

Water Injection in Heterogeneous Deformable Porous Media

In order to evaluate the effect of mesh construction in a multiscale model, water injection in heterogeneous porous media is presented. The permeability field belongs to Tarbert formation and is a part of 20th layer of SPE 10 [40] as shown in Figure 7. The size of the problem was selected in a way to evaluate four different mesh sizes, namely (15×15-3×3), (9×9-5×5), (5×5-3×3), and (3×3-15×15). The first set of numbers are related to the numbers of coarse cells in each direction, and the second set shows the number of fine cells encapsulated in each spatial direction of a coarse cell. The porous media is initially filled with 85% oil and 15% water. The initial pressure is 10 kPa. In order to increase the pressure, water is injected in the lower left corner, i.e. fine cell (1, 1), and production occurs from the upper right corner, i.e. fine cell (54, 54) at a constant pressure. The oil in place has a viscosity value 4 times more than that of the injected water. The pressure and saturation contours are depicted in Figure 8 after 0.11 PVI. As it is clear from this figure, the injected water front follows the lower permeability field and deviates from the quarter circle shape in a homogenous permeability pattern, which was reported in a homogenous case. However, in order to evaluate the effect of coarse grid construction and related mesh, the pressure values at three locations of the domain with coordinates of A (2, 2), B (12, 16), and C (4, 28) are assessed.

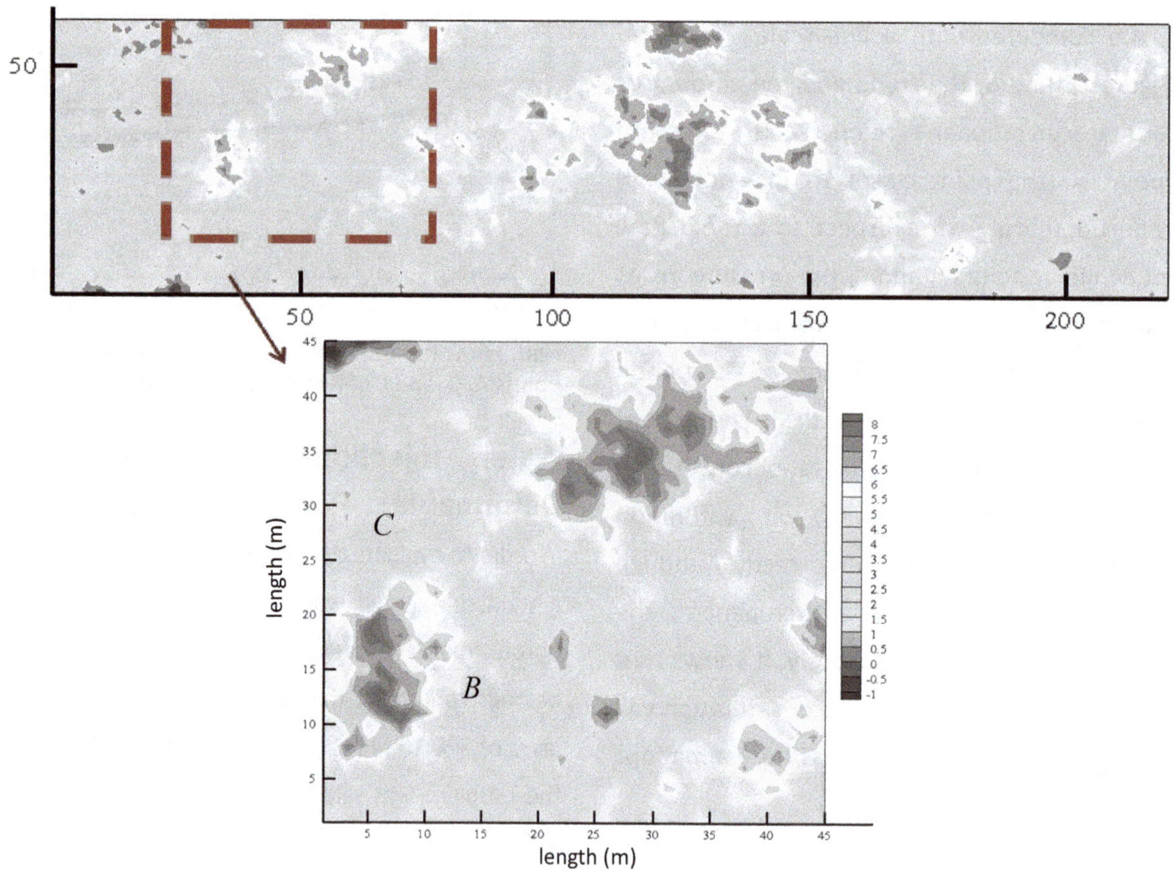

Figure 7: Permeability field belonging to Tarbert formation is obtained from 20th layer of SPE 10.

The pressure values at these locations are compared with the four sets of meshes of those locations mentioned earlier.

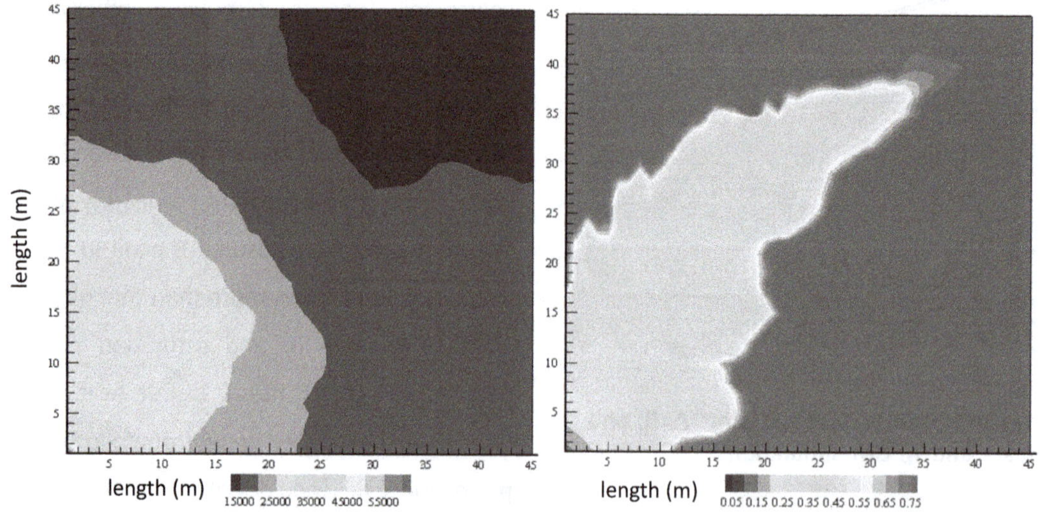

All Dimension are in meter.

Figure 8: Pressure and oil saturation contours after 0.11 PVI.

As shown in Figure 9, reasonable agreement is observed in mesh (15×15-3×3), (9×9-5×5), and (5×5-3×3). However, in a coarser cell, i.e. (3×3-15×15), the result oscillates and do not converge, which is due to the fact that the flow front fills just one coarse cell and affects the hardness matrix. Also, considering high contrast permeability value in the construction of coarse grid and avoiding constructing a dual cell border in this region are recommended.

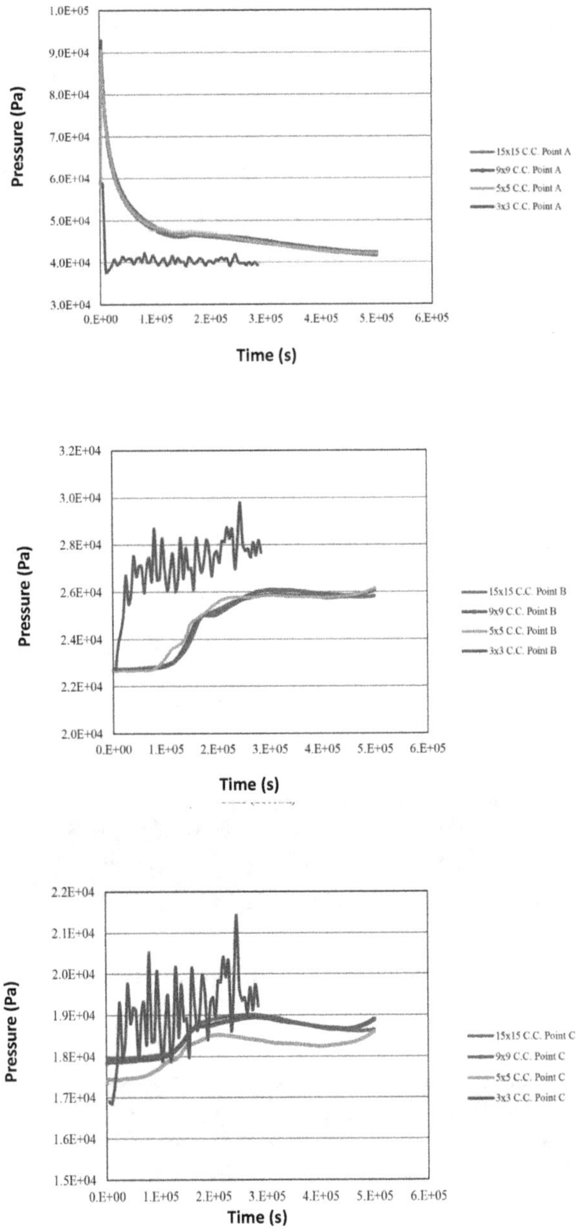

Figure 9: Pressure history at points A, B, and C through different mesh sizes.

On the other hand, the deformations of reservoir formation are important not only to preserve the uppermost facilities, but also to predict fluid flow and production rare correctly. Figure 10 shows the deformation of the reservoir domain through water injection. The elastic modules and Poisson's ratio are considered 25 MPa and 0.25 respectively. The mechanical boundary conditions and mesh sizes are also shown in Figure 10. As it is clear from this figure, the maximum deformations take place on the top of the injection well, where the maximum pressure changes are observed.

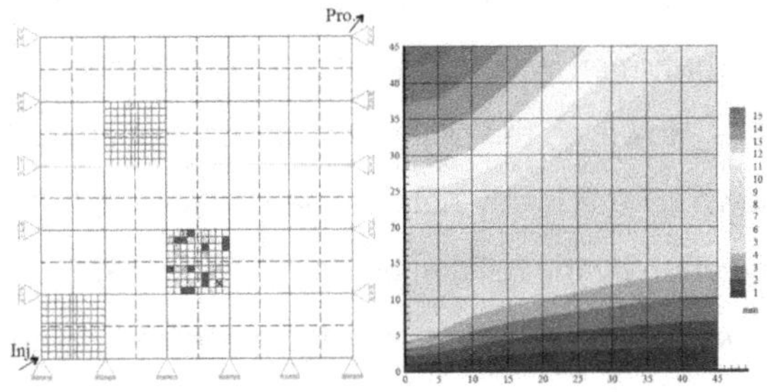

Figure 10: Left: A schematic of water injection and mechanical boundary conditions' right: the deformation distribution of the soil domain in a five-spot water flood problem after 0.085 PVI.

Moreover, production rate through injection is shown in Figure 11 in fine scale and multiscale frameworks. As shown in this figure, the multiscale results are in good agreement with the fine scale solution.

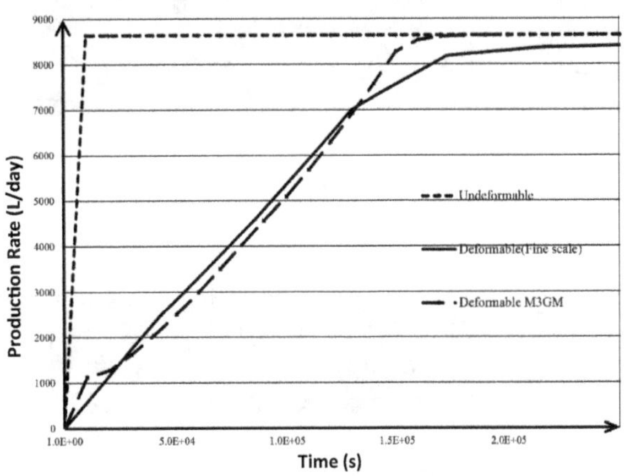

Figure 11: Production rate in a five-spot problem within the reservoir in the rigid and deformable media.

It worth mention that in Multiscale frame work 81 coarse element is utilize in compare with 2025 fine scale model and 25 times savings is obtained. Moreover, as it is clear from Fig. 11 the production rate in the porous media with consideration of deformation is lower than in rigid media that is more realistic. Since in the real field the reservoir formations are not rigid, some energy of injection is consumed through the deformation of rock media. While the injection proceeds, the deformations of rock media have been occurred and productions rate is asymptotes to the threshold value. As explained heretofore, the validity and efficiency of the model are confirmed with respect to aforementioned test cases. However, the robustness of the model will be shown by the highly heterogeneous layer of Tarbert formation by the full domain of 20th layer of SPE 10 as shown in Fig. 7. This layer has the mean and variance of the natural logarithm mean and variance of 2.35 and 5.66, respectively. The permeability varies over 6 orders of magnitude. The injection is imposed by the 10 times less viscos fluid from the fine cell (1,1) and the production with constant pressure of 0 at fine cell (220, 55).

No-flow boundary conditions are considered at the domain borders. The physical properties of fluids and mechanical parameters of porous media are shown in Table. 1. Fine scale results are obtained from Hajibeygi (2011). In the present Multiscale model, two different scales (coarse and fine) for the pressure deformation and saturation equation are applied respectively. The M3GM utilize 20 × 5 coarse grid and 220 × 55 fine grid cells, with respect to pressure deformation and saturation field resolution, respectively. The pressure field and saturation font contours are shown in Fig. 12 After 0.132 PVI.

Table 1: Physical properties of fluids and mechanical parameters of porous media.

μ_o	μ_w	ρ_o	ρ_w	ϕ	S_o	Δx	Δy	E	υ
$10^{-2}\,Pa.s$	$10^{-3}\,Pa.s$	$950\frac{kg}{m^3}$	$1000\frac{kg}{m^3}$	0.2	1	1m	1m	350 *kPa*	0.2

As it is clear from Fig. 11, M3GM results are in the same pattern as the fine scale solution. Figure 13 also shows the deformation of the formation after 0.132 PVI. The maximum deformations take place above the injection well where the pressure and effective stress changes is on the peak.

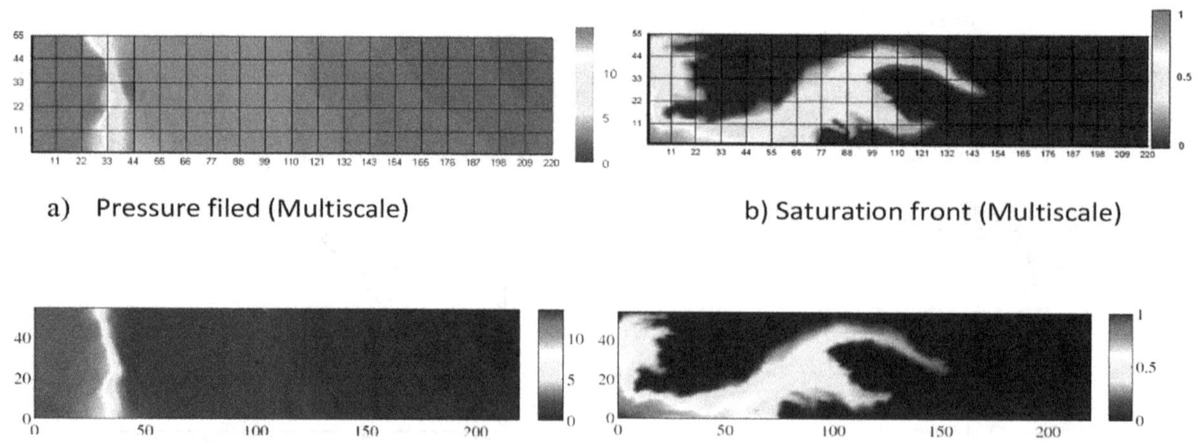

a) Pressure filed (Multiscale) b) Saturation front (Multiscale)

c) Pressure filed (Fine scale) d) Saturation front (Fine scale)

All Dimension are in meter.

Figure 12: comparison of pressure filed saturation front between M3GM and fine scale solution after 0.132 PVI

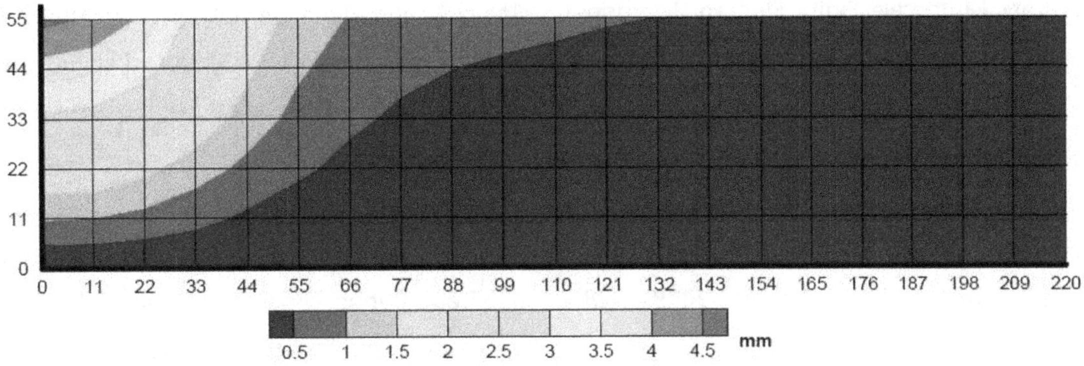

Figure 13: M3GM deformation fields of SPE 10 Tarbert formation after 0.132 PVI

It is important that during injection and depletion, deformation of reservoir also could have undesirable effects on the sensitive uppermost facilities with respect to settlement or heave. In this regard, instrumentations such as shut down valves, High-integrity pressure protection system (HIPPS), mechanical equipment such as floating storage tank that are easily affected by deformation and other deformation sensitive equipment should be located with reliable distance from this region especially in onshore area.

CONCLUSIONS

The new model namely, M^{3}GM, is presented by considering the multiscale and multiphysic nature of reservoir porous media. Its application in the petroleum reservoir through water injection in depleted reservoirs is validated. The M^{3}GM model confirms the results of other methods at high computational efficiency. The present model not only simulates fluid flow in porous media, but also evaluates the deformation of the reservoir domain.

REFERENCES

1. Aarnes J. E., "On the Use of a Mixed Multiscale Finite Element Method for Greater Flexibility and Increased Speed or Improved Accuracy in Reservoir Simulation," *Multiscale Model Simul.*, **2004**, *2*, 421-439.
2. Arbogast T., "Implementation of a Locally Conservative Numerical Sub-grid Upscaling Scheme for Two-phase Darcy Flow," *Comput. Geosci.*, **2002**, *6*, 453-481.
3. Durlofsky L. J., "Numerical Calculation of Equivalent Grid Block Permeability Tensors for Heterogeneous Porous Media," *Water Resource Research*, **1991**, *27*, 699-708.
4. Durlofsky L. J., "Upscaling of Geocellular Models for Reservoir Flow Simulation: A Review of Recent Progress," *International Forum on Reservoir Simulation. Buhl/Baden-Baden*, Germany, **2003**, 1-58.
5. Farmer C. L., "Upscaling: A review," *Int. J. Numer. Meth Fluids*, **2002**, *40*, 63-78.
6. Chen Z. and Hou T. Y., "A mixed multiscale finite element method for elliptic problems with oscillating coefficients," *Math. Comput.*, **2002**, *72*, 541–576.
7. Efendiev Y., Ginting V., Hou T. Y., and Ewing R.,

"Accurate Multiscale Finite Element Methods for Two-phase Flow Simulations," *J. Comp. Phys.*, **2006**, *220*, 155–174.

8. Hajibeygi H. and Jenny P., "Multiscale Finite-volume Method for Parabolic Problems Arising from Compressible Multiphase Flow in Porous Media," *Computational Physics*, **2009**, *228*, 5129-5147.
9. Sadrnejad S. A., Ghasemzadeh H., and Taheri E., "Multiscale Multiphysic Mixed Geomechanical Model in Deformable Porous Media," *Journal for Multiscale Computational Engineering*, **2014**, *12*(6), 529-547.
10. Taheri E., "Multiscale Modeling of Oil Transport in Deformable Porous Media," Ph.D. Thesis, K. N. Toosi University of Technology, **2015**.
11. Lewis R. W. and Schrefler B. A., "Finite Element Method in the Static and Dynamics Deformation and Consolidation of Porous Media," (2nd Ed.) England, Wiley, **1998.**
12. Bruno M. S. "Subsidence-induced Well Failure," In SPE (Chevron Oil Field Research, Society of Petroleum Engineers), SPE 20058-PA, **1992**.
13. Rhett D. W. and Revisited E., "A New Model of Ekofisk Reservoir Geomechanical Behavior," In SPE/ISRM Rock Mechanics in Petroleum Engineering, (Trondheim, Norway 1998), **1990**.
14. Carbognin L., Gatto P., Mozzi G., and Gambolati G., "Land subsidence of Ravenna and its similarities with the Venice case. Proceedings of the Engineering Foundation Conference on Evaluation and Prediction of Subsidence," New York: ASCE, **1978**, 254-266.
15. Lewis R. W. and Sukirman Y., "Finite Element Modelling for Simulating the Surface Subsidence above a Compacting Hydrocarbon Reservoir," *Int. J. Analytic. Numer. Meth. Geomech.*, **1993**, *18*, 619-639.
16. Li X. and Zienkiewicz O. C., "Multiphase Flow in Deforming Porous Media and Finite Element Solutions," *Computers & Structures*, **1992**, *45*(2), 211-227.
17. Minkoff S. E., Stone C. M., Bryant S., Peszynska M., et al., "Coupled Fluid Flow and Geomechanical Deformation Modeling," *Journal of Petroleum Science and Engineering*, **2003,** *38*, 37-56.
18. Phillips P. J. and Wheeler M. F. "A Coupling of Mixed and Continuous Galerkin Finite Element Methods for Poroelasticity I: the Continuous in Time Case," *Comput. Geosci.*, **2007**, *11*, 131-144.
19. Phillips P. J. and Wheeler M. F., "A Coupling of Mixed and Continuous Galerkin Finite Element Methods for Poroelasticity II: The Discrete-in-time Case," *Comput. Geosci.*, **2007**, *11*, 145-158.
20. Settari A. and Walters D. A., "Advances in Coupled Geomechanical and Reservoir Modeling With Applications to Reservoir Compaction," SPE Reservoir Simulation Symposium, Houston, Texas, **1999**.
21. Settari A. and Mourits F. M., "Coupling of Geomechanics and Reservoir Simulation Models. Proc. Of the 8th Int. Conf. on Computer Methods and Advances in Geomechanics," Morgantown, West Virginia, **1994**, 2097-2158.
22. Dean R. H., Gai X., Stone C. M., and Minkoff S. E., "A Comparison of Techniques for Coupling Porous Flow and Geomechanics," In SPE Reservoir Simulation Symposium (Houston 2003), SPE 79709, **2003**.
23. Gutierrez M. and Lewis R. W. "The Role of Geomechanics in Reservoir Smulation," In Proceedings of the SPE/ISRME; ROCK198, (SPE paper 47392) (Trondheim, Norway, **1998**.
24. Gutierrez M., Lewis R. W., and Masters I. "Petroleum Reservoir Simulation Coupling Fluid

and Geomechanics," SPE Reservoir Evaluation & Engineering, **2001**, *4*, 164-171.

25. Wan J., Durlofsky L. J., Hughes T. J. R., and Aziz K. "Stabilized Finite Element Methods for Coupled Geomechanics Reservoir Flow simulations," In SPE Res Simul Sym (SPE 79694) (Houston 2003).
26. Jeannin L., Mainguy M., Masson R., and Gilbert S. V. "Accelerating the Convergence of Coupled Geomechanical-reservoir Simulations," *Int. J. Numer. Anal. Meth. Geomech.*, **2007**, *31*, 1163–1181.
27. Prevost J. H. "Partitioned Solution Procedure for Simultaneous Integration of Coupled-field Problems," *Commun. Numer. Meth. Engng.*, **1997**, *13*, 239-247.
28. Wheeler M. F. and Gai X., "Iteratively Coupled Mixed and Galerkin Finite Element Methods for Poroelasticity," *Numer Meth for PDEs*, **2007**, *23*, 785–797.
29. Kim J., Tchelepi H. A., and Juanes R., "Rigorous Coupling of Geomechanics and Multiphase Flow with Strong Capillarity," *SPE Journal*, **2013**, 1123-1139.
30. Aarnes J. E., Kippe V., Lie K. A., and Rustad A. B. "Modelling of Multiscale Structures in Flow Simulations for Petroleum Reservoirs in Geometrical Modeling," *Numerical Simulation and Optimization Applied Mathematics at SINTEF. Springer*, Verlag, **2007**.
31. Sadrnejad S. A., Ghasemzadeh H., and Taheri E., "Multiscale Advance Features in Modeling Oil Transport in Porous Media," In 21st Annual International Conference on Mechanical Engineering-ISME, Tehran, Iran, **2013**.
32. Wen X. H., Durlofsky L. J., and Edwards M. G., "Use of Border Regions for Improved Permeability Upscaling," *Mathematical Geology*, **2003**, *35*, 521-547.
33. Jenny P., Lee S. H., and Tchelepi H. A., "Multi-scale Finite-volume Method for Elliptic Problems in Subsurface Flow Simulation," *Computational Physics*, **2003**, *187*, 47-67.
34. Jenny P., Lee S. H., and Tchelepi H. A. "Adaptive Fully Implicit Multi-scale Finite-volume Method for Multi-phase Flow and Transport in Heterogeneous Porous Media," *Journal of Computational Physics*, **2006**, *217*, 627-641.
35. Lee S. H., Wolfsteiner C., and Tchelep H. A., "Multiscale Finite-volume Formulation for Multiphase Flow in Porous Media: Black Oil Formulation of Compressible, Three-phase Flow with Gravity," *Computational Geoscience*, **2008**, *12*, 351-366.
36. King M. J., MacDonald D. G., Todd S. P., and Leung, H., "Application of Novel Upscaling Approaches to the Magnus and Andrew Reservoirs," SPE European Petroleum Conference. The Hague, The Netherlands, SPE 50643, **1998**.
37. Chen Y., Durlofsky L. J., Gerritsen M., and Wen X. H., "A Coupled Local–global Upscaling Approach for Simulating Flow in Highly Heterogeneous Formations," *Advances in Water Resources*, **2003**, *26*, 1041-1060.
38. Wen X. H., Durlofsky L. J., and Edwards M. G., "Upscaling of Channel Systems in Two Dimensions Using Flow-based Grids," *Transport in Porous Media*, **2003**, *51*, 343–366.
39. Peszy´nska M., Wheeler M. F., and Yotov I., "Mortar upscaling for multiphase flow in porous media," *Comput. Geosci.*, **2002**, *6*, 73-100.
40. Christie M. and Blunt M. , "Tenth SPE Comparative Solution Project: a Comparison of Upscaling Techniques," *SPE Reserv. Evaluat. Eng.*, **2001**, *4*, 308–317.
41. Durlofsky L. J. "Upscaling of Geocellular Models for Reservoir Flow Simulation: A Review of Recent

Progress," In International Forum on Reservoir Simulation (Buhl/Baden-Baden), Germany, **2003**, 1-58.

42. Jenny P., Lee S. H., and Tchelepi. H. A., "Adaptive Multiscale Finite-Volume Method For Multiphase Flow And Transport In Porous Media," *Multiscale Model. Simul.*, **2004**, *3*, 50-64.
43. Lee S. H., Wolfsteiner C., and Tchelep H. A., "Multiscale Finite-volume Formulation for Multiphase Flow in Porous Media: Black Oil Formulation of Compressible, Three-phase Flow with Gravity," *Computational Geoscience*, **2008**, *12*, 351-366.
44. Jenny P. and Lunati I. "Modeling Complex Wells with the Multi-scale Finite-volume Method," *Journal of Computational Physics*, **2009**, *228*, 687–702.
45. Lee S. H., Zhou H., and Tchelepi H. A. "Adaptive Multiscale Finite-volume Method for Nonlinear Multiphase Transport in Heterogeneous Formations," *Journal of Computational Physics*, **2009**, *228*, 9036–9058.
46. Lunati I. and Jenny P., "Multiscale Finite-volume Method for Density-driven Flow in Porous Media," *Computational Geoscience*, **2008**, *12*, 337-350.
47. Hajibeygi H., Bonfigli G., Hesse M. A., and Jenny P., "Iterative Multiscale Finite-volume Method," *Journal of Computational Physic*, **2008**, *227*, 8604–8621.
48. Hajibeygi H. and Jenny P. "Adaptive Iterative Multiscale Fnite Volume Method," *J. Comput. Phys.*, **2011**, *230*, 628–643.
49. Aarnes J. E. and Efendiev Y. "Mixed Multiscale Finite Element for Stochastic Porous Media Flows," *SIAM Sci. Comp.*, **2008**, *30*, 2319–2339.
50. Efendiev Y., Ginting V., Hou T. Y., and Ewing R. "Accurate Multiscale Finite Element Methods for Two-phase Flow Simulations," *J. Comp. Phys.*, **2006**, *220*, 155–174.
51. Efendiev Y., Hou T. Y., and Wu X. H., "Convergence of a Nonconforming Multiscale Finite Element Method," *SIAM J. Numer. Anal.*, **2000**, *37*, 888–910.
52. Hou T. Y., Wu X. H., and Cai Z. "Convergence of a Multiscale Finite Element Method for Elliptic Problems With Rapidly Oscillating Coefficients," *Math. Comput.*, **1999**, *68*, 913–943.
53. Efendiev Y. and Hou T. Y. "Multiscale Finite Element Methods Theory and Applications," Series: Surveys and Tutorials in the Applied Mathematical Sciences, Springer, **2009**.
54. Lunati I. and Jenny P., "Treating Highly Anisotropic Subsurface Flow With The Multiscale Finite-Volume Method," *Multiscale Model. Simul.*, **2007**, *6*, 308-318.
55. Samier P., Quettier L., and Thiele M. "Applications of Streamline Simulations to Reservoir Studies," *SPE Res. Eval. & Eng.*, SPE-78883-PA, **2002**, *5*(4): 324–332.
56. Baker R. O., Kuppe F., and Chug S., et al., "Full-Field Modeling Using Streamline-Based Simulation: 4 Case Studies," Presented at *the SPE Reservoir Simulation Symposium*, SPE-66405-MS, Houston, Texas, **2001**.
57. White I. R. and Lewis R. W., "The Numerical Simulation of Multiphase Flow Through a Porous Medium and its Application to Reservoir Engineering," *Applied Mathematical Modelling*, **1981**, *5*(3), 165–172.
58. Geiger S., Roberts S., Mattha S. K., Zoppou C., et al., "Combining Finite Element and Finite Volume Methods for Efficient Multiphase Flow Simulations in Highly Heterogeneous and Structurally Complex Geologic Media," *Geofluids*, **2004**, *4*, 284–299.

11

A Study of Flow Properties of Kuwaiti Crude Oil Obtained from Different Sources

Bader H. Albusairi* and Adam Al-Mulla

Department of Chemical Engineering, Kuwait University, Kuwait

ABSTRACT

The rheological properties of four different Kuwaiti crude oils in the presence of polyacrylamide (PAA) were investigated. PAA was varied at three different concentrations of 25, 50, and 75 ppm to study the viscoelastic properties in the untreated samples. Four different temperatures (25, 30, 35, and 40 °C) were used to evaluate different viscoelastic parameters of the treated and untreated Kuwaiti crude oils. The sulfur content and physical properties; such as, density and refractive index of Kuwaiti crude oils were also measured.

An important observation noted from the rheological analysis is that the dynamic viscoelastic parameters like complex modulus, elastic modulus, and storage modulus were found to be a function of the amount of sulfur present in the crude oil. The most important observation noted is that PAA acts as a drag reducing agent in lowering the viscosity, shear stress, and all the viscoelastic parameters of the Kuwaiti crude oil. Carreau model was used in this investigation to determine the relaxation time and the shear thinning characteristics of the treated and untreated crude oil samples at different temperatures.

Keywords: Crude oil, Rheology, Polyacrylamide, Elemental Analysis, Complex Modulus, Storage Modulus

INTRODUCTION

N-paraffins and asphaltenes form the main constituents of crude oil. They are a cause of crystallite formation when a temperature drop in crude oil occurs [1, 2]. International energy sources indicate that around 52% of the international heavy crude oil are recoverable, but it is not economically feasible for processing due to its high viscosity, causing refining and transportation problems. The heavy crude oil possesses many technological problems related to production, refining, and transportation, but the need for energy outweighs these issues which are inherently flow related. Thus there is a need to probe for innovative and amicable solutions which are mainly related to the production, refining, and transportation of the high viscosity crude oil. For a smooth flow in a pipeline, the viscosity of crude oil should be lower than 400 MPa, but the viscosity of the heavy crude oil is normally greater than 10^5 MPa. This is due to the complex nature of asphaltenes, maltenes, and paraffinic waxes [3], which contain carbon (83-87%), hydrogen (10-15%), and trace amounts of metals like nickel, vanadium, and

***Corresponding author**
Bader H. Albusairi
Email: bader.albusairi@ku.edu.kw

cobalt existing in the heavy crude oil [4]. The amount of carbon in the paraffinic wax ranges between C_{18}-C_{70} and it mainly consists of saturates and the asphaltene. The asphaltene composition consists of polycyclic aromatic hydrocarbons [5]. Petroleum crude from varying wells and sources have properties which are different [6]. The heavy crude oil having very high viscosity values [7] has large amounts of asphaltenes and low molecular weight hydrocarbons [8- 10]. Wax precipitation in crude oil is a thermodynamic process. As low temperatures are encountered by the oil, the wax precipitates, and these waxes show shear thinning. The wax built up at the interphase between the borewell and the reservoir can cause skin effect, leading to increased roughness and a pressure drop in pipelines which are carrying crude oil. The build-up of wax can cause malfunctioning and plugging of equipment during the transportation of crude oil.

Various methods are employed to reduce the viscosity of crude oils, the various methods include forming oil in water emulsions [11] and preheating the crude oils or adding light oil fractions to reduce the viscosity to facilitate pumping [12]. Additive help changes the wax crystal structure or act as a dispersant to enhance the flow properties of crude oil at temperatures below 20°C [13]. However, the drawbacks of the emulsion method is that it involves huge capital costs in terms of using costly surfactants to stabilize the oil in water emulsion during transportation and also separation of crude oil from water once the destination is reached. Also, huge capital costs which are involved in heating pipelines include covering large distances and altering the chemical composition of the original crude oil by heating or adding external substances

Knowing the rheology of crude oil is important for reducing the cost involved in transportation. The rheological study would help know the viscosity of crude oil during its production, transportation from reservoir to gathering centers, and transportation from the gathering centers to refineries or crude oil filling stations. A limited number of research papers exist on the study of rheological property of crude oil, containing large amounts of maltenes [10]. The size distribution of isoparaffinic waxes, asphaltenes, and maltenes determines the role of temperature and rheology in the flow properties of crude oil (14]. The rheological properties are affected by the shear conditions, the temperature conditions, and the external dilution of parent crude oil using mineral oils [15]. Rasha et al. [16] has also observed that incorporation of copolymers of maleic anhydride and octadecane have altered the flow characteristics of heavy crude oil. Yang et al. [17] have used copolymers of lauric amine and dodecylbenzenesulfonic acid to alter the crystal structure of waxes, which could change the rheological characteristics of crude oil. Different types of flow improvers have been reported in the literature. The most commonly used approaches are comb-shaped copolymers and linear polymers; such as, ethylene-butylene [11, 12] and ethylene vinyl acetate [8-10] copolymers. These polymers interact with wax crystallization, breaking their crystal structures and decreasing the viscosity of crude oils. The main objective of this work was to study the amount of sulfur presented in different Kuwaiti crude oils obtained from different locations, and relate the sulfur content to rheology. Another important objective was to evaluate the effect of drag reducing agent on its rheological properties.

Methodology

In this work, four different types of Kuwaiti crude

oils have been used for studying the effect of a drag reducing polymer on its rheological properties at four different temperatures. The samples have been obtained from different wells which are located approximately 10 km away from each other. The samples were analyzed for its sulfur content using Antek by PAC instrumentation (Multitek), density using density meter and refractive index using Mettler Toledo RM40/RM50 refractometer at 25°C.

Emulsion Preparation

Polyacrylamide (PAA), a synthetic polymer was used as the surfactant in the present study. PAA was dissolved in 100 mL of crude oil by vigorously stirring it for a period of 15 minutes at room temperature, and this solution was further subjected to high speed shearing at 3000 rpm using a Silverson SL-2 machine for a time period of 15 minutes. Three different concentrations (25, 50, and 75 ppm) of PAA were used in the drag reducing analysis. 50 ppm was chosen as the optimum concentration after carrying out viscosity measurements.

Rheological Analysis

Gemini 200 rheometer supplied by Bohlin Instrument Co., UK using CP 180 cylinder peltier was used for measuring the rheology of fluids. The rheological measurements were carried out at 25, 30, 35, and 40°C. The temperatures in the rheometer system were controlled using a Julabo F 25 water bath. A linear viscoelastic region was obtained by plotting storage modulus (*G′*) as a function of stress; moreover, the linear viscoelastic region was found to be in the range of 0.01-0.75 Pa. The stress value was kept constant at 0.5 Pa, and it was obtained from the linear region, and it was kept constant for further rheological studies carried out in the oscillation test mode. The uniform frictionless transmission of applied stress to the crude oil was controlled by an air bearing mounted in the middle of the shaft. A digital encoder was used to process the deformations occurring in the crude oil at 106 impulses per revolution. These impulses help to measure the minute strains or shear rates or small yield stress values. The gap which was kept between the spindle and the cylinder peltier was 0.15 mm; also, the cone angle was 4°. The samples were brought to equilibrium at the experimental temperature before the test. A set of three experiments was conducted for each run to obtain reproducibility, and the estimated error between each run was approximately 3%. In the viscometry experiments, a cutoff point of 100 s^{-1} for the shear rate values was chosen due to inconsistent trend observed in the viscosity and shear stress values. Similarly a cutoff point of 0.7 s^{-1} was kept in the frequency values in the oscillatory mode because an inconsistent trend was noted for complex modulus (*G**), storage modulus (*G′*), and loss modulus (*G″*).

RESULTS AND DISCUSSION

Four different Kuwaiti crude oils were characterized for sulfur content, density, and refractive index, and the values are indicated in Table 1.

Table 1: The sulfur content, density, and the refractive index values of Kuwaiti crude oil samples.

Sample	Sulfur Content	Density	Refractive Index
MG 23T	2.021	0.896	1.504
MG 29T	1.977	0.895	1.503
MG 108T	1.811	0.871	1.488
MG 34T	1.027	0.867	1.487

Variation of viscosity with a shear rate for a typical Kuwaiti crude oil (MG 23T) at four different temperatures are displayed in Figure 1. The majority of the crude oils show a Newtonian behavior. The flow becomes complex and non-Newtonian due to wax precipitation at lower temperatures [18]. As seen in Figure 1, the viscosity values are found to decrease with a rise in the temperature, showing a drop by 30% in viscosity between the lowest and highest temperature. This indicates a strong dependence of temperature on viscosity for all the different types of crude oil. This behavior is due to the structure breakdown of complex asphaltenes and maltenes in the presence of temperature. The viscosity values are also found to be high at low shear rates because at low shear rates, more time is provided to break the complex chemical structures of the crude oil sample at the corresponding temperature.

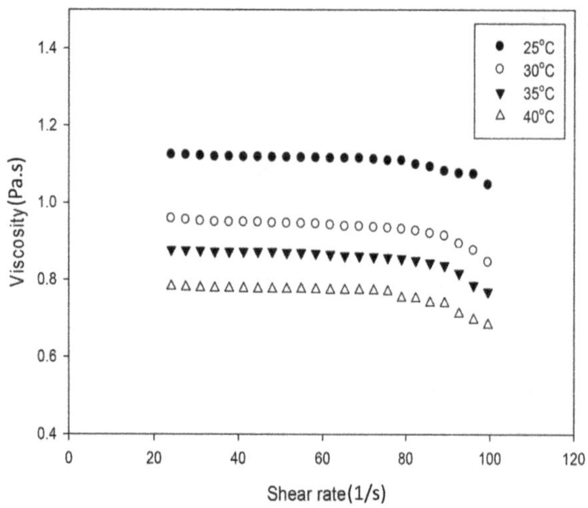

Figure 1: Plot comparing the viscosity versus shear rate for MG 23T at four different temperatures.

Figure 2 depicts the variation of viscosity with the shear rate for four different Kuwaiti crude oils at 30°C. It is seen that the crude oil MG 23T has the highest amount of sulfur (Table 1) and the crude oil MG 23T exhibits the maximum viscosity. The viscosity values are found to decrease as the percentage of sulfur decreases. All the crude oil samples show a non-Newtonian behavior.

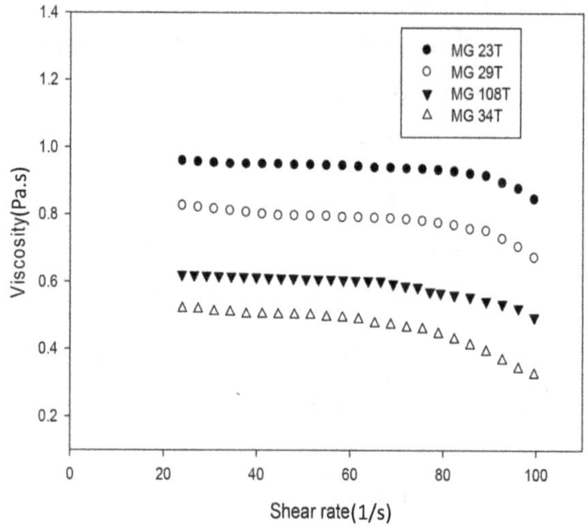

Figure 2: Comparing viscosity versus shear rate for four different Kuwaiti crude oils at 30°C.

Figure 3 depicts a plot of viscosity versus a shear rate for a typical sample MG 108T which is treated with PAA of three different concentrations of 30, 50, and 75 ppm. The addition of 75% of DRA polymer decreases the viscosity by 15%, while the difference noted between 50 and 75 ppm was hardly 2%. This is a clear observation that the addition of PAA leads to a decrease in viscosity. Therefore, the concentration of 50 ppm was chosen as the optimum concentration of PAA for all the further studies.

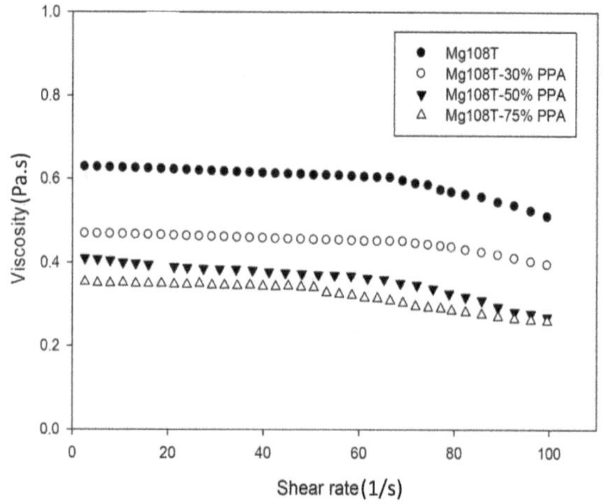

Figure 3: Variation of polyacrylamide (PAA) concentrations 30, 50, and 75 ppm at 30°C for MG 108T.

Figure 4 shows the comparison which is made on the study of variation of viscosity versus the shear rate for untreated and PAA-treated MG 34T crude oil sample at 30°C. From this figure, it is noted that untreated crude oil has higher viscosity than the treated crude oil sample possibly due to the breakdown of chemical complex structures by PAA in the crude oil sample.

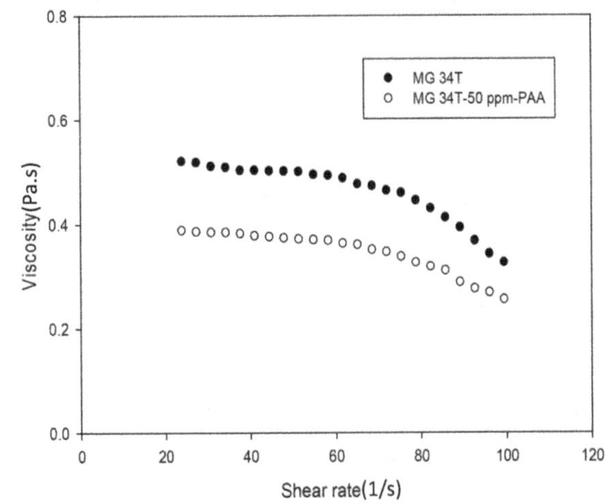

Figure 4: Comparison of viscosity versus shear rate for the untreated and treated MG 34T sample (50ppm) at 30°C.

Figure 5 indicates the plot of viscosity versus the shear rate at 40°C for four different crude oils containing 50 ppm PAA. It is clear from Figure 5 that a decrease in the crude oil viscosity occurs when the amount of sulfur content in the oil (Table 1) decreases.

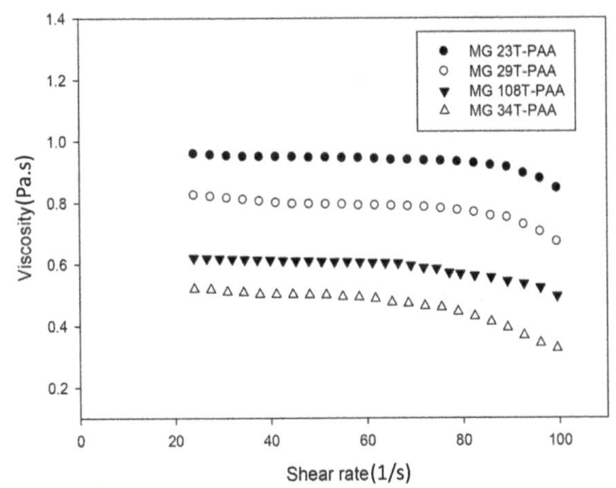

Figure 5: Comparison of viscosity versus shear rate for four different crude oils containing PAA at 40 °C and 50 ppm.

Figure 6 shows the variation of the shear stress as a function of the shear rate for four different untreated crude oils at 30°C. The shear stress behavior is found to be dependent on the amount of sulfur which presents in the crude oil. High shear stress values are noted for the crude oil which is containing the maximum amount of sulfur.

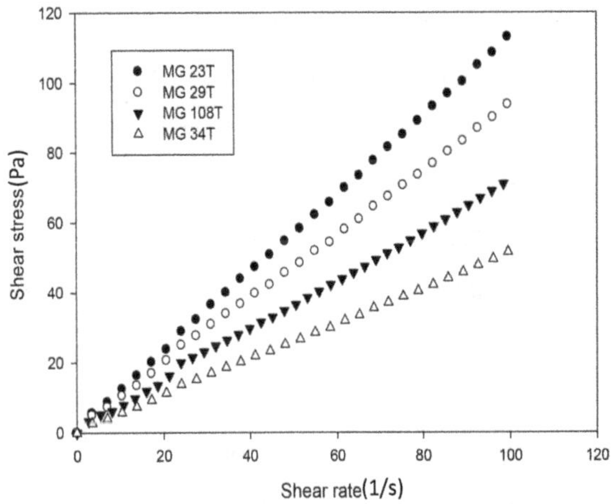

Figure 6: Comparison of shear stress versus the shear rate at 30°C for four different untreated crude oils.

Figure 7 is comparing the change in shear stress versus the shear rate for the untreated and treated MG 34T at 30°C. It is obvious that the untreated crude oil sample has higher shear stress compared to the treated sample and PAA is found to decrease the shear stress of the crude oil sample.

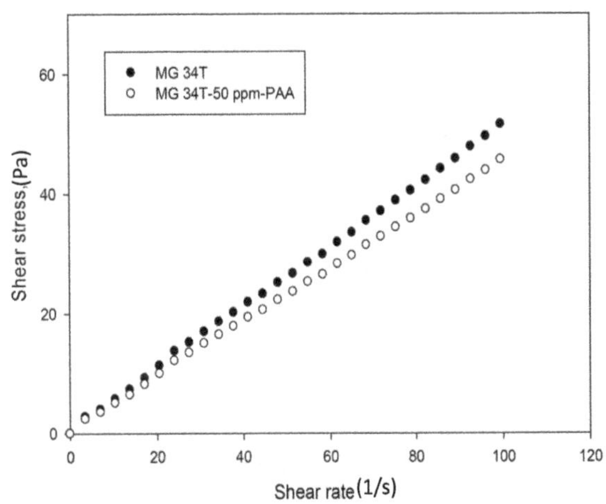

Figure 7: Comparison of shear stress versus shear rate of MG 34T untreated and treated (50 ppm) at 30°C.

The values of yield stress for the neat crude oil is 5.2 Pa, while that of the PAA treated sample is 4.1 Pa, indicating that the breakdown of the complex wax structure has occurred. Figure 8 compares the shear stress as a function of shear rate for different crude oils containing 50 ppm PAA at 40°C. This figure clearly indicates that shear stress values are higher for crude oil containing the maximum amount of sulfur in a system containing 50 ppm PAA. Systems (Figures 6 to 8) which contain microstructure domains display thixotropic phenomenon which implies the extent of micro-structural changes occurring from the initial state to the final state and reverting back to the initial state due to the application of predefined stress. The micro-structural changes under the influence of stress and time determine the viscosity or shear stress of the sample at the relevant shear rate. Thixotropic hysteresis curve was unnoticeable for treated and untreated crude oil possibly due to the fact that large microstructures under the influence of progressively increasing shear rate break down into small flocks and immediately convert back into the parent structures in a very short time interval. The dynamic behavior (varied oscillating stress or strain) of the treated and the untreated samples determines its viscoelastic nature. The total resistance of a material under a given strain represents the complex modulus of a material.

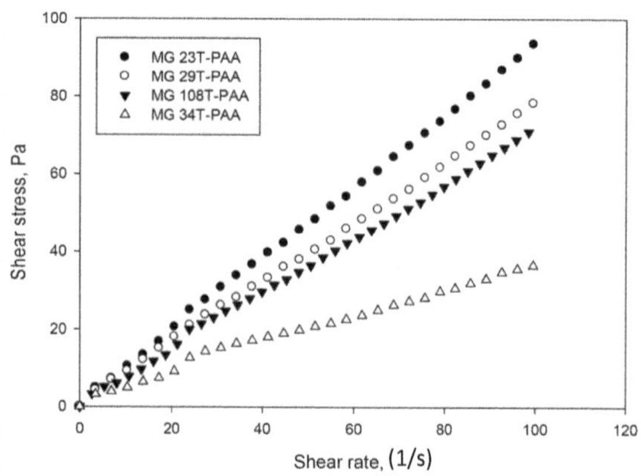

Figure 8: Comparison of shear stress versus shear rate for four different crude oils containing 50 ppm PAA at 40°C.

A dynamic test is important to determine the linear viscoelastic (LV) properties of the crude oil under varying temperatures and varying concentrations of PAA. When the analysis of the crude oil is carried out in the LV region, the chemical nature of the crude oil remains unchanged with the application of stress. The only parameter which can change the flow property in the LV region is temperature. In stress-sweep experiments, the LV region corresponds to complex modulus (G^*) which denotes whether the chemical structure of the sample is broken or is left intact. The stress value was kept constant at 0.5 Pa, which is in the linear viscoelastic range (0.01-0.75) of the sample, where G^* would show a constant response.

Figure 9 depicts the variations of complex modulus versus frequency (ω) for the untreated crude oil (MG 29T) at different temperatures. Figure 9 shows that the complex modulus is strongly dependent on temperature, and it decreases with an increase in temperature. This observation implies that resistance is a function of strain and decreases by raising temperature.

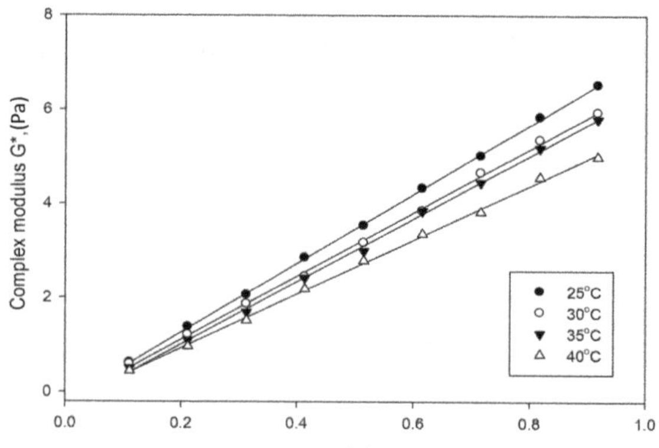

Figure 9: Complex modulus as a function of frequency for untreated crude oil MG 29T at different temperatures.

Figure 10 shows complex modulus versus frequency for different untreated Kuwaiti crude oils at 30°C. The complex modulus is found to increase linearly with increasing frequency. As noted from the previous graphs, the samples with the maximum amount of sulfur content show the maximum value of complex modulus. At high frequencies, the complex modulus values increase as the sulfur content rises.

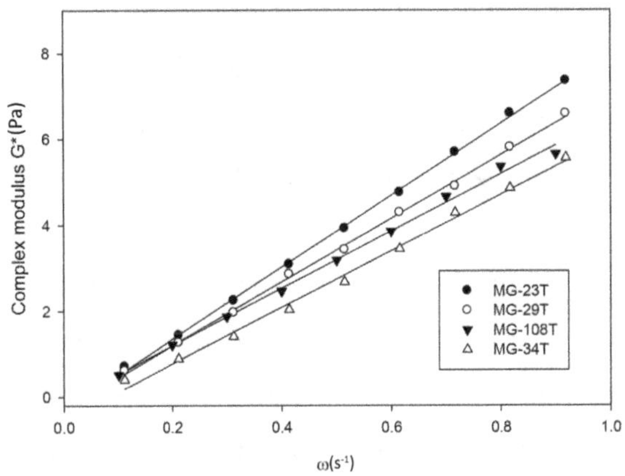

Figure 10: Comparison of complex modulus versus frequency at 30°C for four different untreated Kuwaiti crude oils.

Figure 11 displays complex modulus versus frequency for untreated and treated MG 34T at 30°C. This figure indicates that the addition of PAA decreases the complex modulus, which implies that less resistance is offered by the treated sample compared to the untreated sample.

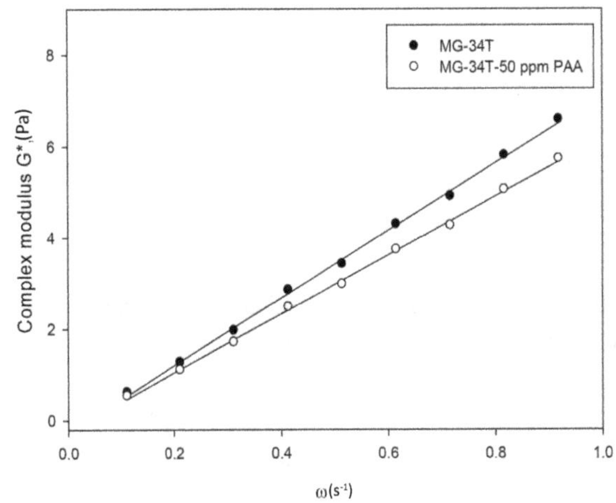

Figure 11: Comparison of complex modulus versus frequency for MG 34T untreated and treated (50 ppm) at 30°C.

Plotting storage modulus (*G'*) versus frequency (Figure 12) shows that the high G' values are noted for untreated MG23T sample containing high sulfur content. A conclusion that can be obtained from this plot is that crude oil with high sulfur content has a high amount of energy per unit cycle stored in it, and it decreases with decreasing sulfur content at high frequencies.

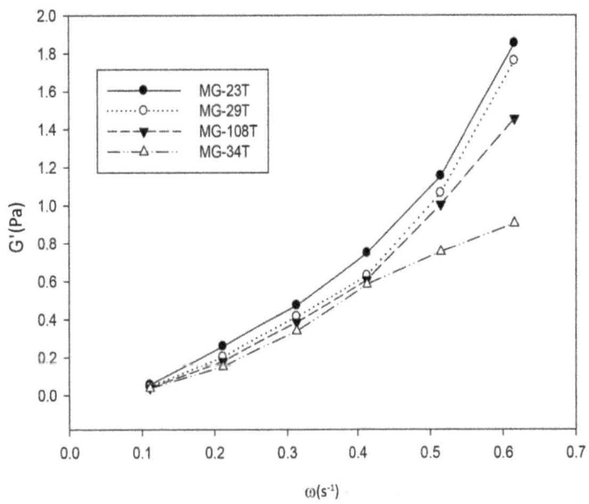

Figure 12: Storage modulus versus frequency at 30°C for four different untreated Kuwaiti crude oils.

Figure 13 indicates that low *G'* values are noted for PAA treated crude oil compared to untreated crude oil implying that energy stored per unit cycle in the material decreases with the addition of PAA.

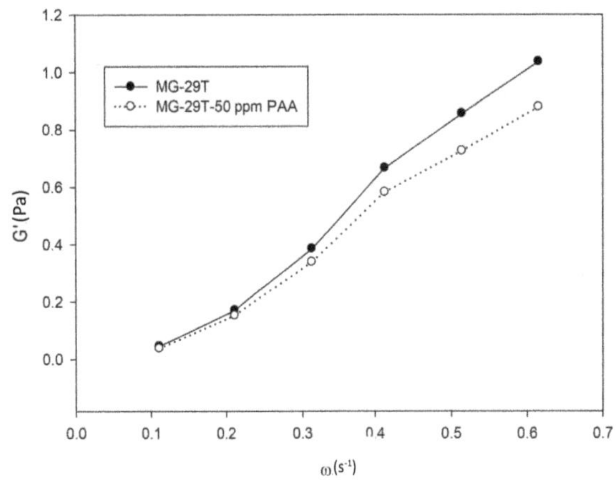

Figure 13: Storage modulus versus frequency for MG 29T for untreated and treated crude oil at 50 ppm and 30°C.

Figure 14 depicts loss modulus (*G″*) versus frequency; moreover, according to this figure, the amount of energy lost per unit cycle in the untreated samples containing high amount of sulfur is higher. The loss modulus values show linear relationships over the entire range of frequencies

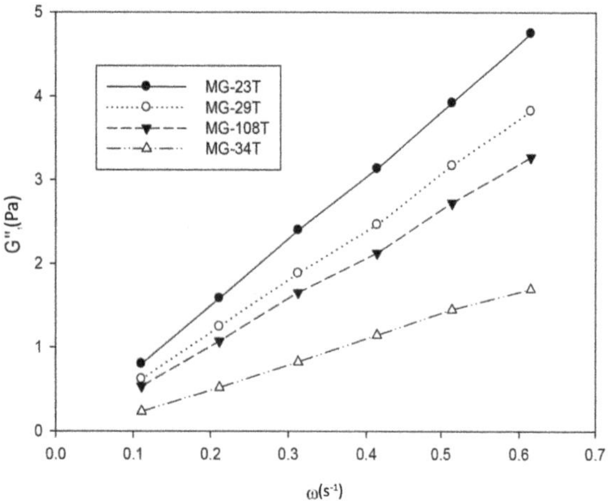

Figure 14: Loss modulus versus frequency at 30°C for four different untreated Kuwaiti crude oils.

Figure 15 compares the loss modulus values for treated (50 ppm) and untreated samples at 30°C. The plot indicates that the energy loss per unit cycle of the treated sample is lower than that of the untreated sample over the complete frequency range.

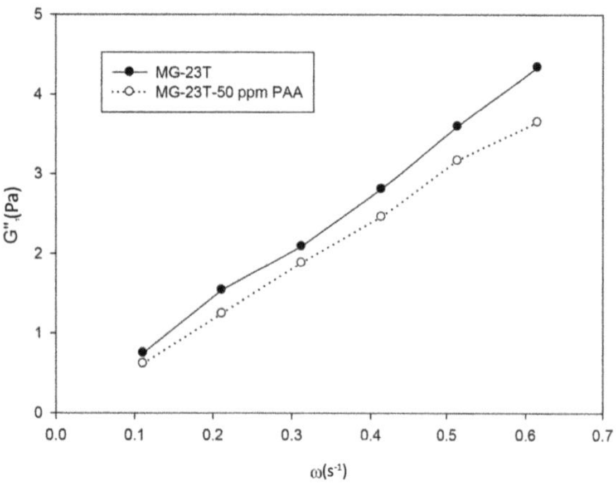

Figure 15: Loss modulus versus frequency for untreated and treated (50 ppm) crude oil at 30°C.

The *G′* and *G″* are constant terms in the LV region of the crude oil. It is affected by varying sinusoidal stresses. The effect of frequency in the crude oil sample can be determined by subjecting the sample to a frequency scan. *G′* and *G″* are dependent on frequency. The waxy viscous nature of crude oil (*G″*) is found to be greater the *G′* (Figure 16).

As seen in Figures 14 and 15, *G″* shows a linear relationship, while *G′*s do not show linearity over the entire frequency range. It is also noted that the G″ values are higher for treated and untreated samples compared to the storage modulus values, which indirectly implies that the amount of energy stored in the crude oil per unit cycle is lower than the amount of energy lost as heat per unit cycle; this causes the crude oil samples to depict the solid characteristics lower than their viscous nature over the entire frequency range.

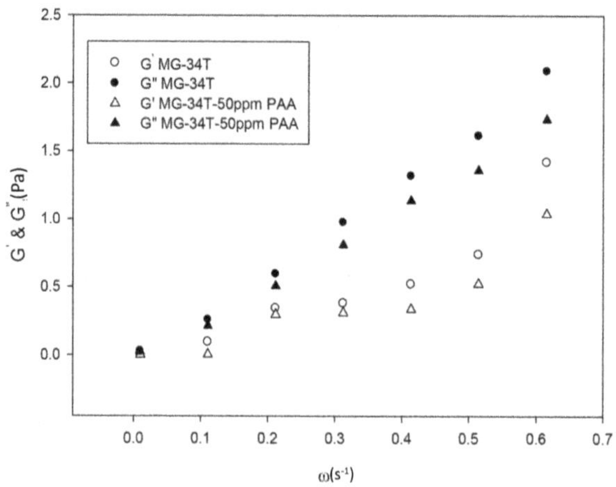

Figure 16: Comparison of *G′* and *G″* versus frequency for untreated and treated (50 ppm) MG 34T at 25°C.

The loss tangent is the ratio of *G″* (energy dissipation stress) and *G′* (energy preserving stress), which defines the viscous and the elastic components of the material respectively. The loss tangent data at different frequencies are indicated in Table 2. Table 2 shows high tan δ values at low frequencies and the value decreases with increasing frequencies. High tan δ values indicate an ability to dissipate deformation energy and relax stresses which would possibly lead to breaking of wax crystals.

The energy stored and recovered per cycle which corresponds to the elastic compliance is found to decrease by reducing time (Table 2). With the addition of PAA to the crude oil, the elastic compliance values are found to decrease further as a function of time, indicating that the elastic properties of the crude oil has decreased due to the disruption of the wax structure. The amount of heat or energy which is given out as the quantity of heat per sinusoidal cycle is found to be less on the addition of PAA to the crude oil. This is a clear indication that the wax present in crude oil after the addition of PAA does not have a large complex structure similar to the crude oil, and thus the amount of heat dissipated is less compared to the neat crude oil.

Table 2: Tan δ and compliance values for the treated and untreated crude oil samples at 25°C.

Frequency (s^{-1})	Tan δ	Tan δ with PAA	Elastic compliance (Pa^{-1})	Elastic compliance with PAA (Pa^{-1})	Viscous compliance (Pa^{-1})	Viscous compliance with PAA (Pa^{-1})
1.019	1.328	1.178	0.1616	0.1596	0.1216	0.1196
2.028	1.042	0.892	0.05217	0.05017	0.05009	0.04809
3.037	0.6816	0.5316	0.03267	0.03067	0.04792	0.04592
4.046	0.4748	0.3248	0.008812	0.006812	0.01856	0.01656
5.056	0.1688	0.0188	0.002141	0.000141	0.01268	0.01068
6.064	0.769	0.619	0.004296	0.002296	0.005585	0.003585
7.074	0.5239	0.3739	0.002662	0.000662	0.005129	0.003129
8.083	0.4192	0.2692	0.0026669	0.0006669	0.005594	0.003594
9.092	0.3241	0.1741	0.003245	0.001245	0.006194	0.004194
10	0.2136	0.0636	0.003032	0.001032	0.002671	0.000671

Table 3: Carreau model constants and coefficient of determination (r^2).

Temperature (°C)	25	30	35	40
λ	0.11237893	0.12034426	0.12073778	0.12981699
N	0.17212747	0.1811687	0.18215563	0.19354373
r^2	0.993	0.996	0.995	0.996

The shear thinning characteristics (*N*) and the relaxation time (λ) for the Kuwaiti crude oil was estimated (Table 3) using the Carreau model, which is applicable to both Newtonian and non-Newtonian fluids and is given by the following equation:

$$\eta = \eta_0 \left(1+\lambda^2\gamma^2\right)^{-N} a \quad (1)$$

where, η corresponds to viscosity; γ corresponds to shear rate; η_0 is the zero shear viscosity at zero shear rate, and λ and N represent the characteristics of the fluid.

The built in solver function in Microsoft Excel 2013 was used to determine the values of λ, N, and the coefficient of determination (r^2) for the fit between the experimental viscosity values and those predicted theoretically using Equation 1.

It is noted that the relaxation time (λ) and the shear thinning constant (N) increase as the temperature rises. Thus a longer relaxation time is needed for the molecular aggregation of the sheared complex substances in the crude oil to reach equilibrium at the temperature in this investigation. The r^2 values are found to be above 0.99 indicating a good fit between the theoretically and experimentally predicted values.

CONCLUSIONS

The present work was performed to evaluate the rheological characteristics of PAA treated Kuwaiti crude oils. The following conclusions are drawn:

1. Non-Newtonian (shear thinning characteristics) behavior is exhibited by the treated and untreated crude oil. With increasing the shear rate and temperature, a decrease in the viscosity of crude oil was observed.
2. The addition of three different concentrations of PAA (25, 50, and 75 ppm) showed that the concentration of 50 ppm PAA is the optimum concentration for decreasing the viscoelastic properties of untreated Kuwaiti crude oils.
3. Plots of viscosity versus the shear rate at varying temperatures indicated that the viscosity values decreased with a decrease in the percentage of sulfur. All the crude oil samples showed a non-Newtonian behavior.
4. Thixotropic behavior was unnoticeable for the treated and untreated crude oil samples because the time required for breakdown and re-flocculation was minimal.
5. With increasing temperature, the complex modulus of the crude oil was found to decrease because the total resistance to flow decreased with a rise in temperature.
6. It is noticed that tan δ values are high at low frequencies and these values decrease by the addition of PAA to the sample, indicating that the wax crystals have a non-cross-linked structure compared to the neat crude oil.
7. The lower elastic compliance values of PAA with respect to the neat crude oil indicate the disruption of wax crystal structure.
8. The loss of energy as heat per cycle with the addition of PAA was found to be lower than that of neat crude oil due to the breakdown of complex structures in wax in the presence of PAA.
9. Carreau model was used to determine the relaxation time (λ) and the shear thinning constant (N) at different temperatures. It was found that these constants were temperature dependent in the case of both treated and untreated samples.

ACKNOWLEDGEMENTS

The authors would like to thank the Research Administration of Kuwait University for providing funds from project RE 01/13. The authors would like to acknowledge Johnson Mathew, Joyson John, and Bader Masier for the technical support provided during the course of this work. The authors would also like to acknowledge the PRP center GE 03/08 for analyzing the crude oil samples.

NOMENCLATURE

PAA	: Polyacrylamide
G′	: Elastic modulus (Pa)
G″	: Loss modulus (Pa)
G*	: Complex modulus (Pa)
DRA	: Drag reducing agent
LV	: Linear viscoelastic
Ω	: Frequency (s^{-1})
δ	: Phase angle (Tan δ= G″/G′)
η	: Viscosity (Pas)
η_0	: Zero shear viscosity
λ	: Relaxation time (s)
γ	: Shear rate (1/s)
N	: Shear thinning constant
r^2	: Coefficient of determination

REFERENCES

1. Herh P. K., Mai B. C., Roye N., and Hedmann K., "Rheology of Crude Oil at High Pressure," *Am. Lab.*, **2003**, *9*, 28-30.
2. Asma E., "Microbial Treatment of Waxy Crude Oil for Migration of Wax Precipitation," *Journal of Petroleum Science & Engineering*, **2007**, *55*, 111-121.
3. Johnsen E. E. and Ronningsen H. P., "Viscosity of Live Water-in-Crude-Oil Emulsions: Experimental Work and Validation of Correlations," *Journal of Petroleum Science & Engineering*, **2003**, *38*, 23–36.
4. Sjoblom J., Aske N., Auflem I. H., Brandal, O., and et al., "Our Current Understanding of Water in Crude Oil Emulsions, Recent Characterization Techniques and High Pressure Performance," *Journal of Colloid Interface Science*, **2002**, 100-102, 399-473.
5. Groffe D., Takhar S., Anderson S. I., Stenby E. H., et al., "A Waxy Inhibition Solution to Problematic Fields: A Chemical Remediation Process," *Petroleum Science & Technology*, **2001**, *19*, 205-217.
6. Yaghi B. M., and Al-Bemani A., "Heavy Crude Oil Viscosity Reduction for Pipeline Transportation," *Ener. Sources*, **2002**, *24*, 93-102.
7. Centeno G., Sanchez-Reyna G., Ancheyta J., Munoz J., et al., "Testing Various Mixing Rules for Calculation of Viscosity of Petroleum Blends," *Fuel*, **2011**, *90*, 3561–3570.
8. Castro L. V. and Vazquez F., "Fractionation and Characterization of Mexican Crude Oils," Energy and Fuel, **2009**, *23*, 1603–1609.
9. Murillo-Hernandez J., Garcia-Cruz I., Lopez-Ramirez S., Duran-Valencia C. et al., "Aggregation Behavior of Heavy Crude Oil-Ionic Liquid Solutions by Fluorescence Spectroscopy," *Energy and Fuel*, **2009**, *23*, 4584–4592.
10. Evdokimov I. N., Eliseev N. Y., and Eliseev D. Y., "Rheological Evidence of Structural Phase Transitions in Asphaltene-containing Petroleum Fluids," *Journal of Petroleum Science & Engineering*, **2001**, *30*, 199–211.
11. Pal R. and Rhodes E., "Viscosity/Concentration Relationships for Emulsions," *J. Rheol.*, **1989**, *33*, 1021–1045.
12. Ghannam M. T. and Esmail N., "Flow Enhancement of Medium-Viscosity Crude Oil," *J. Pet. Sci. Technol.*, **2006**, *24*, 985-999.
13. Feustel M., Oshmann H. J., and Kentsche U., "Additives Based on Component Present in Petroleum for Improving the Cold Flow Properties of Crude and Distillate Oils," **2004**, U.S. Patent No. 4728412.
14. Banerjee S., Kumar R., Mandal A., and Naiya T. K., "Use of Novel Natural Surfactant for Improving Flowability of Indian Heavy Crude Oil," *Pet. Sci. Technol.*, **2015**, *33*, 819-826.
15. Kumar R., Mohapatra S., Mandal A., and Naiya T. K., "Studies on the Effects of Surfactants on Rheology of Synthetic Crude," *J. Pet. Sci. Res.*, **2014**, *3*, 90-99.

16. Rasha A., Atta A., and Kabel K., "Modified Maleic Anhydride-Co-Octadecane as Flow Improver for Waxy Egyptian Crude Oil," *Journal of Petroleum Science & Engineering*, **2014**, *122*, 411-419.
17. Yang F., Li C., Yang S., Zhang Q. et al., "Effect of Dodecyl Benzene Sulfonic Acid (DBSA) and Lauric Amine (LA) on the Associating State and Rheology of Heavy Oils," *Journal of Petroleum Science & Engineering*, **2014**, *124*, 19-26.
18. Hasan S. W., Ghannam M. T., and Nabil E., "Heavy Crude Oil Viscosity Reduction and Rheology for Pipe-Line Transportation," *Fuel*, **2010**, *89*, 1095-1100.

Two-phase Simulation of Well Control Methods for Gas Kicks in Case of Water and Oil-based Muds

Yaser Jahanpeyma and Saeid Jamshidi*

Chemical and Petroleum Engineering Department, Sharif University of Technology, Tehran, Iran

ABSTRACT

Kick occurrence is a possible event during a drilling process. It is required to be handled immediately using a well control method to avoid blowout, financial losses and damages to the drilling crew. Several methods including driller, wait and weight, and concurrent are applicable in the drilling industry to control a well during a kick incident. In this study, typical well control methods were simulated for both cases of water and oil-based muds, and essential parameters such as the required time were calculated. Additionally, for each well control approach, a mathematical algorithm was proposed to simulate the process. In case of oil-based mud, the flash calculation was utilized in each depth and time by considering the effect of kick fluid dissolution in drilling mud to improve the accuracy of control parameters. Based on the results, when oil-based mud is used for drilling, extra time is required to control the well due to kick fluid dissolution in the mud and extensive changes in the mud density. In order to improve the accuracy of the calculations, critical parameters including temperature changes in the well column, dynamic drilling hydraulics, and pressure drop were considered during a well control process. In addition, the simulation of the concurrent method is one of the study innovations because of mud density alternations especially when the mud becomes heavier by a non-linear or complicated mathematical function during the process.

Keywords: Well control, Two-Phase, Simulation, Water and Oil Based Muds.

INTRODUCTION

Kick is a prevalent problem during a large number of drilling operations in which the formation fluid enters the wellbore as a result of higher pressure of formation relative to hydrostatic pressure. Gas type kicks are more commonplace than liquid types. In order to specify kick type, kick fluid density should be estimated using drilling parameters such as Shut-in Drill Pipe Pressure (SIDPP), Shut-in Casing Pressure (SICP) and depth. Due to the fact that gas kicks have more complicated behavior than the liquids, and are more challenging to be controlled, usually, a gas type is assumed for kick fluid in a well control simulation.

Well control methods enable drilling crew to remove the kick fluid from the well, and fill it with a heavier mud proportional to the formation pressure gradient. The driller's method, the wait and weight method and the concurrent method are popular methods for controlling the wells.

***Corresponding author**

Saeid Jamshidi

Email: jamshidi@sharif.edu

In all the mentioned methods, bottom-hole pressure should remain constant during the operation by means of alternating pump speed and choke opening and closing. Furthermore, it is possible for the crew to close the choke, and survey the well conditions in any step of the operation. The final pressure should remain constant when the new mud reaches to the drill bit at the bottom-hole. Choe [1] analyzed well control process applying mass and momentum balance and auxiliary equations for liquid and gas phases in order to calculate density and gas speed values in any region of the well. The analysis contained only water-based mud condition based on the assumption of non-solubility of kick fluid in the drilling mud. He analyzed kick occurrence condition in the Gulf of Mexico.

Avelar et al [2] simulated well control process for newly discovered fields in Brazil. They simulated the process for gas kick type and water-based mud using mass and momentum balance and auxiliary equations. Omosebi et al. [3] analytically predicted annulus pressure during a kick occurrence. They used flash calculations for oil-based muds and assumed that the kick fluid may move as a slug or may be mixed with the drilling fluid. However, they did not consider the effects of gas dissolution in the mud. Marbun and Shidiq [4] calculated annulus and casing shoe pressure during a well control process. They supposed that the gas expands when rising from bottom-hole to the surface with the assumption of non-solubility of gas in the kick fluid. An et al. [5] simulated kick behavior in high-pressure high-temperature conditions in offshore wells filled with oil-based muds. They realistically modeled wellbore pressure profile and kick behavior using their proposed method. Results showed that mud density decreases as well depth increases due to the fact that the temperature effect is more dominant relative to the pressure.

An advanced simulator for predicting standpipe and choking pressure in deep water horizontal well killing based on dynamic bottom-hole pressure was developed by Feng et al [6]. The simulator is able to simulate driller's method considering circulation temperature, gas expansion and choke line friction losses. A gas kick incident was simulated by Sun et al, and a well killing was conducted a dynamic hydraulic simulator using transient multiphase flow model for a high-pressure high-temperature well in western China by Sun et al [7]. The simulation revealed underlying physics and causes of several intriguing phenomena and calculated an appropriate pump rate in order to kill the well successfully.

An additional safety margin before starting the actual circulation by pumping a few strokes into the shut-in well to hand over excess pressure was applied by Mosti et al [8]. This margin can reduce the risk of operation in a well control process.In this paper, three well control methods were simulated using proposed algorithms for both cases of water and oil-based muds, in contrast to other researches in which usually one mud type has been studied. In case of water-based muds, it was supposed that the fluid moves as slugs and the kick fluid is not soluble in the drilling mud. In case of oil-based muds, the effect of kick fluid dissolution in drilling mud during the well control process was considered using flash calculations. It was checked by flash calculations that the fluid includes one or two phases at any time of the process. Furthermore, the effects of temperature changes in well column were considered. In both cases of water and oil-based muds, the pressure drop was calculated using experimental correlations [9-11].

EXPERIMENTAL PROCEDURES

Materials and Methods

Operational approach of the three well control methods contains running new mud (which is heavier than the initial one) into the well. In driller's method, the prepared mud is pumped into the well after running out the initial (two circulations). In the wait & weight method, the new mud is pumped into the well simultaneously as the initial mud is running out (one circulation). In the concurrent method, the mud weight is increased stepwise in contrast to other methods; the heavier mud is pumped into the well at each step of mud preparation until reaching to the surface [12-14].

The Driller's Method

Driller's method, whose procedure has been represented in Figure 1, is a practical well control approach in the drilling industry because of uncomplicated operation [12].

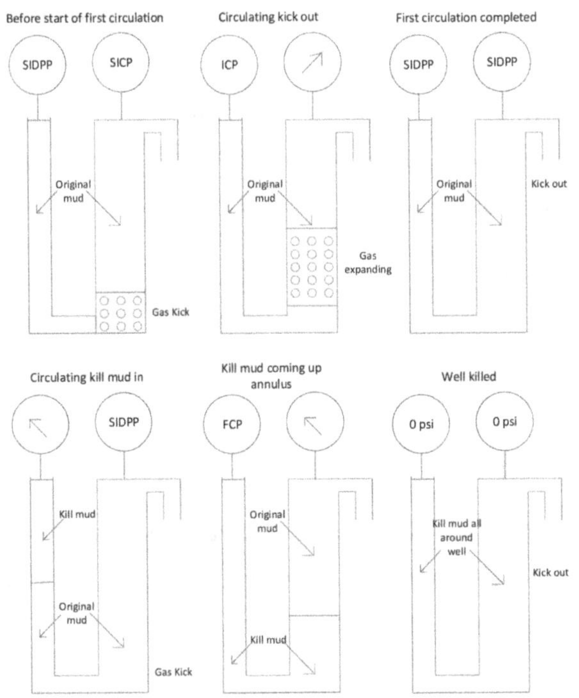

Figure 1: Driller's method procedure.

The Wait & Weight Method

In this method, the heavier mud preparation is started immediately after the kick occurrence and then pumped into the well. The initial mud is withdrawn from the mud circulation system and simultaneously is substituted by a heavier mud until reaching to the surface [12]. Figure 2 illustrates the stepwise procedure of wait & weight method.

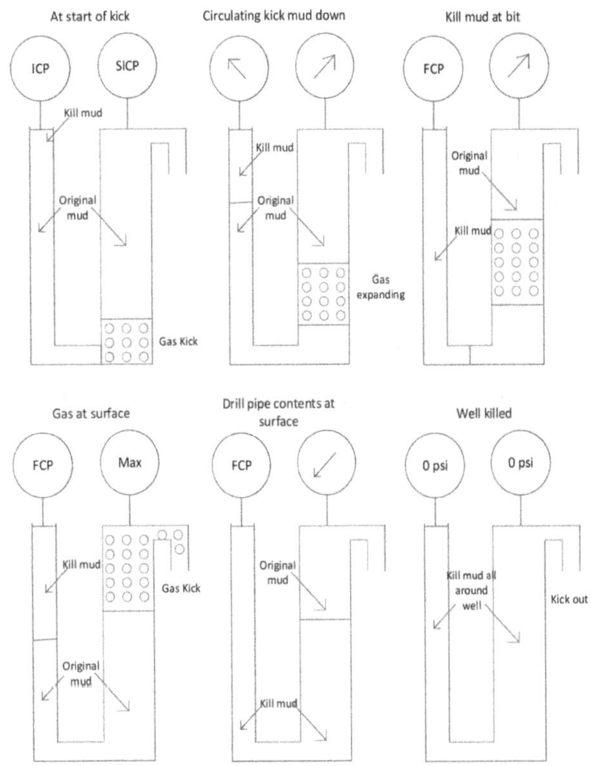

Figure 2: Wait & weight method procedure.

The wait and weight method is faster than the drillers' method, and more desirable for employees because of the limited exerted pressure to the surface facilities. Nonetheless, the method has a few disadvantages including gas migration, pipe sticking and well shutting due to small surface pressure and needing additional surface facilities in the operation.

The Concurrent Method

This method is utilized whenever strong kicks or mud loss take place in the well, in which pump speed should remain at the minimum rate, and mud density is gradually increased.

The major advantage of the concurrent method is that casing pressure is lower compared to the other two methods due to fewer damages to the equipment arising from incremental mud weight increasing. However, the method is more complicated than the others and requires further facilities on the rig.

The procedure of concurrent methods includes:

1- After closing the well and data recording, required parameters such as ICP, FCP are calculated.

2- The pump is turned on and brought up to kill rate speed while casing pressure is held constant. When the pump reaches the kill rate, drill pipe pressure is adjusted to the calculated value. Circulation should be started as soon as ICP is determined.

3- Mud pit crew should increase mud weight during circulation. Whenever mud weight increases to a value at the bottom of the chart, choke operator adjusts circulating pressure to the drill pipe pressure.

4- Circulating is continued until kill weight mud comes back to the surface. [12].

Well Control Simulation for Water-Based Muds

In this study, kick fluid and water-based mud are not supposed to mix or dissolve in each other. Hence, drill string fluid column consists of two layers including old and new mud. Annulus fluid column is composed of four layers: (from bottom to top) new mud, old mud, kick fluid and old mud, schematically displayed in Figure 3. Due to fluid displacement in the well column, the height and boundaries of the layers are changing during the process [15-16].

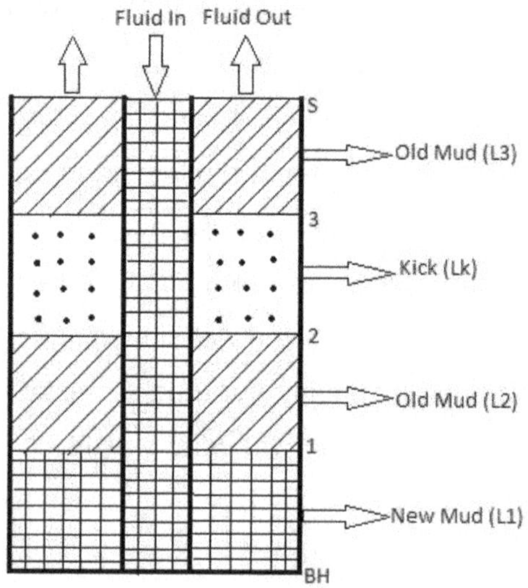

Figure 3: Fluid layers in the well during the well control process.

Equations 1-33 were utilized in order to calculate well control parameters to be used in the flowchart of each method (Figure 4 to Figure 9). The definitions of the parameters are available in the nomenclature section.

Firstly, the basic parameters specially required time of each method is calculated. The required time for passing through the string and annulus are respectively calculated by Equations 1 and 2, and their summation in Equation 3.

$$t_s = \frac{V_s}{SCR} \tag{1}$$

$$t_a = \frac{V_a}{SCR} \tag{2}$$

$$t_3 = t_0 + t_a \tag{3}$$

In Equations 4, 5 and 6, the elapsed time of each well control method has been calculated using Equations 1, 2 and 3.

$$t_{driller} = t_a + t_3 \tag{4}$$

$$t_{wait\,\&weight} = t_w + t_3 \tag{5}$$

$$t_{concurrent} = t_{min} + t_3 \tag{6}$$

The lengths of some sections of Figure 3, which separate different parts of the well versus fluid type, are calculated by Equations 7, 8 and 9.

$$L_n = \frac{SCR(t - t_a)CF}{A_s} \tag{7}$$

$$L_2 = \frac{SCR \times t \times CF}{A_a} \tag{8}$$

$$L_1 = \frac{(SCR(t - t_a) - V_s)CF}{A_a} \tag{9}$$

Pressure values at critical points especially boundaries of fluid layers and surface are calculated using Equations 10, 11 and 12.

$$P_{ss} = P_{BH} - 0.052 \times MW \times L_o - 0.052 \times KMW \times L_n + \Delta P_s + \Delta P_b - SCP \tag{10}$$

$$P_1 = P_{BH} - 0.052 \times KMW \times L_1 \tag{11}$$

$$P_2 = P_1 - 0.052 \times MW \times L_2 \tag{12}$$

Kick fluid properties such as volume, length and top pressure are calculated using trial and error procedures based on the Equations 13 to 20 due to the variable volume of gaseous kick fluid and its expansion when rising toward the surface. Afterward, the constant surface pressure could be calculated by Equation 21. The procedure flowchart has been illustrated in Figures 8 and 9.

$$V_k = \frac{P_{BH} z_2 V_{kBH}}{P_2 z_{BH}} \tag{13}$$

$$L_k = \frac{V_k CF}{A_a} \tag{14}$$

$$P_3 = P_2 EXP\left(\frac{-ML_k}{1544 z_2 T_2}\right) \tag{15}$$

$$P_m = \frac{P_2 + P_3}{2} \tag{16}$$

$$T_m = \frac{T_2 + T_3}{2} \tag{17}$$

$$V_{knew} = \frac{P_{BH} z_m V_{kBH}}{P_m z_{BH}} \tag{18}$$

$$L_{knew} = \frac{V_{knew} CF}{A_a} \tag{19}$$

$$P_{3new} = P_2 \cdot EXP\left(\frac{-ML_{knew}}{1544 z_m T_m}\right) \tag{20}$$

$$P_s = P_3 - 0.052 \times MW \times L_3 \tag{21}$$

$$KTP = D - (L_1 + L_2 + L_k) \tag{22}$$

In the wait and weight method, gas rising velocity during the waiting period of the procedure is calculated using Equation 24. Other formulations in Equations 23 to 28 can be utilized in order to calculate the lengths of fluid sections in Figure 3.

$$L_n = \frac{SCR(t - t_w)CF}{A_s} \tag{23}$$

$$V_{gr} = 0.101\sqrt{gr_t \frac{\rho_L - \rho_G}{\rho_L}} \tag{24}$$

$$L_2 = V_{gr} t \tag{25}$$

$$L_2 = \frac{SCR(t - t_w)CF}{A_a} + V_{gr} t_w \tag{26}$$

$$L_1 = \frac{(SCR(t - t_w) - V_s)CF}{A_a} \tag{27}$$

$$L_2 = \frac{V_s CF}{A_a} + V_{gr} t_w \tag{28}$$

Equations 29 to 33 are used to calculate the lengths of sections in Figure 3 for the concurrent method. The expression g(t) in Equation 29 represents the mathematical function of mud weight changes and its average value. The pressure parameter in the boundaries and kick fluid behavior are the same as the driller and wait & weight methods.

$$f(t_i, t_f) = \frac{\int_{t_i}^{t_f} g(t)dt}{t_f - t_i} \tag{29}$$

$$L_n = \frac{SCR \times t \times CF}{A_s} \tag{30}$$

$$L_2 = \frac{SCR \times t \times CF}{A_a} \tag{31}$$

$$L_1 = \frac{SCR(t - t_s)CF}{A_a} \quad (32)$$

$$L_2 = \frac{V_s CF}{A_a} \quad (33)$$

Figures 4, 5 and 6 exhibit the procedure of principal flowcharts of the three methods by which required parameters could be computed at any desired time during well control process. In order to complete the simulation, calculations should be performed at different time intervals from zero to maximum elapsed time of the process. For instance, the required time for driller's method was calculated 226.63 minutes in this study. The interval (0-226.63) could be divided by 100 time steps, in addition, at any of them, these calculations could be performed to compute major parameters including annular surface pressure, casing shoe pressure, etc.

The procedures of several critical parameters such as casing shoe pressure, kick top depth, kick region length, surface pressure, etc. are the same for the three methods of well control which are stated in Figure 7 along with the general section of the methods.

Based on the computational procedure of driller's method in Figure 4, there are three time intervals whose formulations are different from each other.

-$t<t_a$: the period before bottom-hole fluid reaches to the surface

-$t<t_a+t_s$: the period before heavier mud reaches to the end of the string

-else (other times): the period that heavier mud moving from string bottom to the surface through the annulus

For each of the time intervals, the lengths of the fluid sections and their pressure in boundaries and internal sections have been calculated.

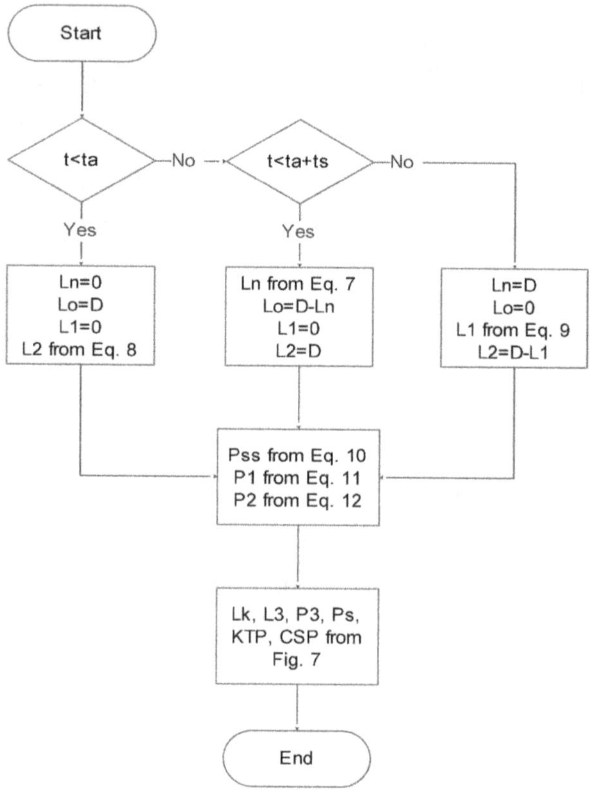

Figure 4: the Computational procedure of driller's method.

Similar to the driller's method, a computational procedure has been proposed for the wait & weight method in Figure 5. There are three time intervals for simulation based on the mechanism of the method:

-$t<t_w$: the period that the crew are preparing the new mud

-$t<t_w+t_s$: the period in which new mud is prepared and is pumped to the string, although has not reached to the bottom

-else (other times): the period in which new mud entering the annulus, before reaching to the surface.

The formulation of each time interval has been referenced in the flowchart of Figure 5.

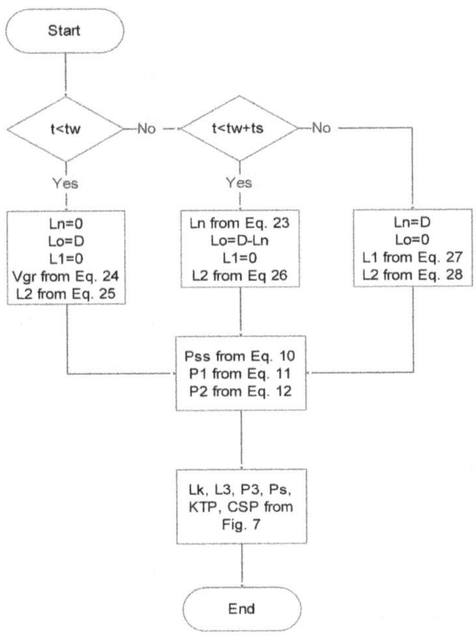

Figure 5: The Computational procedure of wait & weight method.

A computational procedure has been proposed for the concurrent method in Figure 6 based on which three time intervals exist:

-$t<t_0$: the period before mud weight reaches to the final weight from the initial weight.

-$t<t_3$: the period before the complete exit of old mud (which is under kick fluid)

-else (other times): before the final mud completely entering the annulus

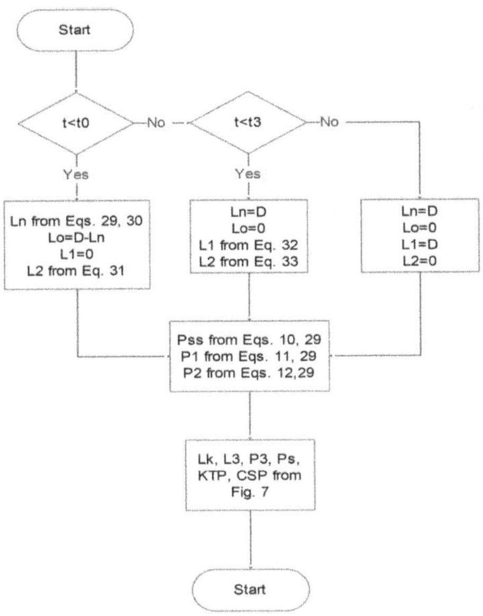

Figure 6: the Computational procedure of the concurrent method.

The procedure of calculating the major parameters such as kick fluid properties and pressure at critical points of the well, especially boundary of fluid layers is the same in all the three well control methods. Figure 7 illustrates the general section in which different procedures and equations have been proposed for both water and oil-based muds because kick region properties such as length, volume and pressure are different depending on the mud type.

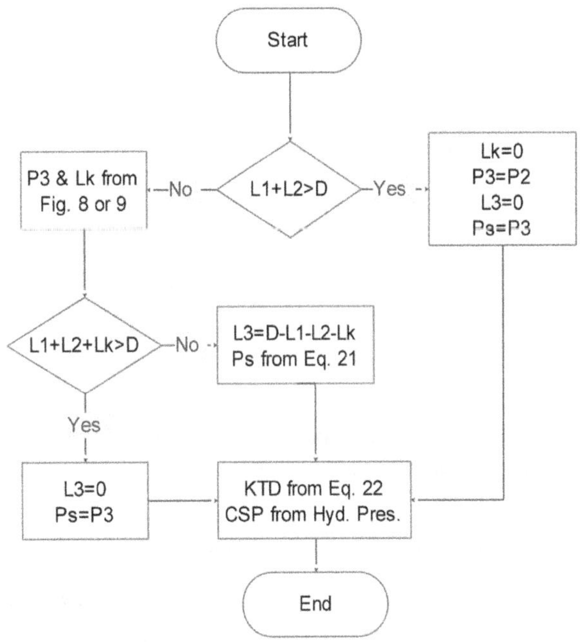

Figure 7: Computational procedure for the general part of the three methods.

Based on the proposed algorithm for water-based muds in Figure 8, the volume, length and top pressure of kick fluid should be calculated by numerical approaches. After an initial guess for kick fluid volume, the procedure is repeated until pressure converges to a constant value. Afterward, the pressure could be calculated at different sections of the well by hydrostatic pressure equations [9].

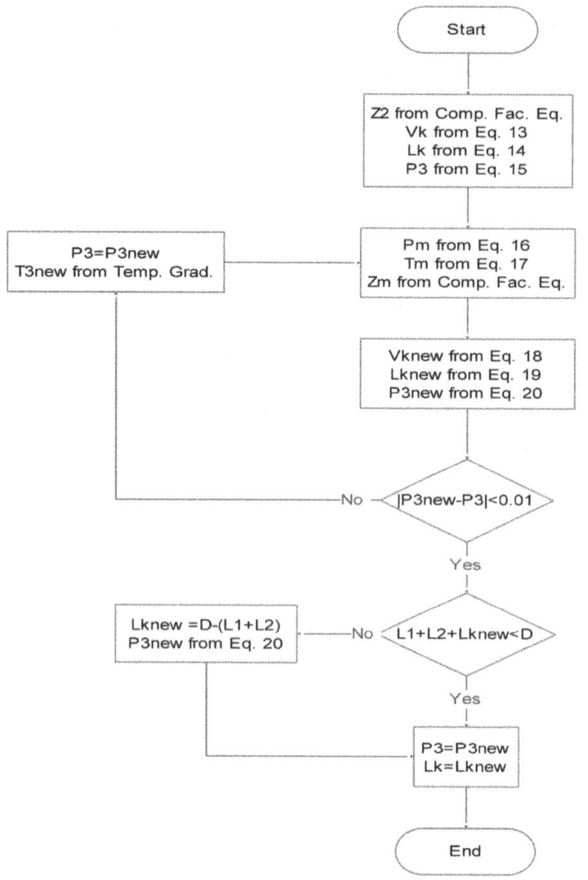

Figure 8: the Computational procedure of kick fluid region for water-based muds.

Oil-based muds consist of diesel as a continuous phase and water as a discontinuous phase. A chemical emulsifier is added to the mixture in order to prevent water droplets from joining and sticking together [9]. Diesel density at 60 F is 820-860 kg/m^3, and it approximately contains hydrocarbons in the range of C_{10} to C_{15} [17].

Figure 9 illustrates well control stages for oil-based muds. Due to the fact that kick fluid mostly contains hydrocarbon gas (methane) and oil-based muds contain hydrocarbon fluids, it could be supposed that the entered gas is soluble in the drilling fluid [18]. Therefore, there is a new kick fluid combination whose composition is lighter than the original diesel and is moving toward the surface. Thus, it may be converted into two phases during the process, and so, flash calculations are required for computing its properties. PR78 was utilized as the equation of state, and Simultaneous Solution Method (one of the flash calculation methods) for two-phase analysis [19-21]. Also, the mud includes diesel and solid particles which are calculated by Equations 34 to 36 have been considered in the calculations.

Based on the procedure details in Figure 9, the kick fluid is discretized to 100 equal sections. Gas and liquid volumes and hydrostatic pressure of each section are calculated using a try and error procedure combined with flash calculation.

$$V_w = x_w V_l \tag{34}$$

$$V_d = x_d V_l \tag{35}$$

$$V_s = x_s V_l \tag{36}$$

The new molar amounts of each component and their fractions in the new kick fluid are calculated using Equations 37 and 38.

$$n_t = n_{c1} + \sum_{i=14}^{20} n_{ci} \tag{37}$$

$$z_i = \frac{n_i}{n_t}(i = 1,\ 14-20) \tag{38}$$

Pressure, volume and length of kick fluid in case of oil-based muds are calculated using Equations 39 to 43.

$$P_{ki} = P_{k(i-1)} - \frac{gCF}{g_c A_a}(\rho_{kgi} V_{kgi} - \rho_{kli} V_{kli}) \tag{39}$$

$$P_{mi} = \frac{P_{k(i-1)} + P_{ki}}{2} \tag{40}$$

$$T_{mi} = \frac{T_{k(i-1)} + T_{ki}}{2} \tag{41}$$

$$V_k = \sum_{i=1}^{N} V_{kgi} + \sum_{i=1}^{N} V_{kli} + V_w + V_s \tag{42}$$

$$L_k = \frac{V_k CF}{A_a} \tag{43}$$

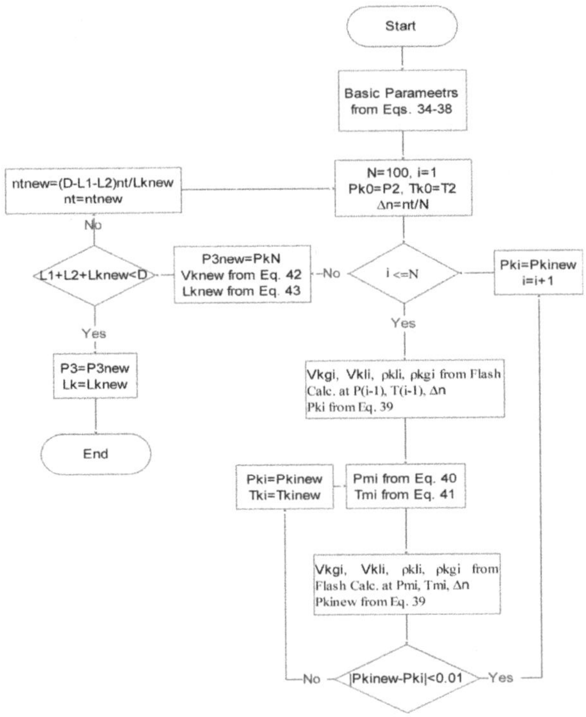

Figure 9: The Computational procedure of kick fluid region for oil-based mud.

RESULTS AND DISCUSSIONS

In this section, the well control process has been simulated for water and oil-based muds. Tables 1 and 2 contain the input data used in the simulation.

Table 1: Input data for water-based mud.

Input parameter (unit)	Value
Depth (ft)	10000
Mud Weight (ppg)	9.6
Casing Shoe Depth (ft)	3500
Shut In Drill Pipe Pressure (psig)	720
Shut In Casing Pressure (psig)	520
Input Kick Volume (bbl)	20

Table 2: Input data for oil-based mud.

Input parameter (unit)	Value
Depth (ft)	5000
Mud Weight (ppg)	7.2
Casing Shoe Depth (ft)	3500
Shut In Drill Pipe Pressure (psig)	250
Shut-In Casing Pressure (psig)	150
Input Kick Volume (bbl)	20

Simulation Results for Water-Based Muds

According to Table 3, the required time for controlling the well by the concurrent method is shorter than wait & weight, and wait & weight's method time is shorter than the driller's method time. Therefore, the concurrent method is more economical relative to the other methods for controlling the well.

Table 3: Input data for oil-based mud.

Well Control Method	Time (min)
Driller	226.63
Wait & weight	159.26
Concurrent	140.88

In Figure 10, annular surface pressure versus time has been displayed for the three methods. In drillers' method, due to the expansion of rising gas in well column, the pressure increases until the kick fluid reaches the surface. During gradual gas withdrawal, the annular pressure is reducing until the gas exit completely from the well, and after that, the new mud is entering into the well. The pressure remains constant during new mud entrance to the annulus, and after which, the pressure is reducing until the annulus eventually occupied by the new mud.

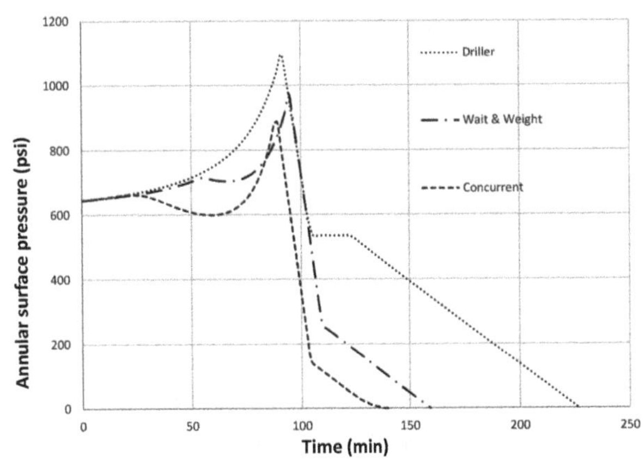

Figure 10: Annular pressure vs. time for water-based muds.

In wait & weight method, the pressure increases during the waiting period as a consequence of gas rising in the well column especially due to extensive expansion near the surface. When heavy mud enters the annular space, the pressure starts to reduce. During gradual gas withdrawal, the pressure is quickly decreasing until complete replacement of heavy mud.

The procedure of concurrent method is similar to the wait & weight except that due to the gradual change of mud weight, the trend of the graph is not linear in some areas. Figure 11 shows casing shoe pressure versus time for the three methods. As long as the gas is moving upward and expanding, the casing shoe pressure is decreasing. When kick fluid passes through the casing shoe, the pressure is decreasing. Until the new mud enters the annulus, the casing shoe pressure remains constant in driller's method and decreases in the others because of new mud weight effects. The casing shoe pressure, which is a critical parameter during the drilling process, remains constant at the end of the process and complete substitution of the new mud.

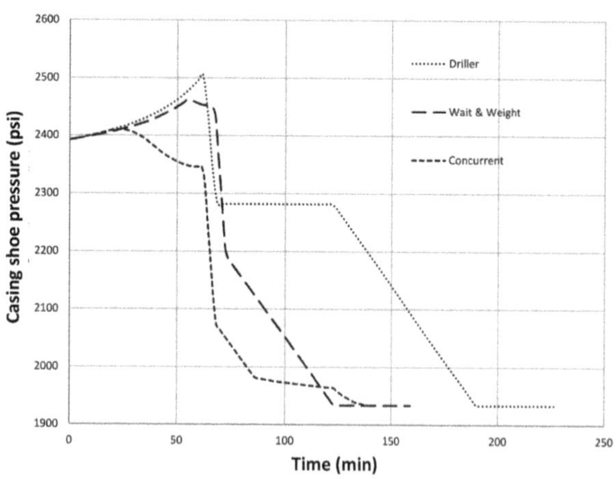

Figure 11: Casing shoe pressure vs. time for water-based muds.

Figure 12 illustrates surface drill pipe pressure versus time for driller's and wait & weight methods in which the pressure remains constant until the arrival of new mud. In this stage, firstly the pressure is reducing and after complete displacement, remains constant. In the concurrent method, since the new mud enters from the start of the process, the pressure is reducing. It should be noted that the immediate alternations of the chart slope are due to the arrival of new mud to the annulus.

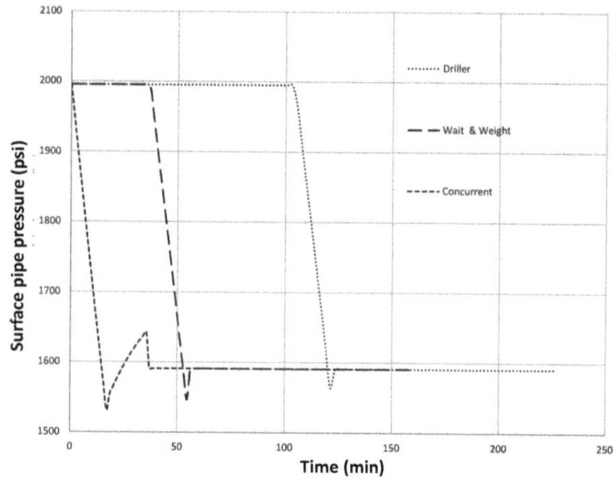

Figure 12: Surface drill pipe pressure vs. time for water-based muds.

The three methods do not have considerable differences in the period of kick fluid arriving at the surface based on Figure 13 in which kick top depth has been plotted versus time. After kick fluid reaching to the surface, the annulus pressure is reducing and consequently, well control process could be run with more safety and confidence.

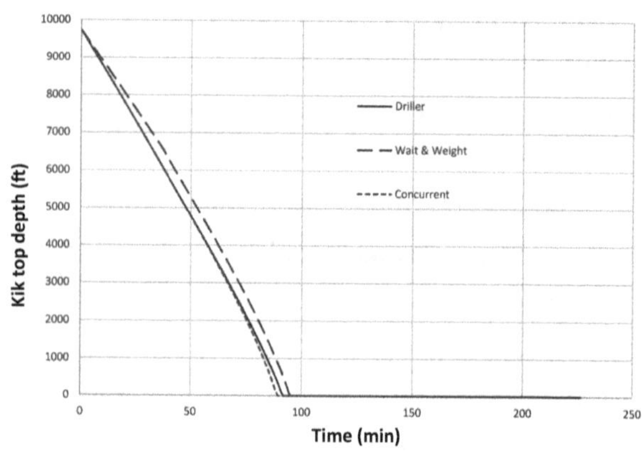

Figure 13: Kick top depth vs. time for water-based muds.

Simulation Results for Oil-Based Muds

The simulation results of oil-based muds have been presented in Table 4. The Table compares well control required times for the three methods. Similar to the water-based muds, the period of the concurrent method is shorter than wait & weight, and wait & weight than the drillers'. Therefore, the concurrent method is more economical relative to the other methods.

Table 4: Well control times for oil-based muds.

Well Control Method	Time (min)
Driller	215.80
Wait & weight	192.31
Concurrent	169.34

Figure 14 illustrates annular surface pressure versus time for the three methods. In the driller's method, the pressure remains constant whenever the kick fluid, composed of initial methane gas mixed with drilling fluid, contains only one phase. When the fluid contains two phases, the annular pressure is reducing until the kick fluid reaching to the surface. The pressure remains constant as long as new mud has not entered to the annulus. It should be noted that during gradual gas withdrawal of the well, the pressure starts to decrease to a final value of zero.

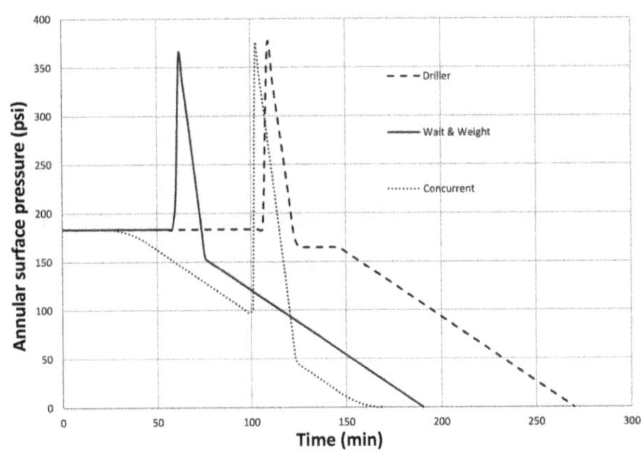

Figure 14: Annular pressure vs. time for oil based muds.

Wait & weight method procedure is similar to drillers' except no constant area in the graph due to the new mud arrival. In the concurrent method, after the arrival of heavy mud into the annular space, the initial constant pressure is reducing. While the kick fluid contains two phases, the pressure decreases until kick fluid completely exit from the well. During the new mud entrance to the annulus, the pressure is decreasing to zeros. Figure 15 shows casing shoe pressure versus time for the three methods whose general trend is similar to Figure 14. The difference is in rapid changes in the graph when the kick fluid contains two phases.

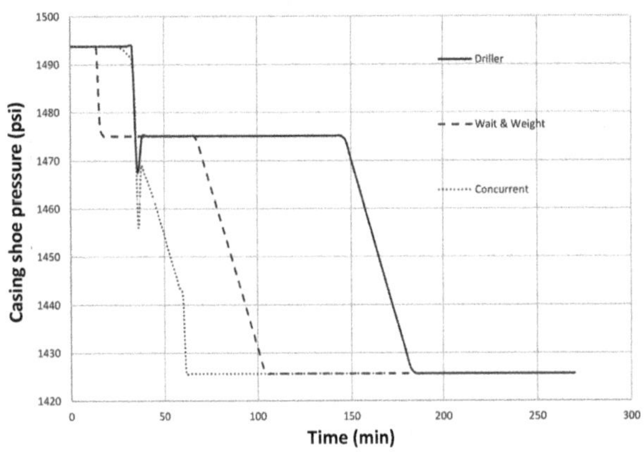

Figure 15: Casing shoe pressure vs. time for oil based muds.

Figure 16 displays drill pipe pressure versus time whose general trend is similar to Figure 15. Two-phase fluid effect of annular fluid on drill pipe fluid is negligible, and consequently on the surface pipe pressure.

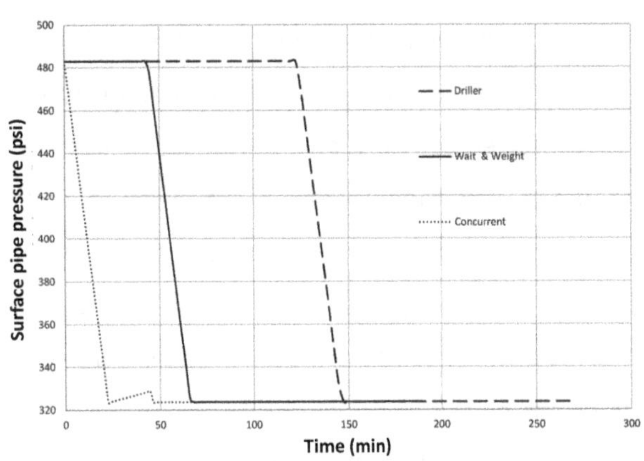

Figure 16: Surface drill pipe pressure vs. time for oil based muds.

Figure 17 displays kick top depth versus time for oil based muds whose procedure for driller's and concurrent methods are the same. However, the concurrent period for kick fluid reaching to the surface is shorter than the drillers' due to single stages procedure. In wait & weight method, the kick fluid period for reaching to the surface is shorter than the others since during initial wait time when the well is shut in, the kick fluid is moving toward the surface. This difference is more obvious in oil-based muds since the wait time for preparing the oil-based muds are more considerable than the water-based muds.

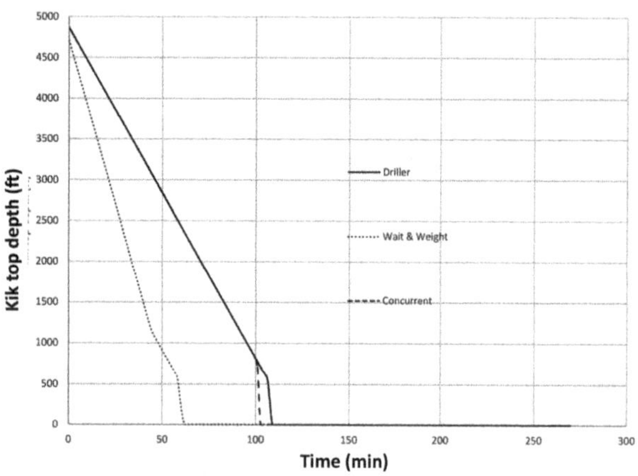

Figure 17: Kick top depth vs. time for oil based muds.

RESULTS AND DISCUSSION

The major differences between the three methods are the approach of heavier mud substitution in the well and bringing out the kick fluid during which exerted pressure to the annulus at the well column is increasing. If the surface pressure could remain constant at a high value by BOP or other equipment, lower pressure will be exerted on the annulus and consequently, the risks and danger of operation will be decreased.

In the concurrent method, it is essential for the drilling crew to increase mud weight from initial to the desired value at a short time, for example by an exponential function. However, this function is technically hard to be run because of equipment and technology limitations. This simulator is able to run the concurrent method using any desired function. It should be noted that the three methods have not considerable differences at the time of arriving the kick fluid to the surface.

It is important for a well control simulator to be run at a short time due to the fact that there is no time for the crew to wait for running the simulator in the actual condition. Maybe combining mass, momentum and energy balance equations could be another appropriate method to simulate the process. Nevertheless, needing extra running time is a disadvantage because of the finite difference or finite element method usage. In this study, simulation has been run at a shorter time relative to the others by flash calculation which is practical to be used in the actual conditions.

CONCLUSIONS

In the concurrent method, required time and exerted pressure to the equipment are lower than wait & weight and wait & weight relative to driller's method. Therefore, the concurrent method is faster and more economical than the others.

In wait & weight method, during the waiting period, gas bubbles are rising toward the surface and expanding which have been considered in this study. Results show that this movement increases the surface pressure, and consequently affects the life and proper operation of the equipment. Thus, it is necessary for the crew to prepare the new mud in a short time to reduce the wait time. Additionally, the drilling crew should be sure about the precise and stable installation of casing shoe and the other equipment due to the fact that until complete kick fluid exit from the well, the casing shoe pressure has a large value, sometimes at a dangerous and destructive range.

In the case of oil-based muds, any changes in pressure and temperature have significant effects on the kick fluid behavior. Dissolution of kick fluid in drilling mud changes the mud density from bottom-hole to the surface, and consequently, more surface choke pressure will be required to maintain the bottom-hole pressure.

Another point about oil-based mud is that due to gas solubility in the kick fluid, the mud reaches later to the surface relative to water-based muds.

NOMENCLATURES

SIDPP: Shut in Drill Pipe Pressure, psi

SICP: Shut in Casing Pressure, psi

BOP: Blowout Preventer

ICP: Initial Circulation Pressure, psi

FCP: Final Circulation Pressure, pi

SCR: Slow Circulation Rate, bbl/min

SCP: Slow Circulation Pressure, psi

MW: Mud Weight, ppg

KMW: Kill Mud Weight, ppg

KTD: Kick Top Depth, ft

CSP: Casing Shoe Pressure, psi

CF: Conversion Factor (bbl to cubic ft), 5.615

L_n: Height of new mud in the drill string, ft

L_o: Height of old mud in the drill string, ft

L_1: Height of new mud in the bottom of the annulus, ft

L_2: Height of old mud in annulus under the kick fluid, ft

L_k: Height of kick fluid in the annulus, ft

L_3: Height of new mud in the annulus above the kick fluid, ft

t_s: Required time for new mud to reach from the surface to bottom of the drill string, min

t_a: Required time for new mud to reach from bottom of the annulus to surface, min

t_1: Required time for old mud which is above kick fluid to exit completely, min

t_2: Required time for the kick fluid to exit completely, min

t_3: Required time for old mud which is under kick fluid to exit completely, min

t_w: Wait time in wait & weight method (the time which is needed for mud to be prepared), min

t_{min}: Minimum time in concurrent method (the time which is needed for mud to reach from initial weight to final weight), ppg

$t_{driller}$: Required time to control the well using driller's method, min

$t_{wait\&weight}$: Required time to control the well using wait & weight method, min

$t_{concurrent}$: Required time to control the well using the concurrent method, min

P_{BH}: Bottom-hole pressure, psi

P_1: Top pressure of new mud layer in the annulus, psi

P_2: Top pressure of old mud layer which is under kick fluid layer, psi

P_3: Top pressure of kick fluid layer, psi

P_s: Annulus surface pressure, psi

P_{ss}: Annulus drill string pressure, psi

D: Well Depth, ft

ΔP_s: Pressure drop in the drill string, psi

ΔP_b: Pressure drop in the bit, psi

A_s: Area of the drill string, ft^2

A_a: Area of annulus areal, ft^2

V_s: Volume of the drill string, bbl

V_a: Volume of the annulus, bbl

V_{kBH}: Kick volume at bottom-hole, bbl

V_k: Kick volume at the desired condition, bbl

V_{gr}: Gas rise velocity, ft/min

g(t): A function that returns mud weight at the desired time for driller's method

$f(t_i,t_f)$: A function that returns average values of mud weight at a time interval

M: Gas molecular weight, lb/lbmol

z_{BH}: Gas compressibility factor at the bottom-hole condition

z_2: Gas compressibility factor at point 2 in Fig 6

V_l: Pit gain volume which is the amount of mud combines with kick fluid, and consist of water, diesel and solid particles, bbl

V_w: Volume of water in pit gain, bbl

V_d: Volume of diesel in pit gain, bbl

V_s: Volume of solid particles in pit gain, bbl

x_w: Volume fraction of water in pit gain

x_d: Volume fraction of diesel in pit gain

x_s: Volume fraction of solid particles in pit gain

n_{c1}: Number of methane moles which enters the well as a kick, mole

n_{c14}-n_{c20}: Mole numbers of each component of diesel, for example, nc14 is mole numbers of C14, mole

n_t= Total mole numbers of hydrocarbon part of the new kick fluid, mole

z_i: Mole fraction of each component in new kick fluid, that is consist of kick fluid and initial mud

P_{ki}: Pressure at the top point of the ith segment, psi

V_{kli}: Volume of kick fluid in segment ith, that its phase is liquid, bbl

V_{kgi}: Volume of kick fluid in segment ith, that its phase is gas, bbl

ρ_{kli}: Density of kick fluid in segment ith, that its phase is liquid, lb/ft^3

ρ_{kgi}: Density of kick fluid in segment ith, that its phase is gas, lb/ft^3

N: Number of segments that are considered for calculation of kick parameters.

REFERENCES

1. Choe J., "Advanced Two-Phase Well Control Analysis," *Journal of Canadian Petroleum Technology*, 2001, 40(05).
2. Avelar C. S., Ribeiro P. R., and Sepehrnoori K., "Deepwater Gas Kick Simulation," *Journal of Petroleum Science and Engineering*, **2009**, *67(*1), 13-22.
3. Omosebi A. O., Osisanya S. O., Chukwu G. A., and Egbon F., "Annular Pressure Prediction during Well Control," In *Nigeria Annual International Conference and Exhibition*, Society of Petroleum Engineers, **2012**.
4. Marbun B. H. and Shidiq A. M., "Estimation of Annulus Pressure Fluids Kick for Vertical Well Using Moore Method and Integrated Numerical Simulation," In *North Africa Technical Conference and Exhibition*, Society of Petroleum Engineers, **2013.**
5. An J., Lee K., and Choe J., "Well Control Simulation Model of Oil-based Muds for HPHT Wells," In *SPE/IATMI Asia Pacific Oil & Gas Conference and Exhibition*, Society of Petroleum Engineers, **2015**.
6. Feng J., Fu J., Chen P., Luo J., and et al., "An Advanced Driller›s Method Simulator for Deepwater Well Control," *Journal of Loss Prevention in the Process Industries,* **2016**, *39*, 131-140.
7. Sun G., Li S., Yu Z., Wang K., and et al., "Dynamic Simulation of Major Kick for an HP/HT Well in Western China," In *Abu Dhabi International Petroleum Exhibition & Conference*, Society of Petroleum Engineers, **2016.**
8. Mosti I., Morrell D., Anfinsen B. T., Vielma W. E., and et al., "Kick Tolerance and Frictional Pressure Losses, Added Safety or Added Risk?,"

In *SPE/IADC Drilling Conference and Exhibition*, Society of Petroleum Engineers, **2017**.

9. Bourgoyne A. T., Millheim K. K., Chenevert M. E., and Young F. S., "Applied Drilling Engineering," *TX: Society of Petroleum Engineers*, Richardson, **1991.**
10. Azar J. J. and Samuel G. R., "Drilling Engineering," PennWell Books, **2007.**
11. Austin E. H., "Drilling Engineering Handbook," *Springer Science & Business Media*, **2012.**
12. Baker R. and Fitzpatrick J., "Practical Well Control. Petroleum Extension Service," University of Texas at Austin, **1998**.
13. Watson D., Brittenham T., and Moore P. L., "Advanced Well Control," Society of Petroleum Engineers, **2003.**
14. Grace R. D., "Blowout and Well Control Handbook," Gulf Professional Publishing, **2017.**
15. Talaia M. A., "Terminal Velocity of a Bubble Rise in a Liquid Column," *World Academy of Science, Engineering and Technology*, **2007,** *28*, 264-268.
16. Viana F., Pardo R., Yánez R., Trallero J. L., and et al., "Universal Correlation for the Rise Velocity of Long Gas Bubbles in Round Pipes," *Journal of Fluid Mechanics*, **2003**, *494*, 379-98.
17. Date A. W., "Analytic Combustion: with Thermodynamics," *Chemical Kinetics and Mass Transfer,* Cambridge University Press, **2011.**
18. Kim N. R., Ribeiro P. R., and Pessôa-Filho P. A., "PVT Behavior of Methane and Ester-based Drilling Emulsions," *Journal of Petroleum Science and Engineering*, **2015**, *135*, 360-366.
19. McCain W. D., "The Properties of Petroleum Fluids," PennWell Books, **1990**.
20. Danesh A., "PVT and Phase Behavior of Petroleum Reservoir Fluids," Elsevier, **1998.**
21. Xiang-Qi L. X. and Tao S. X., "Study on Two-phase Gas-liquid Flow during Gas Kick," *Journal of Engineering Thermophysics*, **2004**, *1*, 0-20.

13

A New Method for Predicting Well Pattern Connectivity in a Continental Fluvial-delta Reservoir

Yanfeng Liu[1]*, Yuetian Liu[2], Lu Sun[2], and Jian Liu[3]

[1] Sinopec Petroleum Exploration and Production Research Institute, Beijing, China
[2] China University of Petroleum (Beijing), Beijing, China
[3] PetroChina Research Institute of Petroleum Exploration and Development, Beijing, China

ABSTRACT

The features of bad flow unit continuity and multiple layers emphesize the importance of a well pattern design for the development of a fluvial-delta reservoir. It is proposed a method to predict well pattern connectivity (WPC) based on the exploration and evaluation of wells. Moreover, the method helps evaluate the risk of well placement. This study initially establishes the parameters for characterizing the lateral and vertical flow unit distributions. Then, extensive statistics on the mature oil-field sands of synthetic geological models are obtained to generate the prediction model of WPC which will reveal the correlation among WPC, flow unit distribution, and well spacing (WS). Finally, a case study is conducted to validate the proposed method for predicting WPC. The procedure of the method is comprised of two steps. The first step is to calculate the parameters which characterize the vertical sand body distribution of the target formation by using the well drilling and logging information. The second step is to integrate the calculated parameters and designed WS into the proposed formula to forecast WPC. The new method of WPC prediction has the advantage of integrating the static and dynamic information of similar mature oil fields with each other. By utilizing the model, making an important decision on well pattern design and a reservoir production forecast in the newly discovered continental fluvial-delta reservoir would be reasonable.

Keywords: Well Pattern Connectivity, Flow Unit Distribution, Well Spacing Optimization, Reservoir Performance

INTRODUCTION

Well pattern optimization is the key procedure in the reservoir development scheme design. Discontinuous sand bodies and alternate sand–shale sequences are typical characteristics of continental fluvial-delta reservoirs, in which well pattern connectivity (WPC) significantly affects the production performance [1,3]. The optimal well pattern significantly relies on hydraulic flow unit distribution [4,6], whereas a detailed interpretation of reservoir architecture is derived from the available drilling wells [7]. For this type of reservoir, making an important decision on well pattern deployment is a challenging task. Although infill drilling has been a practical choice to reduce the risk of well placement, it is still often deployed at the late development stage, where well pattern and flow unit distribution are mismatched, and it causes irreversible loss [8,9]. Consequently, flow unit distribution and WPC

***Corresponding author**

Yanfeng Liu
Email: yanfeng1022@126.com

should be predicted as early as possible to derive the optimal well pattern for reservoir development.

Predicting WPC requires the correct understanding of flow unit distribution. A flow unit is defined as a group of reservoir rocks with similar properties, which affects fluid flow. The flow unit provides a means of uniquely subdividing a reservoir into volumes that approximate the reservoir architecture on a consistent scale with well spacing (WS) [10,11]. Based on the definition of flow unit, we observe that wells in one flow unit are connected, whereas wells located in different flow units are not connected. WPC is a measure of well pattern control on flow unit distribution. If reservoir flow unit distribution can be clearly recognized, then a correct prediction of WPC can be achieved.

The flow unit dynamically displays the reservoir internal architecture, which has been the subject of extensive research [12,14]. Outcrop investigation and a seismic survey are the main techniques used to explore the hierarchical heterogeneity of a sandstone reservoir [15,16]. Moreover, the dimensions and scales of individual sand bodies in different types of sedimentary environments are stored as an expert knowledge base [17,18], and they have significant meanings for the optimization of reservoir development. By considering the characteristics of sedimentary elements, the genetic analysis of sand bodies provides an insight into flow unit distribution [19,22]. The scales and shapes of genetic sand bodies significantly control the distribution of flow units. However, within a given facies, the reservoir properties significantly vary, and the flow units do not always coincide with the geologic lithofacies or sedimentary units. Furthermore, a number of techniques identifying flow unit distribution have been proposed in the literature [23,24]. These methods have significance for reservoir characterizations. However, most of these methods rely on detailed geologic and dynamic information of the target reservoir, which can only be used in mature oil fields. The goal of this study is to develop a methodology for the prediction of WPC after the drilling of appraisal wells because only limited data from drilling wells are available.

EXPERIMENTAL PROCEDURES

The Selection of Parameters Characterizing Flow Unit Distribution

The flow unit is a block of reservoir rock within which the geological and petrophysical properties that affect fluid flow are internally consistent and predictably different from the external properties. Each flow unit has a constant ratio of flow capacity and storage capacity, which indicates that the relationship between permeability and porosity has a constant slope. The wells drilled within one flow unit are connected with each other. Flow units are confined by certain quantitative petrophysical properties such as porosity, permeability, capillarity, and fluid saturation. Flow unit distribution in the vertical direction is usually easier to measure than the one in the lateral direction. Flow units do not have a consistent size, shape, or orientation because of the complex internal architecture of sedimentary units. Therefore, parameters which are independent of the detailed dimensions or scales should be generated to characterize the flow unit distribution, with WPC as the target.

We assumed a theoretical multiple layer of heterogeneous sandstone reservoir with stratigraphy thickness denoted by H, and covered area denoted by S. In the statistical sense, the flow unit number in the vertical direction is N, the total thickness in the vertical direction is h, and the total number in the three-dimensional (3D) space is M. Flow units are randomly distributed in the 3D space. Two pairs of parameters are proposed to characterize flow unit distribution.

The first proposed parameters are flow unit laterally drilled ratio (LR) and vertically drilled ratio (VR). LR is the ratio of wells drilled at any flow unit divided by the sum of the wells in the well pattern, and VR is the proportion of total flow unit thickness to strata thickness. However, LR and VR measure the total volume of all flow units, but they are not able to fully describe the local heterogeneity of flow unit. The other two parameters are flow unit vertically distributed frequency (VF) and laterally distributed frequency (LF). LF is the number of flow units in a layer per square kilometer, and VF is the number of vertical sand bodies per meter (Figures 1 and 2). The parameters of global and local flow unit distributions imply the heterogeneity of the reservoir architecture on different scales. Regardless of the shapes or sizes of the flow units, these two pairs of parameters are capable of determining the distribution affecting WPC.

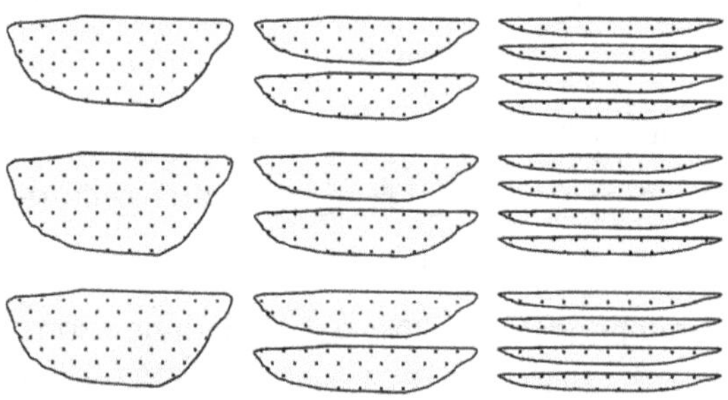

Figure 1: Similar LR and different VF of flow units with a theoretical shape.

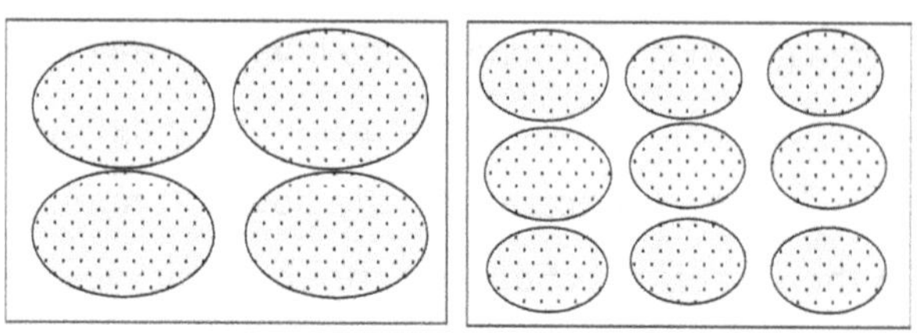

Figure 2: Similar LR and different LF of flow units with a theoretical shape.

WPC is defined as a measure that controls reservoir flow unit distribution. Based on the theoretical model of the continental fluvial-delta reservoir, WPC is expressed as follows:

$$WPC = \frac{\sum_{i=1}^{N}\sum_{j=1}^{NW} C_{ij}}{\sum_{i=1}^{N}\sum_{j=1}^{NW} W_{ij}} \quad (1)$$

where, W_{ij} is the number of wells adjacent to well *j* at layer *i*, and C_{ij} is the number of connected wells in W_{ij}. Then, $W_{ij} \geq C_{ij}$ is a permanent equation (*i*= 1, 2, 3, ..., *N*; *j*=1, 2, 3, ..., *NW*, where *N* is the average number of flow units in the vertical direction, and *NW* is the total number of flow units in a well pattern).

Statistics on the Sampled Mature Oil Fields

As the main type of continental reservoirs, a fluvial-delta reservoir plays a key role in China's oil and gas industry. Not only expert knowledge and experience, but also static and dynaxmic information of old oil fields are extremely valuable for the well pattern optimization of newly discovered similar reservoirs. Thus, the intrinsic relationship between vertical and lateral flow unit distributions is examined on the basis of the extensively sampled mature oil fields. Then, the correlation among WPC, flow unit distribution and WS is determined. A database containing information on more than 40 reservoirs with the fluvial-delta environment located in the eastern and middle parts of China is generated.

In the database, VR, VF, and LR can be easily obtained from the drilling and logging data. By contrast to VR, VF, land LR, the actual LF cannot be detected accurately by any available tools. However, the flow unit distribution of the mature reservoirs which undergo multiple fine characterizations and simulations is presumably clear. We calculate the LF and WPC of each zone based on these previous findings. Then, the relationship between WPC and flow unit distribution is analyzed.

Following the construction of the database of mature oil fields, we analyze the correlation between global flow unit distribution (VR and LR) and local flow unit distribution (LF and VF). LR exhibits a power law correlation with VR (Figure 3), and its mathematical model is expressed in Equation 2. LF varies positively and linearly versus VF according to the relationship of *LF*= 7.62*VF* (Figure 4).

$$AR=1.821\frac{VR^{0.711}}{0.66^{0.711}+VR^{0.711}} \quad (2)$$

Afterward, the intrinsic relationship between lateral and vertical flow unit distributions was examined. The correlation among WPC, flow unit distribution, and WS was determined. Many of the sampled mature reservoirs already have dense WS, the average of which is approximately 200 m. Two histograms are charted to reveal the relationship between WPC and flow unit distribution (Figures 5 and 6). We conclude that WPC has a positive power law function with VR and a negative linear correlation with VF. However, the size of the sample is small because of limited access to the actual reservoirs, indicating unconsolidated discovery. Thus, we attempt to verify and refine the results by other means such as numerical simulation.

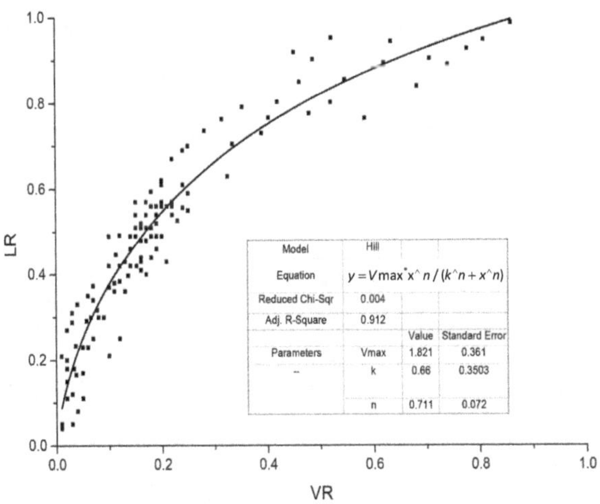

Figure 3: Regression of LR and VR.

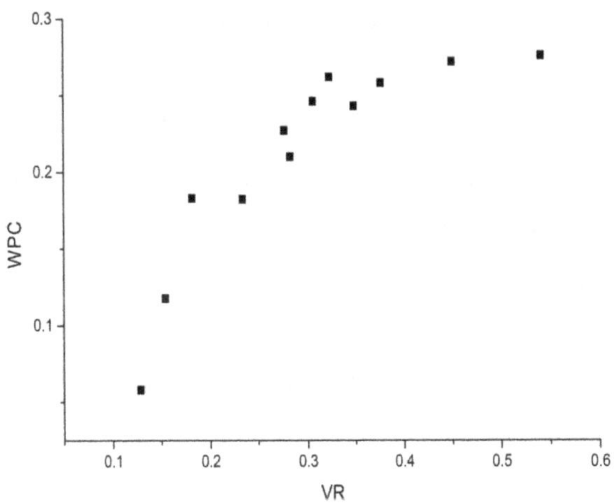

Figure 5: Correlation between WPC and VR.

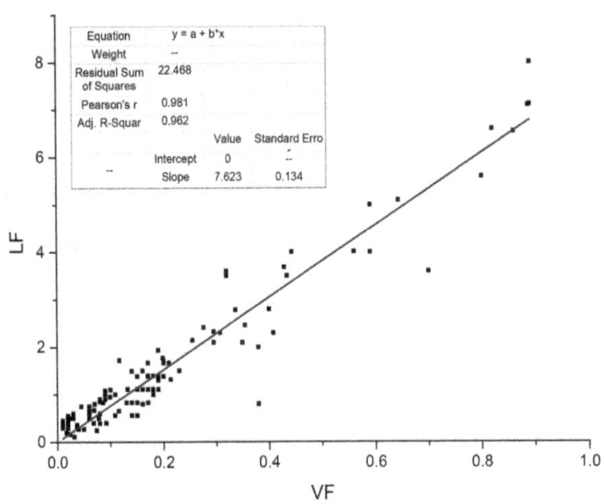

Figure 4: Regression of LF and VF.

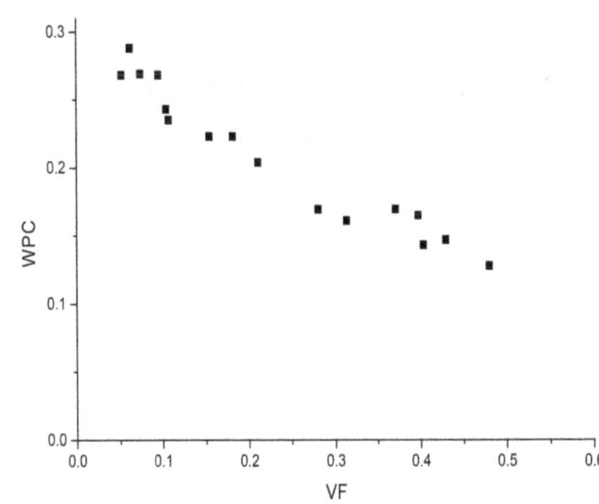

Figure 6: Correlation between WPC and VF.

Regression by Using a Synthetic Geological Model

Although extensive approaches and algorithms have been proposed to characterize discrete reservoir variable distribution [25,29], none of them fulfills the requirement of the aforementioned intrinsic relationship. Relying on the theory of stochastic geometry, a new algorithm combining the philosophies of object-based and pixel-based techniques is created to simulate flow unit distribution. In the procedure, the geometry of the flow unit is similar to the Boolean Simulation, whereas the generating process is analogous to the diffusion-limited aggregation [30]. Flow units with irregular shapes and random sizes are considered the basic volume elements of the model. We call this method the "flow unit growth algorithm" which indicates that a given number of flow units are seeded and then grow thicker and wider simultaneously to conform to the conditions of VR, VF, LR, and AF.

The simulation starts by gridding the reservoir into cells (at a height of *dz* (m) and areas *dx* × *dy* (m^2)). The flow unit generating process contains four steps as follows:

Step 1: The parameters of vertical flow unit distribution (VR and VF) are obtained from drilling data. The lateral flow unit distribution (LR and LF) is determined on the basis of the abovementioned relationship between lateral and vertical flow unit distributions. Flow unit seeds are initiated randomly at the cell grids in 3D, and the seed number is $M = 7.62 \times VF \times H \times S$.

Step 2: Vertical and lateral growth probabilities are calculated and updated to control the direction of flow unit growth in the subsequent step. The real-time average flow unit number (NC) in the vertical direction and the average sand body thickness (hc) control the flow unit generating process. If $\frac{NC}{N} > \frac{hc}{h}$ then the sand body grows vertically; otherwise, the sand body grows laterally.

Step 3: After determining the sand body growth direction, a grid cell is selected randomly and verified whether it is vertically or laterally adjacent to a flow unit. If the growth direction is vertical, then the selected grid cell must be vertically adjacent to the existing flow unit. Otherwise, the selected grid cell must be laterally adjacent to the existing flow unit. A new cell will be selected until it is conditioned.

Step 4: Steps 2 and 3 are repeated to confirm the conditions in which the average flow unit thickness reaches h. The simulation is ended when the average flow unit number in the vertical direction reaches N. The results of the simulation are shown in Figure 7. In the model, each flow unit is an ensemble of grid cells surrounded by impermeable shale.

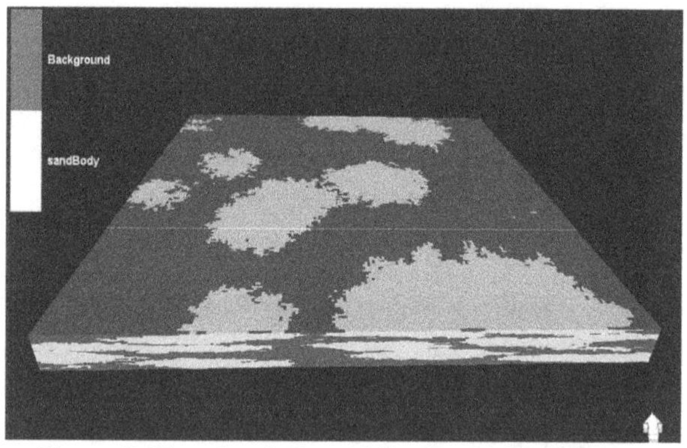

Figure 7: The simulation of flow unit distribution.

The proposed algorithm for 3D flow unit distribution simulation allows us to investigate the influence of flow unit distribution on well pattern. We assumed a reservoir with coverage (S) of 5 km × 5 km, and a stratigraphic thickness (*H*) of 200 m established to forecast flow unit distribution and WPC; the grid cell size is 10 m × 10 m × 1 m. The simulations are conducted in a series of conditions. The flow unit vertically drilled ratios (VR) are 0.2, 0.3, 0.4, 0.5, and 0.6, and the corresponding flow unit vertically distributed frequencies (VF) are 0.02, 0.05, 0.1, 0.2, and 0.3. Each simulation generates 10 realizations to reduce the influence of random error. For every realization, squared well patterns with a WS of 50, 100, 200, 300, 400, 500, 600, and 700 m are arranged to calculate the WPC. Then, an orthogonal experimental design is used to analyze the correlation among WPC, flow unit distribution, and WS.

A single-factor analysis of the simulation results reveals that WPC has a positive power function with VR and a negative linear relationship with VF (Figures 8 and 9). With a given sand body distribution, WPC decreases with an "anti-S" shape, whereas WS increases linearly. The first inflection points on the WPC decrease the curve which denotes the generally recommended reasonable WS (Figure 10). Afterward, we adopt the tools of the multiple nonlinear regression analysis and the orthogonal experimental design to derive the quantitative WPC prediction model. Finally, the well pattern prediction model is expressed in Equation 3, which provides support for well pattern design in the newly discovered fluvial-delta reservoir.

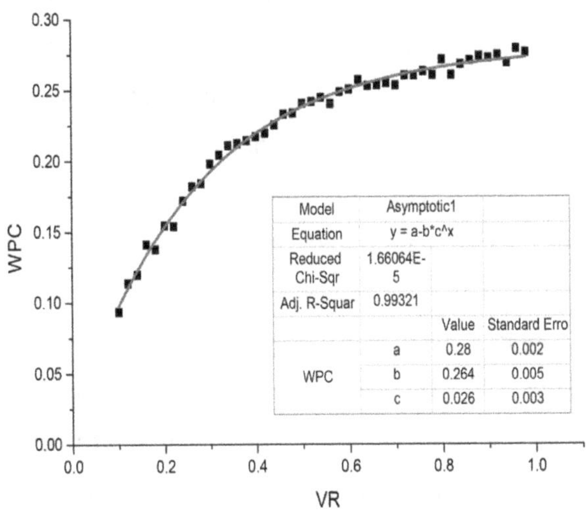

Figure 8: The variation of WPC versus VR.

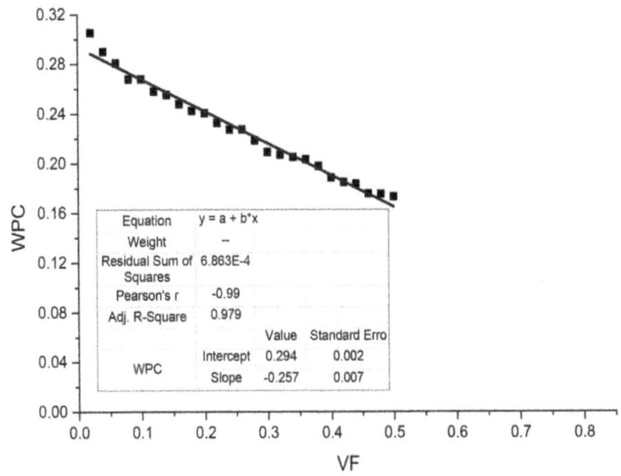

Figure 9: The variation of WPC versus VF.

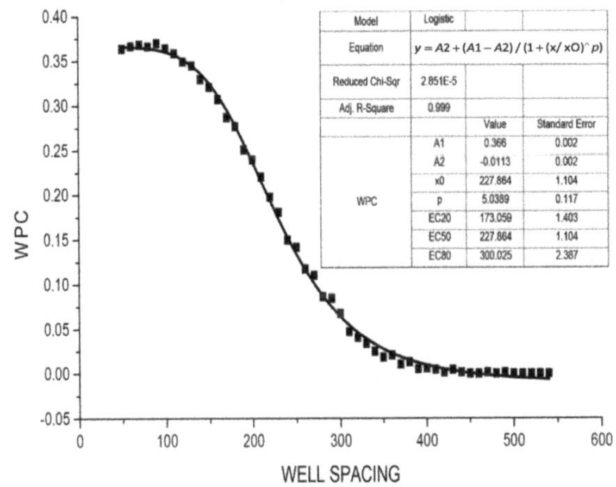

Figure 10: The variation of WPC versus WS.

$$WPC = -1.0226 + 1.1306^{VR} - 0.2012 \times VF + \frac{0.3570}{1+\left(\dfrac{WS}{243.3456}\right)^{4.5337}} \quad (3)$$

RESULTS AND DISCUSSION

Case study

A mature reservoir is required to verify the proposed WPC prediction model. Thus, in this section, a fluvial-delta reservoir is selected to determine whether the prediction result at the early stage coincides with the WPC from the actual dynamic production and the static geological information at the late development stage.

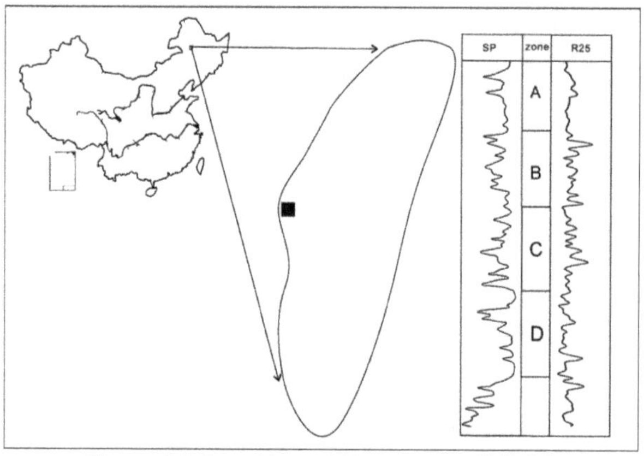

Figure 11: Location of Honggang oil field.

The reservoir is located in the western part of the Honggang oilfield in the Songliao basin of China (Figure 11), which was first discovered in 1965 and utilized in 1973. The block is approximately 4.3 km long and from 3.5 km to 4.2 km wide. Production The reservoir is located in the western part of the Honggang oilfield in the Songliao basin of China (Figure 11), which was first discovered in 1965 and utilized in 1973. The block is approximately 4.3 km long and from 3.5 km to 4.2 km wide. Production information is obtained from Saertu sandstone at 1,860 m. The maximum sand thickness is 9.6 m, and the average thickness is 2.1 m. The average porosity is approximately 30.2%, and the average air permeability is approximately 105 mD. The reservoir contains (1) black oil with an original bubble point pressure of 10.8 MPa and (2) solution gas to the oil ratio of 53.6 std m^3/(stocktank m^3) (standard cubic meters per stock-tank cubic meters) Previous geological studies reported that the reservoir could be categorized as a delta front subfacies.

The lateral distribution of the flow units has an evident influence on the water flooding performance. During the past 42 years, the well pattern has been altered four times, and WS has narrowed from 1,200 m to 100 m. The reservoir is currently in the high water cut development stage, with a current recovery of 46.4% and a water cut of 94.95. The VR and VF obtained from early deployed wells are 0.544 and 0.136 respectively, which are approximately the same as the average values of 0.550 and 0.134 for all wells at the late development stage. After incorporating VR (0.544) and VF (0.134) into the WPC prediction model, WPC varies with WS as shown in Figure 12. If the WS is 100 m, then the predicted WPC is 0.37. The actual WPC obtained from the dynamic performance is 0.374. Therefore, the prediction model has good reliability.

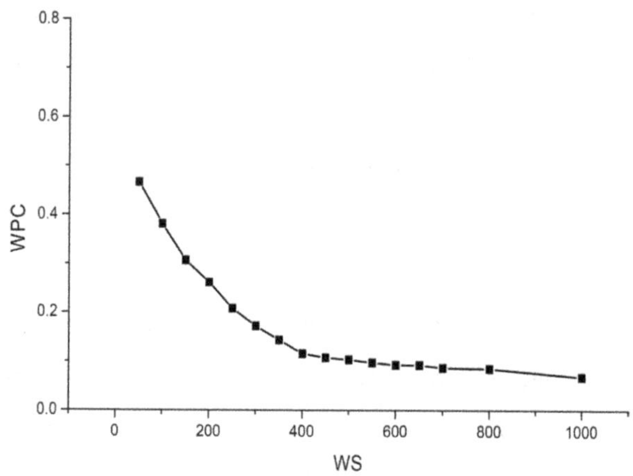

Figure 12: Actual WPC varies with WS.

Volumetric sweep efficiency in the fluvial-delta reservoir is mainly determined by WPC. In this study, we hypothesize that a potential correlation among WPC, flow unit distribution, and WS may exist. In a continental fluvial-delta reservoir, lateral flow unit distribution within sedimentary sequences has a statistical correlation with vertical flow unit distribution. Although bearing certain similarities, flow units and genetic sand bodies essentially differ. The study of a sand body shape or size relies on the hydrodynamic characteristic of the sedimentary process. Meanwhile, the flow unit is a tool used to understand fluid flow during reservoir development. A substantial knowledge of the width to thickness ratio, the length to width ratio, and profile style of sand bodies in braided river, meandering river, and distributed river in the fluvial-delta system has been obtained. Therefore, extensive literature on sand body shape, size, and distribution provides a significant insight into lateral flow unit distribution, but cannot replace the role of flow unit. The irregular shapes of flow units are determined by reservoir architecture and well position, resulting in considerable difficulty when delineating the flow unit distribution. Regardless of the shapes, sizes and variations of flow units, reservoirs would have a similar WPC if their flow unit laterally drilled ratio and laterally distributed frequency are the same. After evaluating the reliability of areal flow units in predicting WPC, we illustrate and compare the methods used to predict the lateral flow unit distribution on the basis of information on the vertical flow unit distribution. Many geostatistical tools such as Boolean Simulation or sequential indicator simulation are used to evaluate the uncertainty of reservoir architecture at the appraisal stage when only several wells are available. The presumed length, width, thickness and variation of the sand bodies as the main parameters are employed as the input to the simulation. However, previous studies conducted by experts on the lateral flow unit distribution mainly concentrate on the late development stage. These approaches require detailed geologic and dynamic information to identify flow unit distribution in the 3D reservoir. The method proposed in this study has the advantage of predicting flow unit distribution and WPC earlier than the regular procedure. The final well pattern prediction model is a nonlinear function of low unit distribution and WS; in addition, the model can be used easily in a newly discovered reservoir.

If the sedimentary facies of a newly discovered reservoir has been confirmed to be a continental fluvial delta during the exploratory and appraisal stages of a reservoir lifecycle, the proposed prediction model of WPC can then be used. The usage of the method implements two steps, as follows:

Step 1: Calculation of the parameters of vertical flow unit distribution;

After revealing the correlation of stratigraphy, the flow units of each drilled well could be interpreted by using the core and logging data within the target zone. The flow unit is a block of sandstone separated by impermeable shale. Then, the parameters characterizing vertical flow unit distribution (VR and VF) are averaged for all wells.

Step 2: Optimization of WS;

By putting parameters of vertical flow unit distribution (VR and VF) into Equation 3, we can evaluate the variation of WPC versus WS. The optimized WS is located at the place where WPC decreases fastest.

CONCLUSIONS

This study proposes a method of WPC prediction by establishing the relationship among WPC, sand body lateral distribution, and sand body vertical distribution. The method can be used when only exploratory and appraisal wells are available.

The results of this study show that WPC increases with the power function of the flow unit vertically drilled ratio; moreover, WPC decreases linearly with the flow unit vertically distributed frequency; in addition, WPC exhibits an "anti-S"-shaped function versus WS.

The WPC integrating the static and dynamic knowledge of many similar mature oil fields can serve as a guide for well pattern design in newly discovered fluvial-delta reservoirs.

ACKNOWLEDGEMENTS

The authors would like to thank China National Natural Science Fund Committee. The paper is sponsored by National Natural Science Foundation Project No. 51374222, National major project No. 2016ZX05013002-001 and No. 2016ZX05032005-002. It is also sponsored by National Key Basic Research & Development Program No. 2015CB2509005.

REFERENCES

1. Robinson, J. W. and McCabe P. J. "Sandstone-body and Shale-body Dimensions in a Braided Fluvial System: Salt Wash Sandstone Member (Morrison Formation), Garfield County, Utah," *AAPG Bulletin*, **1997**, *81*(8), 1267-1291.
2. King P. R., "The Connectivity and Conductivity of Overlapping Sand Bodies." *North Sea Oil and Gas Reservoirs*, Springer Netherlands, **1990**, 353-362.
3. Weber K. J., "Influence of Common Sedimentary Structures on Fluid Flow in Reservoir Models," *Journal of Petroleum Technology*, **1982**, *34*(03), 665-672.
4. Guangming T., Ogbe D. O., Munly W., and Hatzignatiou D. G., "The Use of Flow Units as a Tool for Reservoir Description: a Case Study," *SPE Formation Evaluation*, **1995**, *10*(02), 122-128.
5. Canas J. A., Malik Z. A., and Wu C. H., "Characterization of Flow Units in Sandstone Reservoirs: La Cira Field, Colombia, South America," In *Permian Basin Oil and Gas Recovery Conference*, Society of Petroleum Engineers, **1994.**
6. Pande P. K., Michael B. C., Blasingame T. A., Kelkar M., et al., "Application of Integrated Reservoir Management and Reservoir Characterization to Optimize Infill Drilling," In *Permian Basin Oil and Gas Recovery Conference,* Society of Petroleum Engineers, **1994**.
7. Driscoll V. J., "Recovery Optimization through Infill Drilling Concepts Analysis and Field Results," *Fall Meeting of the Society of Petroleum Engineers of AIME*, Society of Petroleum Engineers, **1974**.
8. Ching H. W., Laughlin B. A., and Jardon M., "Infill Drilling Enhances Waterflood Recovery," *Journal of Petroleum Technology*, **1989**, *41*(10), 1-088.
9. Gulick K. E. and McCain Jr. W. D., "Waterflooding Heterogeneous Reservoirs: An Overview of Industry Experiences and Practices," *International Petroleum Conference and Exhibition of Mexico*, Society of Petroleum Engineers, **1998.**
10. Hearn C. L., Ebanks Jr W. J., Tye R. S., and Ranganathan V., "Geological Factors Influencing Reservoir Performance of the Hartzog Draw Field, Wyoming," *Journal of Petroleum Technology*, **1984**, *36*(08), 1-335.
11. Ebanks Jr W. J. "Flow Unit Concept-integrated Approach to Reservoir Description for Engineering Projects," *AAPG Bulletin*, **1987**, *71*, 551-552.
12. Gunter G. W., Finneran J. M., Hartmann D. J., and Miller J. D., "Early Determination of Reservoir Flow Units Using an Integrated Petrophysical Method," In *SPE Annual Technical Conference and Exhibition*, Society of Petroleum Engineers, **1997**.
13. Rodriguez A., "Facies Modeling and the Flow Unit Concept as a Sedimentological Tool in Reservoir Description: A Case Study," *SPE Annual Technical Conference and Exhibition*, Society of Petroleum Engineers, **1988**.
14. Tyler N., Barton M. D., and Finley R. J., "Outcrop Characterization of Flow Unit and Seal Properties and Geometries, Ferron Sandstone, Utah," *SPE Annual Technical Conference and Exhibition*,

Society of Petroleum Engineers, **1991.**

15. Howell J. A., Martinius A. W., and Timothy R. G., "The Application of Outcrop Analogues in Geological Modelling: a Review, Present Status and Future Outlook," *Geological Society*, London, Special Publications, **2014**, *387*(1), 1-25.
16. Miall A. D., "Reservoir Heterogeneities in Fluvial Sandstones: Lessons from Outcrop Studies," *AAPG Bulletin*, **1988**, *72*(6), 682-697.
17. Robinson J. W. and McCabe P. J., "Sandstone-body and Shale-body Dimensions in a Braided Fluvial System: Salt Wash Sandstone Member (Morrison Formation), Garfield County, Utah," *AAPG Bulletin*, **1997**, *81*(8), 1267-1291.
18. Colombera L., Felletti F., Mountney N. P., and McCaffery W. D., "A Database Approach for Constraining Stochastic Simulations of the Sedimentary Heterogeneity of Fluvial Reservoirs," *AAPG Bulletin*, **2012**, *96*(11), 2143-2166.
19. Miall A. D., "Architectural-element Analysis: A New Method of Facies Analysis Applied to Fluvial Deposits," *Earth-Science Reviews*, **1985**, *22*(4), 261-308.
20. Miall A. D., "Reservoir Heterogeneities in Fluvial Sandstones: Lessons from Outcrop Studies," *AAPG Bulletin*, **1988**, *729*(6), 682-697.
21. Mikes D. and Geel C. R., "Standard Facies Models to Incorporate all Heterogeneity Levels in a Reservoir Model," *Marine and Petroleum Geology*, **2006**, *23*(9), 943-959.
22. Amaefule J. O., Altunbay M., Tiab D., Kersey D. G., et al., "Enhanced Reservoir Description: Using Core and Log Data to Identify Hydraulic (Flow) Units and Predict Permeability in Uncored Intervals/Wells," *SPE Annual Technical Conference and Exhibition*, Society of Petroleum Engineers, **1993**.
23. Yang, L. I., "Flow Unit Mode and Remaining Oil Distribution in Reservoir," *Acta Petrolei Sinica*, **2003**, *3*, 0-10.
24. Aminian K., "Prediction of Flow Units and Permeability Using Artificial Neural Networks," *SPE Western Regional/AAPG Pacific Section Joint Meeting*, Society of Petroleum Engineers, **2003**.
25. Damsleth E. and Henning O., "Geostatistical approaches in reservoir evaluation," *Journal of Petroleum Technology*, **1997**, *49*(05), 498-501.
26. Srivastava R. M., "An Overview of Stochastic Methods for Reservoir Characterization," *AAPG Bulletin Special Publications*, **1994**, 3-16.
27. Chiu S. N., Stoyan D., Kendall W. S., and Mecke J., "Stochastic geometry and its applications," *John Wiley & Sons*, **2013**.
28. Holden L., Hauge R., Skare Q., and Skorstad A., "Modeling of Fluvial Reservoirs with Object Models," *Mathematical Geology*. **1998**, *30*(5), 473-496.
29. Sophie V., Shtuka A., and Mallet J-L., "New Fast, Stochastic, Boolean Simulation of Fluvial Deposits," *SPE Annual Technical Conference,* **1998.**
30. Witten T. T., and Sander L. M. "Diffusion-limited Aggregation," *Physical Review B*, **1983**, *27*(9), 56-86.

14

Static Modeling of Oil Field Mineral Scales: Software Development

Mohammad Parvazdavani*[1], Behjat Kari[2], Mahmood Dinmohammad[3] and S. A. Mousavi Dehghani[3]

[1]Reservoir Studies Research Division, Research Institute of Petroleum Industry (RIPI), Tehran, Iran
[2] Petroleum Engineering Department, Amirkabir University of Technology, Tehran, Iran
[3] Upstream Faculty, Research Institute of Petroleum Industry (RIPI), Tehran, Iran

ABSTRACT

Mineral scale deposition in near wellbore regions of injection wells is one of the main challengeable issues during the water injection process, which magnifies the importance of robust models in predicting the amount of mineral scale deposition such as calcium sulfate. One of the main challenges of $CaSO_4$ scale is in carbonated reservoirs, in which sensitive behavior is observed in related to the contribution of both calcium and sulfate ions in carbonated and sulfated scale reactions. This defect is mirror of wrong procedure and value in the estimation of first kind/value of precipitant contributed in scale deposition reactions (ions competition) as well as inconsistent temperature/pressure dependent coefficients of prediction model. The objective of this study is to develop a model that can accurately predict the formation and amount of $CaSO_4$ scale as the dominant scale in multicomponent aqueous systems by three major tools, namely utilization the best temperature- and pressure-dependent thermodynamic interactive ion coefficients (MSE Model: Pitzer), developing our fine-tuned iterative mathematical solver, and verification of the results of the model by accurate experimental data. The results showed that at the optimum value of precipitant (10%) in scale deposition reactions and by defining the best temperature- and pressure-dependent coefficients, we can attain the best accuracy in the prediction of $CaSO_4$ scale deposited amount (less than 0.06% as a relative error compared to 36% overestimation and 22% underestimation in commercial software). The output of this study is developed software leading to the more accurate prediction of the amount of promising scales in near wellbore regions or pipelines.

Keywords: $CaSO_4$ Scale, Laboratory Static Jar Tests, Iterative Mathematical Solver, Pitzer Thermodynamic Model, Ions Binary Interactive Coefficient

INTRODUCTION

Seawater injection is one of the common methods used in IOR usually used in offshore oil fields to maintain reservoir pressure and improve secondary recovery. Scale deposition is one of the most serious oil field problems that inflict water injection systems primarily when two incompatible waters are involved. Scale deposition can occur from one type of water because of super-saturation with scale forming salts attributable to changes in the physical conditions under which the water exists. Super-saturation can be generated in water by changing

***Corresponding author**
Mohammad Parvazdavani
Email: parvazdavanim@ripi.ir

the pressure and temperature conditions or by mixing two incompatible waters. Scale can develop in the formation pores near the wellbore, which reduces porosity and permeability, and consequently productivity and injectivity decrease. It can block flow by clogging perforations or forming a thick lining in production tubing. Scale is also deposited in down-hole pumps, tubing, casing flow-lines, heater treaters, tanks, and other production equipment and facilities. The consequence could be production-equipment failure, emergency shutdown, increased maintenance cost, and overall decrease in production efficiency [1]. The most common oil field scales are listed in Table 1, along with the primary variables that affect their solubility [2].

Table 1: Common oil field scales.

Name	Chemical Formula	Primary Variables
Calcium Carbonate	$CaCO_3$	Partial pressure of CO_2, T, TDS, and pH
Calcium Sulfate / **Gypsum**	$CaSO_4.2H_2O$	T, TDS, and P
Hemihydrate	$CaSO_4.H_2O$	-
Anhydrate	$CaSO_4$	-
Barium Sulfate	$BaSO_4$	T and P
Strontium Sulfate	$SrSO_4$	T, TDS, and P
Ferrous Carbonate	$FeCO_3$	Corrosion, Dissolved Gases, and pH
Ferrous Sulfide	FeS	Corrosion, Dissolved Gases, and pH
Ferrous Hydroxide	$Fe(OH)_2$	-
Iron(III) oxide-hydroxide	$Fe(OH)_3$	-

Based on the industrial reports, the four dominant scales are sulfates, such as calcium sulfate (anhydrite, gypsum), barium sulfate (barite), and strontium sulfate (celestite), and calcium carbonate. Other less common scales have also been reported such as iron oxides, iron sulfides, and iron carbonate. Lead and zinc sulfide scale have recently become a concern in a number of North Sea oil and gas fields [3].

Calcium sulfate scale poses a unique problem for the salts under consideration because it occurs with one of three different phases. Gypsum, the most common scale formed, occurs at relatively low temperatures. At high temperatures (above 100 °C), the stable phase predicted is anhydrite ($CaSO_4$). However, hemi-hydrate has been known to form in the temperature range of 100 to 121 °C, especially in non-turbulent systems and in high ionic strength brines [4].

According to Oddo et al. [5], calcium sulfate scale formation is somewhat dependent on temperature, but is typically precipitated because of a decrease in pressure or an increase in the relative concentrations of calcium or sulfate. $CaSO_4$ solubility is fairly independent of pH, so it can readily precipitate in an acid environment. The solubility of calcium sulfate is also affected by temperature, salinity, and excess common ions.

The case where water injection (seawater, river, aquifer, or produced water) is used for pressure maintenance and sweep, the mixing of incompatible brines can lead to the formation of sulfate scales when the injection water contains sulfate ions [6]. Since there is sensitive behavior in the prediction of $CaSO_4$ scale deposition due to the contribution of both calcium and sulfate ions in carbonated and sulfated scale reactions, there is a strong motivation for developing accurate modeling tools that can predict the likelihood of $CaSO_4$ scale formation.

Also, other authors such as Bedrikovtsky (2006, 2009) addressed the injectivity decline due to sulfate scales deposition throughout the PWRI [7, 8]. Mackay discussed that sufficient concentrations of scaling ions delivered to the production well can necessitate squeeze treatments (inhibitor) [9]. Therefore, the authentic estimation of mineral scales precipitation is required. The main difficulty in scale prediction modeling lies in distinguishing the ions competition behavior of multicomponent mixtures. One of the main defects in previous studies is related to wrong procedure in the sequence and value of precipitants taken part in mineral scale deposition reactions at each stage of modeling.

The objective of this study is to develop a model that can accurately predict the formation of $CaSO_4$ scale deposition as well as mineral scales in multicomponent aqueous systems by three major tools: huge experimental data bank, including the implemented static tests, gathered temperature and pressure dependent coefficients, and a finely tuned iterative mathematical solver in which the previous described defect was removed. The next parts of this paper are organized in the manner of the following structure. First, we introduce the computational model briefly and determining the used parameters. The mathematical methodology for solving the governing equations has been presented in solving methodology, i.e. section 2. Section 3 contains the results and discussions of the performed Jar tests as the static scale deposition tests as well as the finely tuned mathematical model. In the last section, i.e. section 4, the conclusions are reported to show the accuracy of the proposed mathematical model on the prediction of mineral scale deposition in a specific thermodynamic condition.

EXPERIMENTAL PROCEDURES

Experimental Static Test

To obtain the observed scale precipitation results regarding the anhydrite and compare the results of the model as well as other sources (OLI and StimCade) with the experimental ones, static tests were designed and implemented in laboratory. In order to obtain the anhydrite scales at different ratios of injection to formation water, Na_2SO_4 and $CaCl_2$ was used to provide required divalenets for the water phase (see Table 2 for the compositions of the water phase). After that, the formation water was mixed with injection one at different ratios, and after an optimum time, the solution was filtered with 0.45 micrometer paper filter. The filter was warmed in an oven at optimum time and temperature. The dried filter was weighed and the amount of scales (mg/L) was calculated. Figure 1 shows the stability of the formed scale regarding the time.

Table 2: Composition of synthetic waters for the prediction of individual scale formation.

Formation water	Value (mg/L)	Sea water	Value (mg/L)
Calcium	12244	Sulfate	7000
Chloride (Brine)	129758	Sodium	1627
Sodium (Brine)	84068	-	-
Chloride (Calcium)	21734	-	-

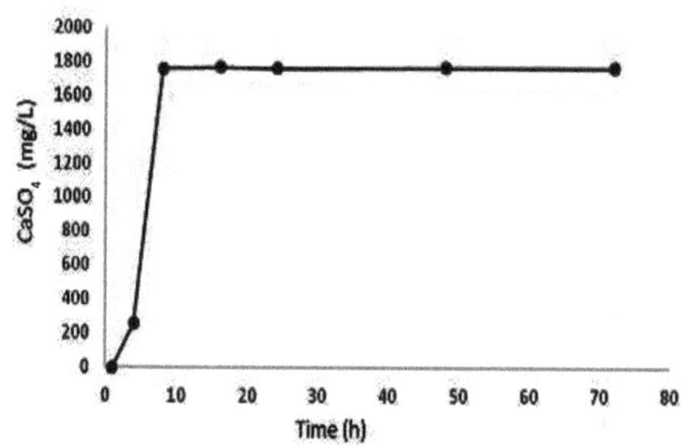

Figure 1: Effect of time on formation of anhydrate scale.

Computational Model, Model Description, Aqueous Chemistry

At the first stage, chemical reactions in the liquid phase have been introduced. The dissolved species in the aqueous phase as well as the satisfaction of the mass-action equations for all the possible association reactions between the master species at every transport node are assumed to be in thermodynamic equilibrium. The association reactions between the master species and the corresponding mass-action equations accompanied by the master species are defined in Equations 1-5.

$$H_2O(l) \leftrightarrow H^+ + OH^-$$
$$K_{H_2O} = [H^+][OH^-]\frac{\gamma_{H^+}\gamma_{OH^-}}{a_{H_2O(l)}} \quad (1)$$

$$H_2O(l) + CO_2(aq) \leftrightarrow H^+ + HCO_3^-$$
$$K_{CO_2\_1} = \frac{[H^+][HCO_3^-]}{[CO_2]}\frac{\gamma_{H^+}\gamma_{HCO_3^-}}{\gamma_{CO_2}a_{H_2O(l)}} \quad (2)$$

$$HCO_3^- \leftrightarrow H^+ + CO_3^{-2}$$
$$K_{CO_2\_2} = \frac{[H^+][CO_3^{-2}]}{[HCO_3^-]}\frac{\gamma_{H^+}\gamma_{CO_3^{-2}}}{\gamma_{HCO_3^-}} \quad (3)$$

$$HA \leftrightarrow H^+ + A^-$$
$$K_{H_2S} = \frac{[H^+][A^-]}{[HA]}\frac{\gamma_{H^+}\gamma_{A^-}}{\gamma_{HA}} \quad (4)$$

$$H_2S(aq) \leftrightarrow H^+ + HS^-$$
$$K_{H_2S} = \frac{[H^+][HS^-]}{[H_2S(aq)]}\frac{\gamma_{H^+}\gamma_{HS^-}}{\gamma_{H_2S(aq)}} \quad (5)$$

Activity coefficients are calculated with the Pitzer model (See Appendix A) [10].

Mineral Dissolution and Precipitation

At the second stage, to consider the effects of mineral dissolution and precipitation, promising mineral phases: calcite ($CaCO_3$), siderite ($SrSO_4$), barite ($BaSO_4$), anhydrite ($CaSO_4$), siderite ($FeCO_3$), and iron sulfide (FeS), which can precipitate from the aqueous solution, were considered (Equations 6-11).

$$Ca^{+2} + CO_3^{-2} \leftrightarrow CaCO_3$$
$$K_{CaCO_3} = [Ca^{+2}][CO_3^{-2}]\gamma_{Ca^{+2}}\gamma_{CO3^{-2}} \quad (6)$$

$$Sr^{+2} + SO_4^{-2} \leftrightarrow SrSO_4$$
$$K_{SrSO_4,} = [Sr^{+2}][SO_4^{-2}]\gamma_{Sr^{+2}}\gamma_{SO_4^{-2}} \quad (7)$$

$$Ba^{+2} + SO_4^{-2} \leftrightarrow BaSO_4$$
$$K_{BaSO_4} = [Ba^{+2}][SO_4^{-2}]\gamma_{Ba^{+2}}\gamma_{SO_4^{-2}} \quad (8)$$

$$Ca^{+2} + SO_4^{-2} \leftrightarrow CaSO_4$$
$$K_{CaSO4} = [Ca^{+2}][SO_4^{-2}]\gamma_{Ca^{+2}}\gamma_{SO_4^{-2}} \quad (9)$$

$$Fe^{+2} + CO_3^{-2} \leftrightarrow FeCO_3(S)$$
$$K_{FeCO_3} = [Fe^{+2}][CO_3^{-2}]\gamma_{Fe^{+2}}\gamma_{CO_3^{-2}} \quad (10)$$

$$Fe^{+2} + HS^- \leftrightarrow H^+ + FeS(S)$$
$$K_{FeS} = \frac{[Fe^{+2}][HS^-]}{[H^+]}\frac{\gamma_{Fe^{+2}} \times \gamma_{S^{-2}}}{\gamma_{H^+}} \quad (11)$$

The temperature and pressure dependence of the equilibrium constants are calculated using Pederson and Christensen correlations (2006) [10].

Determination of Parameters in Model

In case of incompatibility between formation and injection water during the water injection process, it is essential to have accurate data on the thermodynamic properties of the mixed electrolyte solutions. The activity coefficients are of primary importance in accurately describing the thermodynamic behavior of aqueous mixed electrolyte solutions. Among recently developed models of electrolyte solutions, ionic interaction models provide the simplest and most coherent procedures for calculating the thermodynamic properties of electrolyte components. An ion interaction model for predicting the activity coefficients of mixed electrolyte solutions was developed by Pitzer in early 1970s. The Pitzer model extended the Debye-Huckel method, using a virial expansion to account for the ionic strength dependence of the short-range forces in binary and ternary ion interactions [10].

The adjustable parameters for the model are the single electrolyte parameters, $\beta_{0,MX}, \beta_{1,MX}, \beta_{2,MX}$ and $C_{\phi,MX}$ as well as the mixed electrolyte parameters θ_{ij} and ψ_{ijk} for the significant minerals. Researches have previously evaluated most of the electrolyte parameters needed for this system using osmotic coefficient data. Table 3 summarizes the references used to extract the required coefficients and parameters in Pitzer model. Selected θ_{ij} and ψ_{ijk} parameters are consistent with the high concentration solubility data while they slightly reduces the accuracy of the equations in the low concentration range. Since terms in θ_{ij} and ψ_{ijk} are relatively insignificant for low concentrations, the loss of accuracy is minimal in the context of solubility prediction.

Different papers have used different empirical functions describing the variation of Pitzer's parameters with respect to temperature and pressure (11, 12, 13, 14, 15, 16 and 17, 18). For computational ease, these functions can be recast into the following ten parameter expression using simple algebraic transformations (Eq. 12). X(T) can be either Pitzer's parameters $\beta_{0,MX}, \beta_{1,MX}, \beta_{2,MX}$, $C_{\phi,MX}$ θ_{ij} or ψ_{ijk}

$$X(T)=\frac{a_1}{T}+a_2+a_3P+a_4\ln(T)+(a_5+a_6P)T$$
$$+(a_7+a_8P)T^2+\frac{a_9+a_{10}P}{T-227}+\frac{a_{11}+a_{12}P}{680-T} \quad (12)$$

Table 3: References of the required coefficients and parameters in Pitzer model

Mixed electrolyte binary interaction parameters ($\beta_{0,MX}, \beta_{1,MX}, \beta_{2,MX}$ and $C_{\phi,MX}$) at 25°C	Harvie et al., 1984 [11]
Mixed electrolyte ternary interaction parameters and ion-neutral interaction coefficient (θ_{ij} and ψ_{ijk}) at 25 °C	Harvie et al., 1984 [11]
Debye–Huckel slope for the activity coefficient, A^{ϕ}	Moller, 1988 [12]
Values of the fitting constants Eq.12 for the binary interaction parameters for aqueous electrolytes	Christov and Moller, 2004 [13]; Marion, 2001

Solving Methodology

In this paper, an iterative algorithm is developed for modeling and the calculation of the amount of precipitates in a super saturated mineral solution. Iterative calculation for the prediction of mineral scales was introduced in previous studies [10]. Yuan and Todd used an iterative model for the prediction of sulfate scales and did not consider the calcite scale. They performed the modeling by applying the sequence of barite, Celestin, and anhydrate precipitation reactions to the calculation procedure. Pedersen firstly analyzed the ion products of the iron minerals ($FeCO_3$ and FeS) against the solubility products. Afterward, the ion product of calcium carbonate is checked against its solubility product. In case of precipitation possibility, double-loop iteration is employed to calculate simultaneous precipitation of calcium carbonate and iron minerals. In a recent study on scales modeling by iterative solvers, Li performed the modeling of calcite and anhydrate scales solely and did not consider barite and celestite scales. He used the iterative calculation on mass and charge change due to scales precipitation with the conditions of lack of change in activity coefficients.

In the specified studies, the calculation algorithms are based on the specific order of reactions and successive calculations some of which are nested. The main challenge in literature, which is the source of error in the estimation of scale precipitation, is lack of possibility to simultaneously calculate the individual solid precipitation in multicomponent mixtures.

To handle the described challenge in literature, two main assumptions are considered; the first one is the simultaneous competition of ions contributed in mineral scale reactions as the common cations/anions with the different reaction rates. Secondly, the formed mineral scales are proportional to the total concentration. With these assumptions, precipitation ratio coefficient (PRC) would be defined as the ratio of precipitant taken part in reactions at each stage of modeling and is between 0 and 1. The amount of mineral scale for each compound with positive SI is equal to the total potential of precipitation multiplied by PRC. At the end of each stage, the new equilibrium concentrations of ions would be calculated and the updated values of activity coefficients can be obtained by the Pitzer's model (Appendix A). These stages proceed up to the step in which SI changes into zero or negative. The cumulative mineral scale precipitations can be calculated by summation of the experiment steps. In this mathematical solver, convergence speed and calculation error is exactly proportional to the value of PRC. Sensitivity analysis was carried out to obtain the optimum value of PRC in which the modeling results are consistent and matched with the experimental results and are independent of the sequence of scales formation reactions. Figure 2 shows the algorithm applied to the modeling.

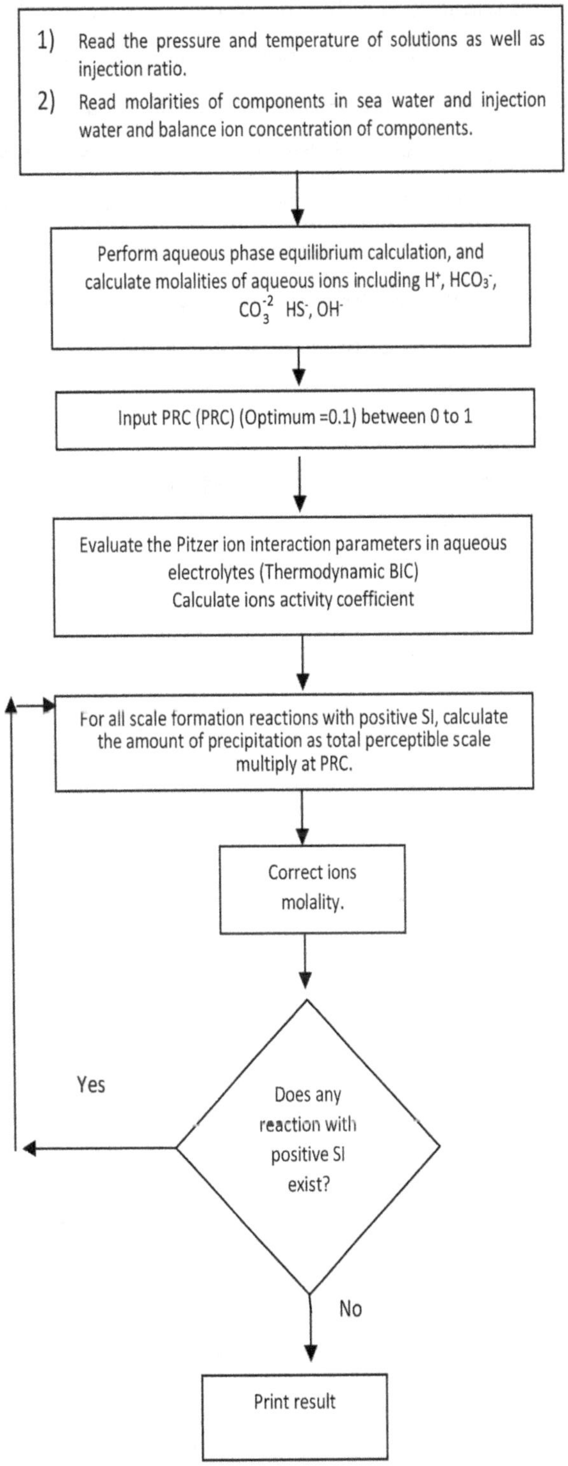

Figure 2: Algorithm applied to mineral scale precipitation modeling.

RESULTS AND DISCUSSION

A specified algorithm (Figure 1) was written in Visual Basic (V.B.) programming language and the following results have been obtained. Figure 3 shows the executive input window of the developed software.

In this window, all of the required parameters for the model, including ions compositions and TDS as well as pH, are inserted and pushing the check data button, the charge balance will be carried out to commence the ions equilibrium reactions calculations. Figure 4 shows the executive output window, in which calculation procedure will be done in a specific thermodynamic condition and different mixing ratios for the case study. Also, the graphical charts will be plotted in the output window simultaneously.

To validate the parameters used in the model, the solubility of one individual solid (anhydrate) was calculated and compared to the literature results at different temperatures and at the vapor pressure by applying the algorithm in Figure 5.

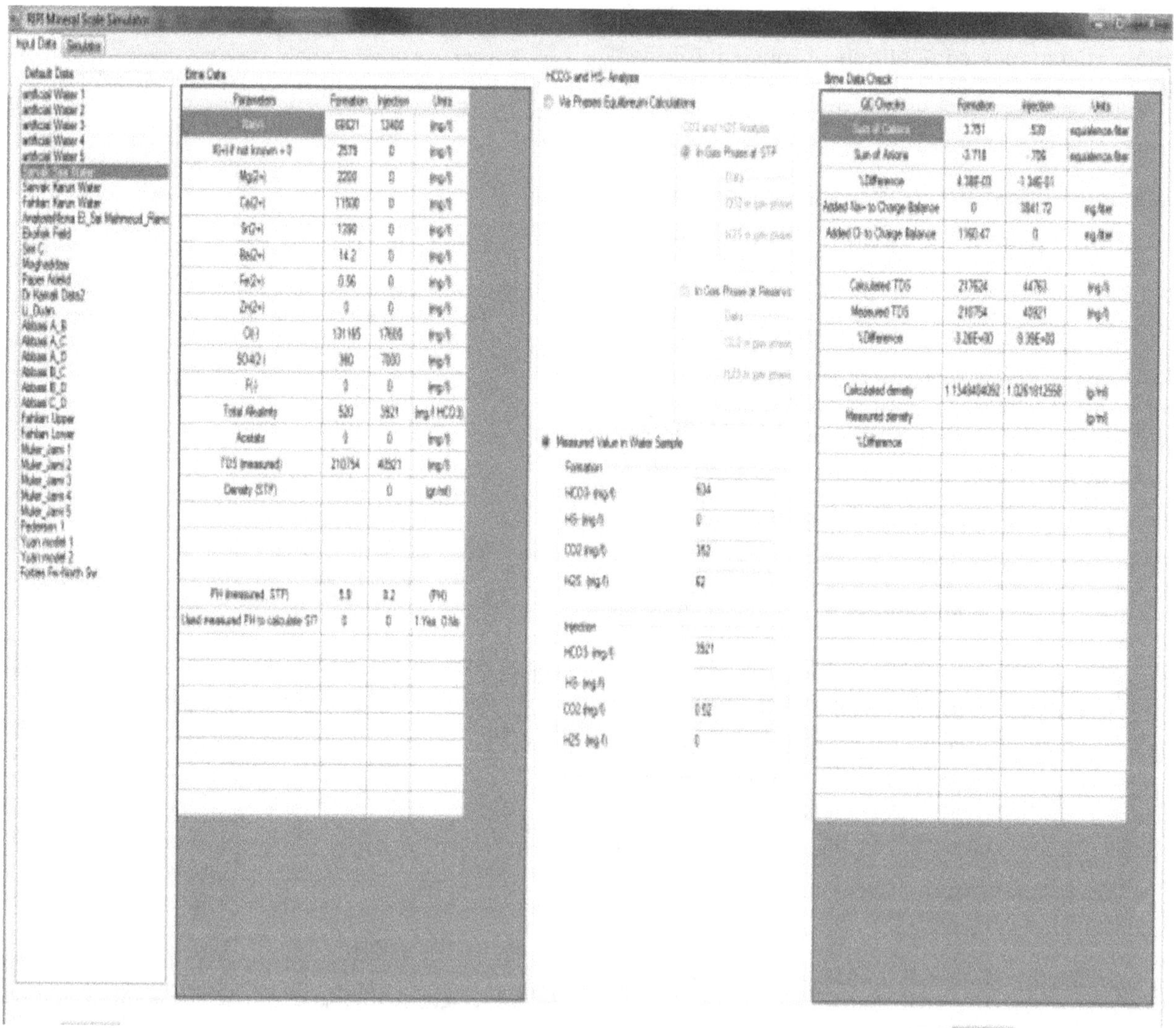

Figure 3: Input window of the developed software.

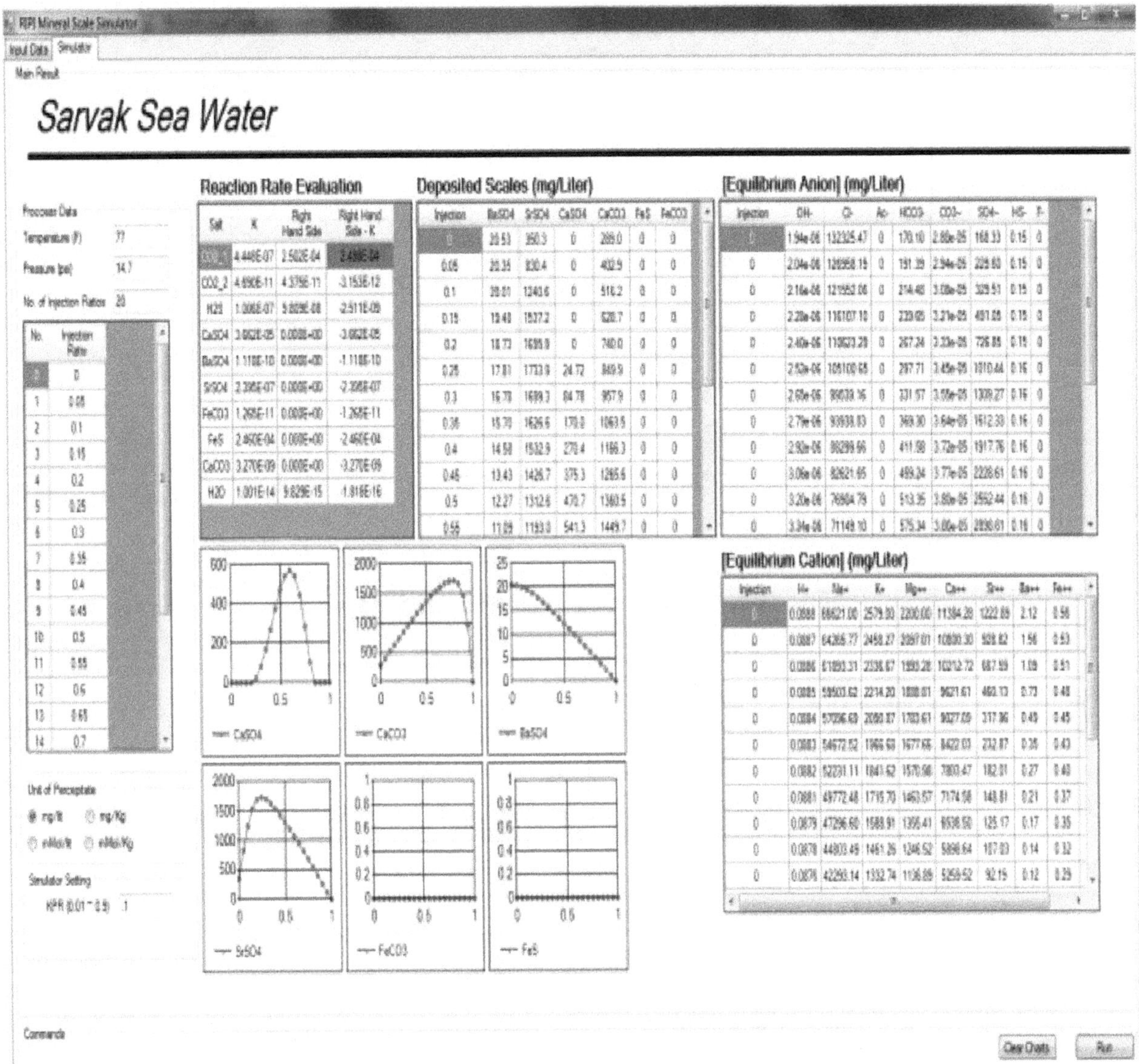

Figure4. Executive output window of developed software

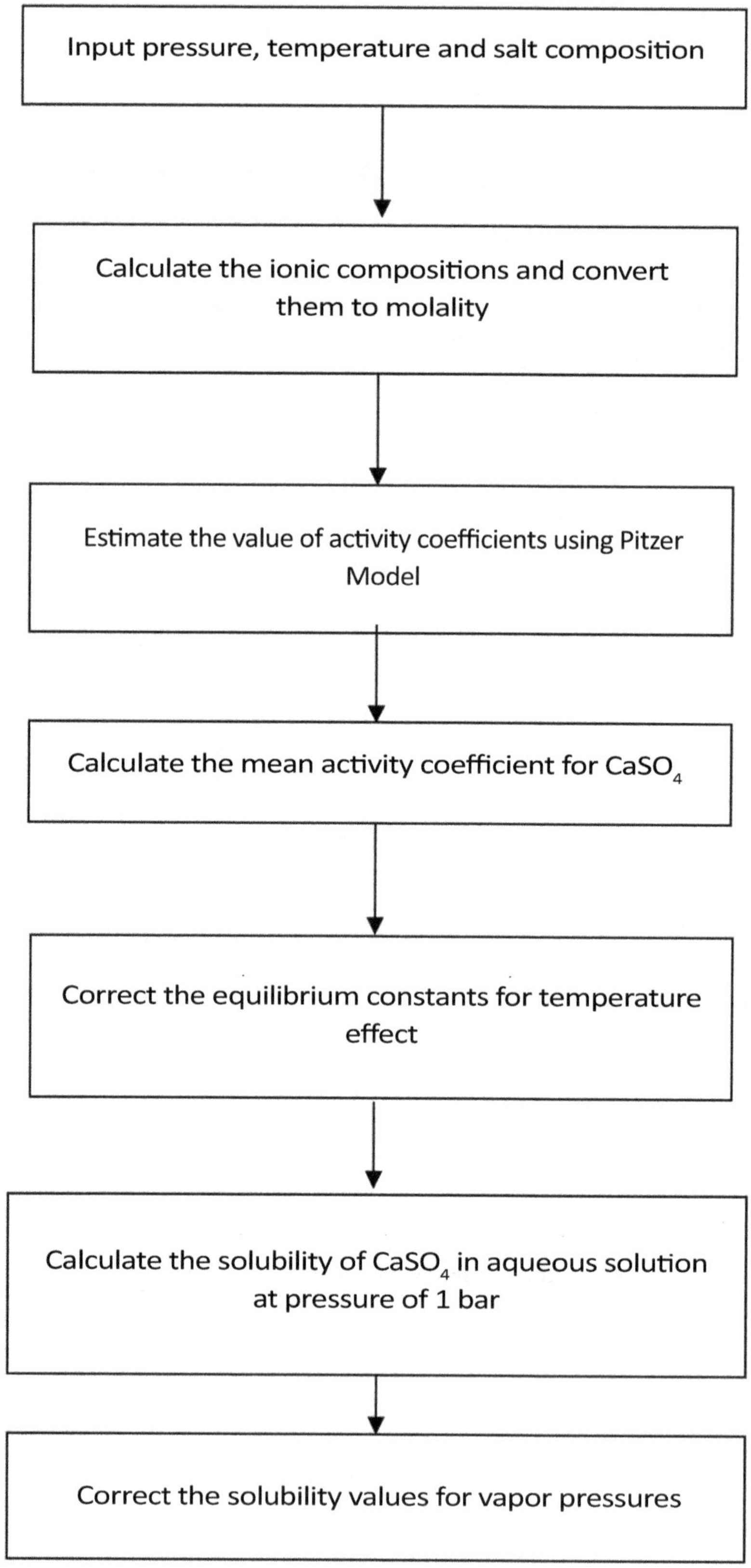

Figure 5: Algorithm Applied to solubility calculation.

The result of the comparison of $CaSO_4$ solubility in pure water is shown in Figure 6. As can be observed, there is a good match between our model and the previously developed models, which reflects using of suitable pressure and temperature dependent parameters as well as adjustable parameters for the model.

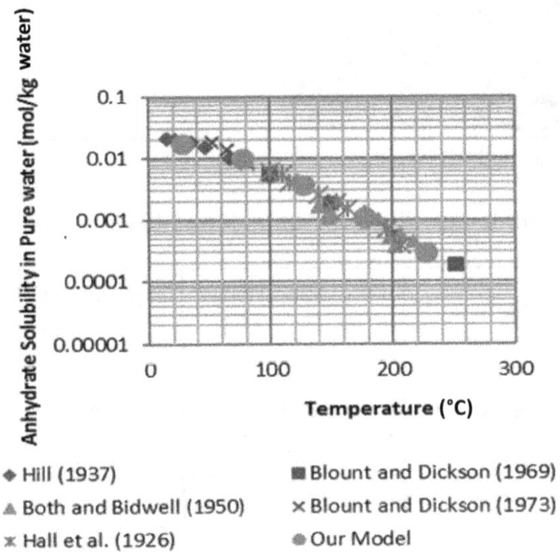

Figure 6: A comparison of the calculated anhydrite solubility in water using our model with the experimental data.

Since there is $CaSO_4$ as one of the dominant scales in blending the used case studied water (Sarvak) with sea water (Table 2), in the second stage, the scale precipitation of individual solid ($CaSO_4$) has been assessed with the valid proposed model. In order to solely monitor the formation of one type of mineral scale, synthetic waters have been used as the formation and injection sea waters (see Table 3 for their compositions). By blending the used waters at different mixing ratios, the results for $CaSO_4$ scales at different ratios are obtained in both modeling and experimental sections. The comparison between the experimental data and the model results as well as commercial software ones is shown in Figure 7.

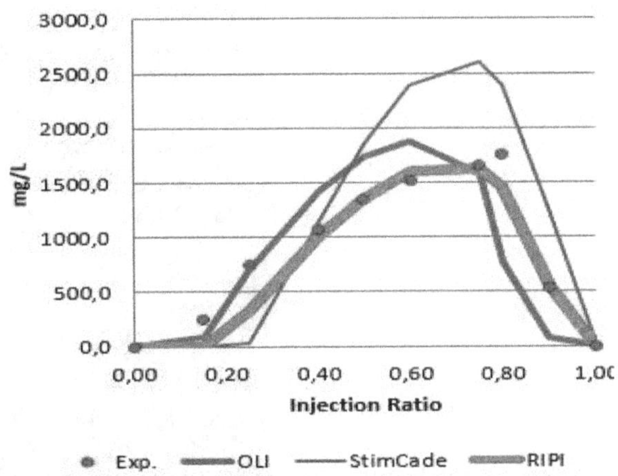

Figure 7: Comparison of mineral scale precipitation in the developed model, different common tools, and laboratory tests (at T=25 °C, P=1 bar).

Different software packages use different solvers but one model, i.e. Pitzer/modified Pitzer, to predict the precipitation and most of the difference in the results is referred to different algorithms defining ions competition through the scaling reactions.

To be sure about the kind of formed scale as individual solid, SEM analysis was performed on each stage of individual solid static Jar tests (Figure 8).

To verify the obtained results in the section of individual scales and use the computational model on an industrial scale, real formation water (Case study; Sarvak) was mixed with sea water in two distinct formation water/sea water mixing ratios of 0.5 and 0.75 (see Table 4 for the compositions of the candidate waters). Modeling was conducted at all the mixing ratios and the results were compared with the experimental data and commercial software results (Figure 9). Mixing ratios in the experimental studies were planned from 0.25 to 0.75 (0.25, 0.5, and 0.75). Since the main difference between commercial software and the developed model was in the scope of mixing ratios between 0.4 and 0.9, two points of experimental data (0.5 and 0.75) were selected to be compared with the available sources.

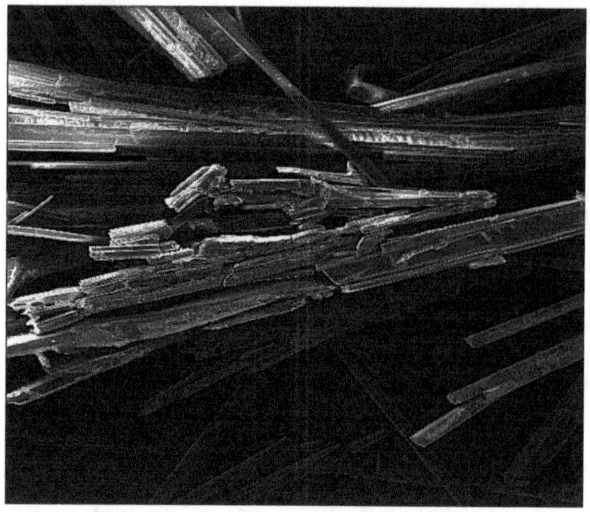

Figure 8: SEM image of anhydrate scale.

Table 4: Composition of Sarvak formation water and sea water for the prediction of scales formation in a real case.

Component	Formation Water Value (mg/L)	Sea Water Value (mg/L)	Component	Formation Water Value (mg/L)	Sea Water Value (mg/L)
Cations			Anions		
Sodium	66621	17600	Chloride	131165	12400
Potassium	2579	-	Sulfate	360	7000
Calcium	28750	-	Sulfite	48	-
Magnesium	5500	-	Bicarbonate	634	3921
Ferrous Iron	0.56	-	Nitrate	63.4	-
Barium	14.2	-	Nitrite	1.00E-02	-
Strontium	1390	-	Neutrals	-	-
Carbon dioxide, aq.	352	-	Hydrogen sulfide, aq.	62	-
Total Dissolved Solids, mg/L	237125	40921	Measured Density, g/cc	1.1441	1

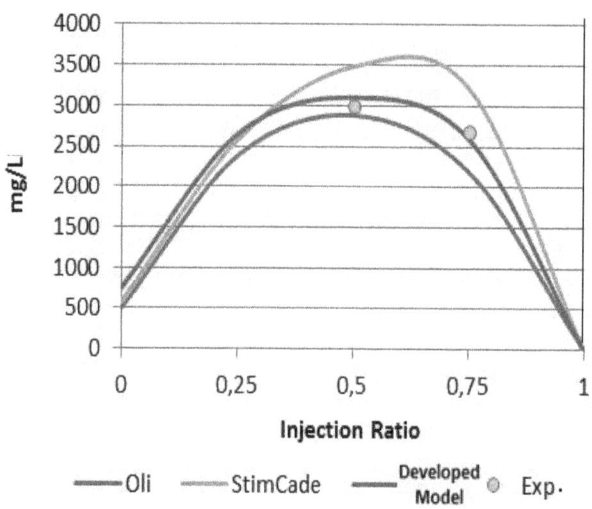

Figure 9: Comparison of total scales precipitation by different sources

To validate the developed model, a sensitivity analysis was performed on two main criteria, namely the sequence of scale precipitation and PRC. Two different patterns of precipitation at different PRC values were planned to check the dependency of model on the sequence of scale precipitation as well as PRC; A: ($CaCO_3$-$SrSO_4$-$CaSO_4$-$BaSO_4$) and B: the opposite sequence of A. Figures 10 and 11 show that the obtained results are independent of the sequence of scale precipitation and PRC in all scales, except $CaSO_4$.

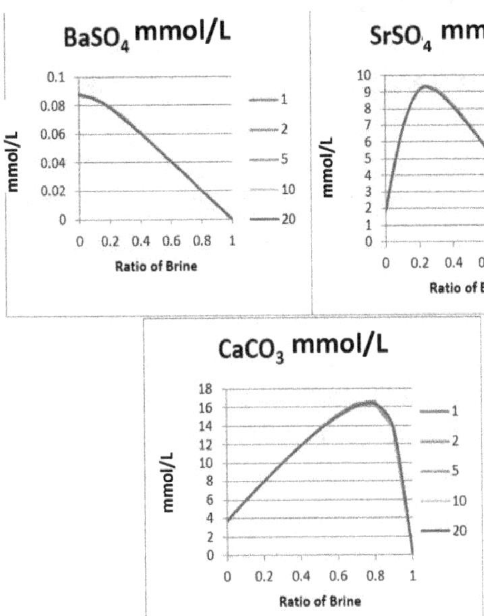

Figure 10: Sensitivity analysis on the sequence of scale precipitation and PRC in pattern A.

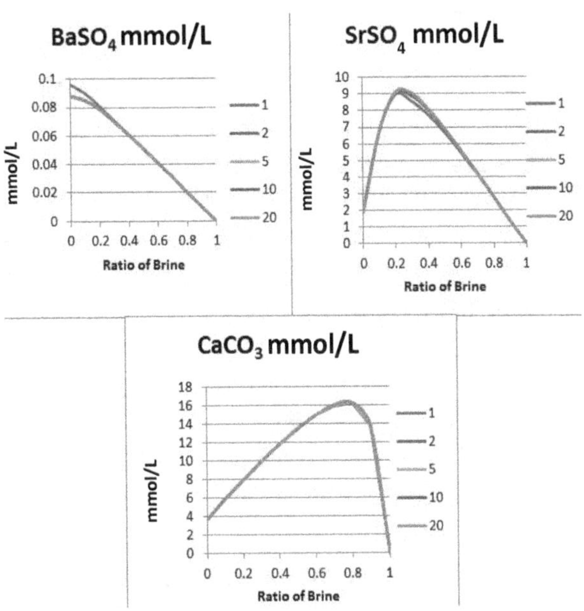

Figure 11: Sensitivity analysis on the sequence of scale precipitation and PRC in pattern B.

In case of $CaSO_4$, a significant change was observed in both aspects of changes in PRC values and the sequence of scale precipitation (Figure 12). The origin of this sensitive behavior in $CaSO_4$ is related to the contribution of both calcium and sulfate ions in carbonated and sulfated scale reactions. However, it was approved that this dependency would be diminished as the PRC value approached its optimum value (10%) and would be omitted approximately for lower values of PRC (Figure 13).

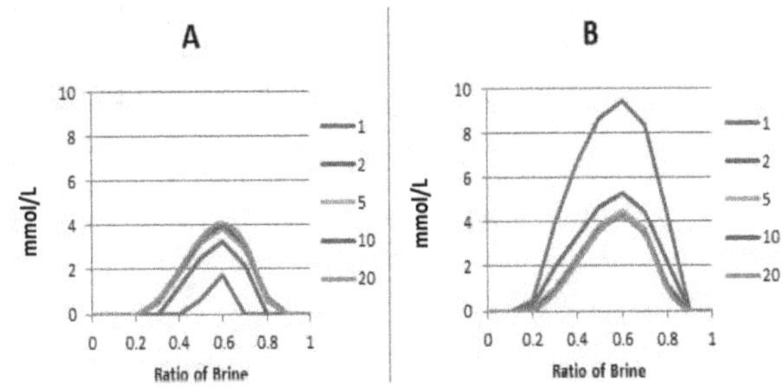

Figure 12: Sensitivity analysis on the sequence of scale precipitation and PRC in $CaSO_4$ Scale; A: Carbonate/sulfate; B: Sulfate/carbonate.

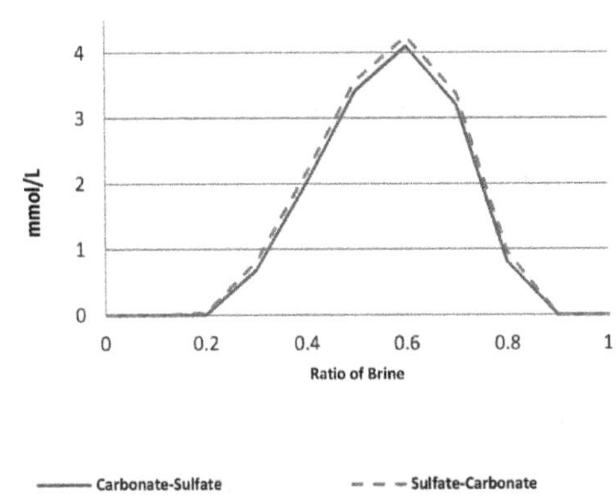

Figure 13: Independency of scale precipitation amount from the sequence of scale precipitation in PRC=5%.

As observed in Figures 10 to 13, the developed model is independent of the sequence of scale precipitation as well as PRC values in each step.

CONCLUSIONS

The results of model developed for predicting the formation of mineral scales in multicomponent aqueous systems caused by mixing incompatible waters (sea water and formation water) and by changing thermodynamic conditions are consistent with experimental results for

$CaSO_4$ and with the previous scale prediction models and commercial software (OLI Scale Chem, StimCade). Briefly, the following conclusions can be drawn according to the results of the developed model:

- The main advantageous of the present model is related to applying a procedure to the sequence and value of precipitants taken part in mineral scale deposition reactions at each stage of modeling.
- At the optimum value of precipitant (10%) in scale deposition reactions and by defining the best temperature and pressure dependent coefficients, a relative error of less than 0.06% was observed compared to commercial software with 36% overestimation and 22% underestimation in the prediction of $CaSO_4$ scale deposited amount.
- The origin of sensitive behavior in $CaSO_4$ scale prediction is related to the contribution of both calcium and sulfate ions in carbonated and sulfated scale reactions (ions competition).
- The sensitivity analysis of our developed model shows that the scale prediction results are independent of the sequence of scale precipitation and PRC in all scales, except $CaSO_4$. In the case of $CaSO_4$, scale precipitation curves approach each other in PRC values lower than 10% for both defined reaction sequences, and the amount of $CaSO_4$ scale become constant.

REFERENCES

1. BinMerdhah B. M., Azam Mohd Yassin A., and Muherei A. M., "Laboratory and Prediction of Barium Sulfate Scaling at High-Barium Formation Water," *Journal of Petroleum Science and Engineering*, **2010**, *70*, 79-88.
2. Moghadasi J., Jamialahmadi M., Muller-Steinhagen H., Sharif A., et al., "Scale Formation in Iranian Oil Reservoir and Production Equipment during Water Injection," *The 5th International Oilfield Scale Symposium and Exhibition. Aberdeen*, UK, SPE 80406, **2003**.
3. Collins R. I. and Jordan M. M., "Occurrence, Prediction and Prevention of Zinc Sulfide Scale within Gulf Coast and North sea High Temperature/High Salinity Production Wells," *The SPE Third International Symposium on Oil field Scale. Aberdeen*, UK, SPE 68317, **2001**.
4. Moghadasi J., Jamialahmadi M., Muller-Steinhagen H., Sharif A., et al., "Scale Formation in Iranian Oil Reservoir and Production Equipment during Water Injection," *The 5th International Oil field Scale Symposium and Exhibition*, Aberdeen, UK, SPE 80406, **2003**.
5. Oddo E. J., Smith P. J., and Tomason B. M., "Analysis of and Solutions to the $CaCO_3$ and $CaSO_4$ Scaling Problems Encountered in Wel ls Offshore Indonesia," *The 66th Annual Technical Conference and Exhibition of the Society of Petroleum Engineering*, Dallas, TX. SPE 22782, **1991**, 1-10.
6. Mackay J. E., "Jordan Impact of Brine Flow and Mixing in the Reservoir on Scale Control Risk Assessment and Subsurface Treatment Options: Case Histories," *Journal of Energy Resources Technology*, **2005**, *127*, 201-213.
7. Bedrikovetsky P., Mackay E. J., Rosario F. F., Monteiro R. P., et al., "Injectivity Impairment Due to Sulfate Scaling During PWRI: Analytical Model, SPE International Oil field Scale Symposium," *Society of Petroleum Engineers*, Aberdeen, UK, **2006**.
8. Bedrikovetsky P. G., Mackay E. J., Silva R. M. P., Patricio F. M. R. et al., "Produced Water Re-Injection with Seawater Treated by Sulfate Reduction Plant: Injectivity Decline, Analytical Model," *Journal of Petroleum Science and Engineering*, **2009**, *68*, 19-28.
9. Mackay E. "Predicting In Situ Sulfate Scale Deposition and the Impact on Produced Ion Concentrations," *Journal of Chemical Engineering and Data*, **2003**, *81*, 326-332.
10. Pedersen K. S., Christensen P. L., and Azeem S. J., "Phase Behavior of Petroleum Reservoir Fluids," Taylor and Francis, **2006**.
11. Roberto T. P. and Pitzer S. K., "Thermodynamics of Concentrated Electrolyte Mixtures and the Prediction of Mineral Solubilities to High Temperatures for Mixtures in the System Na-K-Mg-Cl-SO-OH-$H_2$0," *Geochim Cosmochim Acta*, **1987**, *51*, 2429-2443,

12. Greenberg P. and Moller G. N., "The Prediction of Mineral Solubilities in Natural Waters: A Chemical Equilibrium Model for the Na-K-Ca-Cl-SO_4-H_2O System to High Concentration from 0 to 250 °C," *Geochim. Cosmochim. Acta*, **1989**, *53*, 2503-2518.
13. Harviec E., Mueller N., and Weare J., "The Prediction of Mineral Solubilities in Natural Waters: the Na-K-Mg-Ca-H-Cl- SO-OH-Hcos-Cos-CO_2-H_2O System to High Ionic Strengths at 25 °C," *Geochim. Cosmochim. Acta*, **1984**, *48*, 723-751,
14. Mueller N., "The Prediction of Mineral Solubilities in Natural Waters: A Chemical Equilibrium Model for the Na-Ca-Cl-SO_4-H_2O System, to High Temperature and Concentration," *Geochim. Cosmochim. Acta*, **1988**, *52*, 821-837.
15. Christov C. and Moller N., "A Chemical Equilibrium Model of Solution Behavior and Solubility in the H-Na-K-Ca-OHCl-HSO_4-SO_4-H_2O System to High Concentration and Temperature," *Geochimica et Cosmochimica Acta*, **2004**, *68*, 3717–3739.
16. Hee-Talk K. and J. Frederick W., "Evaluation of Pitzer Ion Interaction Parameters of Aqueous Mixed Electrolyte Solutions at 25 °C. 2. Ternary Mixing Parameters," *J. Chem. Eng. Data*, **1988**, *33*, 278-283,
17. Monnin C., "The Influence of Pressure on the Activity Coefficients of the Solutes and on the Solubility of Minerals in the System Na-Ca-Cl-SO_4-H_2O to 200°C and 1 kbar and to High NaCl Concentration," *Geochimica et Cosmochimica Acta*, **1990**, *54*, 3265-3282.
18. Safari H., Shokrollahi A., Jamialahmadi M., Ghazanfari M. H., et al., "Prediction of the Aqueous Solubility of $BaSO_4$ Using Pitzer Ion-interaction Model and LSSVM Algorithm," *Journal of Fluid Phase Equilibria*, **2014**, *374*, 48-62.

15

Application of Least Square Support Vector Machine as a Mathematical Algorithm for Diagnosing Drilling Effectivity in Shaly Formations

Abdolhamid Sameni[1]* and Ali Chamkalani[2]

[1] Chemical Engineering Faculty, Tarbiat Modares University, Tehran, Iran
[2] Petroleum University of Technology, Ahwaz, Iran

ABSTRACT

The problem of slow drilling in deep shale formations occurs worldwide causing significant expenses to the oil industry. Bit balling which is widely considered as the main cause of poor bit performance in shales, especially deep shales, is being drilled with water-based mud. Therefore, efforts have been made to develop a model to diagnose drilling effectivity. Hence, we arrived at graphical correlations which utilized the rate of penetration, depth of cut, specific energy, and cation exchange capacity in order to provide a tool for the prediction of drilling classes.

This paper describes a robust support vector regression (SVR) methodology that offers superior performance for important drilling engineering problems. Using the amount of cation exchange capacity of the shaly formations and correlating them to drilling parameters such as the normalized rate of penetration, depth of cut, and specific energy, the model was developed. The method incorporates hybrid least square support vector regression into the coupled simulated annealing (CSA) optimization technique (LSSVM-CSA) for the efficient tuning of SVR hyper parameters. Also, we performed receiver operating characteristic as a performance indicator used for the evaluation of classifiers. The performance analysis shows that LSSVM classifier noticeably performs with high accuracy, and adapting such intelligence system will help petroleum industries deal with the well drilling consciously.

Keywords: Support Vector Regression, Shaly Formations, Coupled Simulated Annealing, Drilling Region

INTRODUCTION

The remarkable portion of footage drilled occurs in shales; moreover, the drilling operation is performed at high ROP because shales are not rigid and stiff [10]. Due to chemical and mechanical factors, the cuttings stick together or to the bit. Therefore, the bit performance decreased which subsequently caused the drilling rate to decline [10,37,40,41]; this phenomenon is called "bit balling." The agglomeration and compaction of clay-rich cuttings to the bit, junk slots of a bit, or bottom hole assembly surfaces are regarded as balling, and they deteriorate the overall performance [39,48]. Smith and his coworkers [39,40,41,42] deduced that when clay-rich shale is drilled with water-based mud, the main reason for low rate of

***Corresponding author**
Abdolhamid Sameni
Email: a.sameni@modares.ac.ir

penetration (ROP) is global balling. Therefore, whatever prevents or minimizes the global balling can improve and optimize the drilling performance.

Bit Balling Investigations Review

Numerous research studies have considered the framework and the effect of bit balling, and drilling ineffectivity has subsequently been taken into account when drilling a shale formation. The bit balling study has been divided into mechanical and chemical approaches [5,7,8,10,14,31,34,46, 47]; the mechanical explanations ascribe balling to two problems [3]: (I) the difficulty in preserving fluid between the cutter and the cuttings lead to differential sticking of the cuttings to the bit (cutter), and (II) the dilatancy in the shear zone of the cuttings causes a drop in pore pressure within the cuttings, which leads to differential sticking. In addition, the chemical approach can be explained via two strategies [3]: (I) the tendency of drilling fluid to wet the surface of the bit, and (II) sticking of the cuttings by virtue of swelling, because hydrophilic cuttings try to imbibe water. This water imbibition is due to [3]: (1) cohesion among cuttings, and (2) adhesion to bit surfaces.

Furthermore, the effect of other possible parameters on bit balling was also surveyed by exploring shale properties (such as cation exchange capacity (CEC)) [12,14,31], mud properties [5,7,8, 10,12,15,34,44,46,47,56], down-hole pressure [5, 7, 10, 12, 34, 46], and bit and cutter design [51, 56]. Other studies [1,3,6,8,17,21,22,26,27,29, 35,39,40,43,50,51,56] have been developed and proposed several normalized, dimensionless, and bit operating parameters for characterizing and diagnosing bit performance.

In this paper, in addition to the normalized rate of penetration (ROPn), other parameters such as specific energy (Es) and depth of cut (Dcut) are also used to represent drilling performance. The field data used in the correlation are obtained from the southern Iranian oil field drilling wells at the time of drilling.

EXPERIMENTAL PROCEDURES

The Normalized Rate of Penetration

For making a comparison among the combined data from various drilled intervals, normalizing the ROP has been suggested [25]. The normalization is implemented by the following model[15]:

$$ROP_n = \frac{ROP_{\text{actual}}}{\left(\frac{WOB}{d_b}\right)^2 \left(\frac{rpm}{60}\right)} \tag{1}$$

where, ROP_{actual}, ROP_n, *WOB*, d_b, and *rpm* are the actual rate of penetration, normalized rate of penetration, weight on bit, bit diameter, and rotary speed respectively.

Specific Energy

Pessier and Fear (1992) defined specific energy as the work done per unit volume of rock drilled. In the clean drilling, the specific energy decreases gradually, whereas in the balling region, the specific energy increases gradually. In general, a relatively low value implies efficient drilling and/ or weak rock, and a high value denotes ineffective drilling and/or strong rock [25]. The equation used by Pessier and Fear (992) to express specific energy is defined by:

$$E_s = \frac{WOB}{(Borehole\ Area)} + \frac{120\pi \times rpm \times Torque}{(Borehole\ Area) \times ROP} \tag{2}$$

where *Es*, *WOB*, rpm, *Torque*, and *ROP* are correspondingly specific energy, weight on bit, rotary speed, bottom hole torque, and the rate of penetration.

A balled or worn bit requires higher specific energy than a new and/or clean bit for drilling the same rock under identical conditions. The difference between total torque and torque off-bottom is used to estimate bottom hole torque and to subsequently calculate the specific energy for all bit runs[25].

Depth of Cut

Smith (1998) defined the depth of cut as a supplementary parameter for surveying drilling performance:

$$D_{cut} = \frac{ROP}{5 \times rpm} \quad (3)$$

where D_{cut}, *ROP*, and *rpm* represent the depth of cut, rate of penetration, and rotary speed respectively.

Cation Exchange Capacity

The cation exchange capacity (CEC) is a tool to describe the electrolytic cation absorbability of a clay-rich rock onto pore surfaces. CEC measurement in oil and gas studies is usually used for shaliness evaluation of sedimentary formations [52]. However, the CEC of rocks in bore hole cannot directly and continuously be measured. The measurement of CEC is subject to laboratory or empirical correlations developed from logs. Typically, the oil and gas industries measure the CEC with an API-recommended methylene blue capacity test [33], but by taking the advantage of the logs derived, using correlations is a beneficial way. In the Appendix, the corresponding calculation and relationship for empirical correlations are described.

Support Vector Machines

Support vector machines (SVM) on the basis of statistical learning theory and structural risk minimization were introduced by Vapink in 1955 [48]. The SVM was constructed to maximize the minimum distance between the data which leads to an optimum hyper-plane. Assuming the objective is $f(x) = \omega.x + b$, where the training set is given by $\{(x_i,y_i)\}_{i=1,2,..,l}$, $x_i \in \mathbf{R}$ is the input and $y_i \in \{-1,1\}$ is the output. For the linear separable case, SVM formulations appear to be:

$$\begin{Bmatrix} w^T x_i + b \geq +1 & if \;\; y_i = +1 \\ w^T x_i + b \geq +1 & if \;\; y_i = +1 \end{Bmatrix} \quad (4)$$

And for the non-separable case,

$$\begin{aligned} w^T \varnothing(x_i) + b \geq +1 \quad if \;\; y_i = +1 \\ w^T \varnothing(x_i) + b \geq +1 \quad if \;\; y_i = +1 \end{aligned} \quad (5)$$

where, $\varnothing(x)$ maps the input data into a higher dimensional feature space. Support vector machines for multi-classification have been developed in different supervised classification applications for various issues [4,13,16,19,24,30,32,54,55,57].

A nonlinear kernel function computes the hyper-plane with a maximum margin of $\frac{2}{\|w\|}$ between the classes. Standard SVM model utilizes the quadratic programming problem in the following equation:

$$\begin{aligned} & min \frac{1}{2} w^T w + C \sum_{i=1}^{l} \xi_i \\ & s.t \quad y_i \left(w^T \varnothing(x_i) + b \right) = 1 - \xi_i \, , \; \xi_i \geq 0, i = 1, \ldots, l \end{aligned} \quad (6)$$

where, *C* is the representative of the trade-off parameter.

By implementing Karush-Kuhn-Tucker (KKT) and using Lagrangian multipliers, α_i, the quadratic programming problem can be solved. Consequently, *w* can be obtained by using $w = \sum_{j=1}^{s} \alpha_{t,j} y_{t,j} \varnothing(x_{t,j})$ where, α_i is support vector (SV) used to specify decision boundary.

Let donate t_j (j=1, ..., s)j to the s support vectors. Then, one can rewrite:

$$w = \sum_{j=1}^{S} \alpha_{tj} y_{tj} \varnothing(x_{tj}) \tag{7}$$

By taking the dual problem into account, the quadratic programming problem would be solved.

$$max\, Q(\alpha) = -\frac{1}{2}\sum_{j=1}^{l} \alpha_i \alpha_i y_i y_j K(x_i, x_j) \tag{8}$$

$$s.t. \begin{Bmatrix} 0 \le \alpha \le C, \forall_i \\ \sum_{j=1}^{l} \alpha_i y_i = 0 \end{Bmatrix}$$

According to Mercer theorem, the kernel is similar to the following equation:

$$K(x_i, x_j) = \varnothing(x_i)^T \varnothing(x_j) \tag{9}$$

The general formulation for SVM is represented by:

$$y(x) = sign\left(\sum_{j=1}^{l} \alpha_i y_i K(x_i, x_j) + b\right) \tag{10}$$

Least Square Support Vector Machines

Suykens and Vandewalle proposed a modified version of SVM called least square SVM (LS-SVM) [45]. Like SVM, LS-SVM has application in both regression and classification cases. LS-SVM has reduced the run time and shown more adaptivity. Moreover, LS-SVM draws attention because it employs an equality constraint-based formulation instead of using quadratic programming methods. In LS-SVMs, the regression appears as:

$$min \frac{1}{2} w^T w + C\sum_{i=1}^{l} e_i^2 \tag{11}$$

$$s.t \quad y_i\left(w^T \varnothing(x) + b\right) = 1 - e_i, \quad i \ge 0, i = 1, \ldots, l$$

The Lagrangian equation is defined as follows:

$$L(w,b,e;\alpha) = \frac{1}{2} W^{T+1} + C\sum_{i=1}^{l} e_i^2 - \sum_{i=1}^{l} \alpha_i \left\{ y_i\left(w^T \varnothing(x) + b\right) - 1 + e_i \right\} \tag{12}$$

where, Lagrangian multipliers $\alpha_i \in \mathbf{R}$; if Lagrangian equation is differentiated with respect to w, b, α_i, and e_i, and the conditions are applied, one may obtain:

$$\begin{Bmatrix} \frac{dL}{dw} = 0 \rightarrow w = \sum_{i=1}^{l} \alpha_i y_i \varnothing(x_i) \\ \frac{dL}{db} = 0 \rightarrow \sum_{i=1}^{l} \alpha_i y_i = 0 \\ \frac{dL}{de_i} = 0 \rightarrow \alpha = Ce_i, \quad i = 1, \ldots. l \\ \frac{dL}{\alpha_i} = 0 \rightarrow y_i = y_i\left(w^T \varnothing(x) + b\right) - 1 + e_i, \quad i = 1, \ldots. l \end{Bmatrix} \tag{13}$$

Omitting e and w results in a KKT system:

$$\begin{bmatrix} 0 & Y^T \\ Y & \Omega + C^{-1} I \end{bmatrix} = \begin{bmatrix} 0 \\ 1 \end{bmatrix} \tag{14}$$

where, C is a positive constant, and b is the bias.

$$y = (y_1, \ldots, y_i)^T$$
$$\overset{\vee}{1} = (1, \ldots, 1)$$
$$\alpha = (\alpha_1, \ldots, \alpha_i)$$
$$\Omega_{ij} = y_i y_j \varnothing(x_i)^T \varnothing(x_j) = y_i y_j K(x_i, x_j), \qquad 1 \le i, j \le 1 \tag{15}$$

By applying the Mercer condition, the final result would be:

$$y(x) = sign\left(\sum_{j=1}^{l} \alpha_i y_i K(x_i, x_j) + b\right) \tag{16}$$

The most common kernel functions used in SVM are defined as follows [2,9]:

1. Linear kernel: $K(x_i, x_j) - x_i x_j$ (17)

2. Polynomial kernel:

$$K(x_i, x_j) = (x_i x_j + C)^d \quad for\, d \in \Theta, c \ge 0 \tag{18}$$

3. Gaussian (RBF) kernel:

$$K(x_i, x_j) = \exp\left(\frac{-\| x_i - x_j \|^2}{2\sigma^2}\right) \tag{19}$$

Coupled Simulated Annealing

Coupled simulated annealing (CSA) method was proposed by Xavier de Souza et al. in 2010 [53]. CSA features a new form of acceptance probability function that can be applied to an ensemble of optimizers. This approach considers several current states which are coupled together by their energies in their acceptance probability function. Moreover, as it is distinct from classical simulated annealing (SA) techniques [1,29], parallelism is an inherent characteristic of this class of methods. The objective of creating coupled acceptance probability functions that comprise the energy of many current states, or solutions, is to generate more information when deciding to accept less favorable solutions.

The following equation describes the acceptance probability function A with coupling term ρ:

$$A_\theta(\rho, x_i \to y_i) = \frac{exp\left(\frac{-E(y_i)}{T_k^\infty}\right)}{exp\left(\frac{-E(y_i)}{T_k^\infty}\right) + \rho} \tag{20}$$

where, $A_\vartheta(\rho, x_i \to y_i)$ is the acceptance probability for every $x_i \in \Theta$, $y_i \in \Upsilon$ and $\Upsilon \subset \Theta$. Υ denotes the set of all possible states, and the set $\Theta \equiv \{x_i\}_{i=1}^q$ is presented as the set of current states of q minimizers; moreover, T_k^{ac} is the acceptance temperature parameter at time instant k. The variance σ^2 of A_ϑ is equal to:

$$\sigma^2 = \frac{1}{q}\sum_{\forall x \in \Theta} A_\Theta{}^2 - \frac{1}{q^2} \tag{21}$$

The coupling term ρ is given by:

$$\rho = \sum_{x_j \in \Theta} exp\left(\frac{-E(y_i)}{T_k^\infty}\right) \tag{22}$$

Hence, CSA considers many current states in the set Θ, which is a subset of all possible solutions ϒ, and accepts a probing state y_i based not only on the corresponding current state x_i but by considering also the coupling term, which depends on the energy of all other elements of ϒ.

Performance Evaluation: Receiver Operating Characteristic (ROC)

The ROC curve [18,21] shows the separation abilities of a binary classifier; by setting different possible classifier thresholds, the data set is tested on misclassifications. If the plot has an area under the curve equal to 1 for the test data, a perfectly separating classifier is found (on that particular dataset); if the area is equal to 0.5, the classifier has no discriminative power at all.

The receiver operating characteristic (ROC) curve can be used to measure the performance of a classifier. ROC utilizes sensitivity and specificity analyses as given by:

$$\text{Sensitivity}(\%) = \frac{\text{TP}}{\text{TP} + \text{FN}} \times 100 \tag{23}$$

$$\text{Specificity}(\%) = \frac{\text{TN}}{\text{FP} + \text{TN}} \times 100 \tag{24}$$

where *TP*, *TN*, *FP*, and *FN* denote true positives, true negatives, false positives, and false negatives respectively.

True Positive (TP): An input is detected Effective as diagnosed by field observation, Effective.

True Negative (TN): An input is detected Ineffective as diagnosed by field observation, Ineffective.

False Positive (FP): An input is detected Effective, but it is experimentally labeled Ineffective.

False Negative (FN): An input is detected Ineffective, but it is experimentally labeled Effective.

Cation Exchange Capacity Measurement

The API-recommended methylene blue test (MBT) is carried out by providing one gram of finely ground dried shale. Then, the powdered shale is dispersed in water with a small amount of sulfuric acid and hydrogen peroxide (acting as a dispersant). Afterwards, it boils gently for a few minutes, and then it cools to reach room temperature. Subsequently, it is titrated by a methylene blue solution. When a drop of the sample suspension is placed on a filter paper, and the result is a faint blue halo surrounding the dyed solids, the end point is reached. Also, the CEC analyses can be performed at the well site with a minimum amount of equipment.

Results and Discussion

The data have been gathered from a shaly formation in one of the southern Iranian oilfields. The formations were drilled by a PDC bit using water-based mud. For evaluating and training the proposed model, the data set was divided into two categories. A randomly 12 data sets were selected for the testing stage, and the remaining data sets were used for training the diagnostic model. The method for measuring CEC is described in the preceding section. The input parameters to LS-SVM classifier include cation exchange capacity, normalized rate of penetration, depth of cut, and specific energy, whereas the output parameter represents the drilling effectivity/ ineffectivity. Furthermore, in the present study, it is tried to correlate CEC with the normalized rate of penetration, depth of cut, and specific energy in order to develop such classifications. It means that the effectivity/ineffectivity of drilling operations can be assessed by relating the aforementioned parameters.

In a separate study, the same dataset was exposed to the LS-SVR algorithm only (without the CSA algorithm), and the different parameters were optimized based on an exhaustive search. It was found out that it was not possible to reach the best solutions starting from arbitrary initial conditions. In particular, it is difficult to obtain the optimum choices of σ^2 and γ after starting with some discrete values. The solutions frequently become stuck in sub optimal local minima. These experiments justified the use of a hybrid technique for LS-SVR parameter tuning. Then, these data were exposed to the hybrid CSA-LSSVM model. Figure 2 depicts the flowchart of the developed CSA-based hyper parameter determination approach for SVM.

The *k*-fold method presented by Salzberg was applied to the experiments using a *k* value of 10 [38,20]. Thus, the dataset was split into 10 parts, with each part of the data sharing the same proportion of each class of data.

Due to fewer parameters to set and an excellent overall performance, the RBF is an effective option for kernel function [49,26]. Through an initial experiment without feature selection, the parameter values of the proposed CSA-LS-SVM approach were set as follows: T_0 and T_k^{ac} were set to 1.0, and the range of the parameters of RBF kernel is selected arbitrary: σ^2 = [0.001, 10^3] and C = [0.01, 10^4]. The optimal parameters for each pair of the inputs are listed in Table 1.

Table 1: Optimal values of categories optimized by CSA optimization technique.

	C	σ^2
D_{Cut}-CEC	1.6889	1.1384
1/[E_S]-CEC	2.5756	2.4853
$ROP_{Normalized}$-CEC	1.2435	1.7996

As shown in Figure 1, in the case of D_{cut} versus CEC, the classifier divided the plane area into two parts: effective and ineffective drilling region. One pattern represents minimally effective bit cleaning or reversible balling, and the other pattern represents ineffective drilling or irreversible balling.

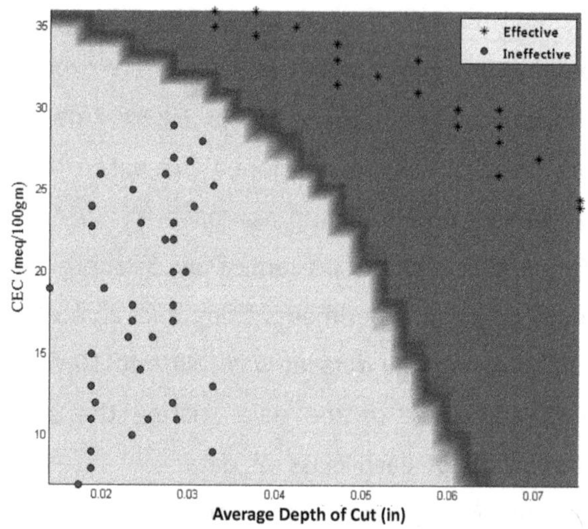

Figure 1: LSSVM model classify the drilling data into effective and ineffective for inputs of CEC-D_{cut}.

As D_{cut} and CEC increase simultaneously, effective drilling can be achieved. In other word, at low depth of cut, when the CEC is high, the drilling will be effective. Correlating CEC versus reciprocal of specific energy (Figure 2) shows a same trend as one described for D_{cut} vs. CEC. Since, specific energy is related to the strength of the rock being drilling and subsequently efficiency of the drilling operation, therefore a low value of specific energy implies an effective drilling operation, while high values shows an ineffective drilling process.

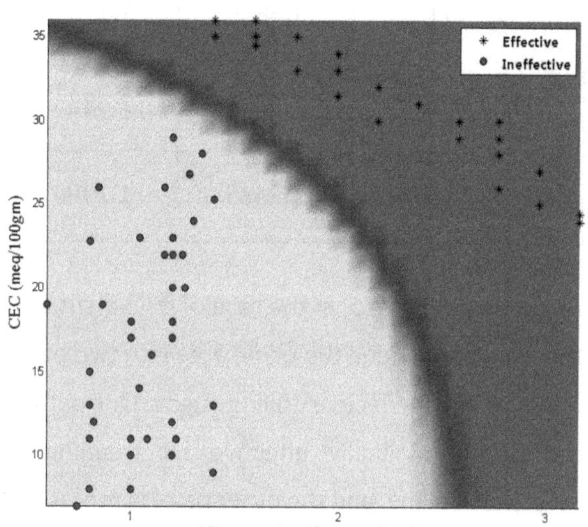

Figure 2: LS-SVM model classifies the drilling data into effective and ineffective for the inputs of CEC-1/Es.

As can be seen in Figure 3, the normalized rate of penetration was employed to distinguish effective bit cleaning from an ineffective drilling pattern.

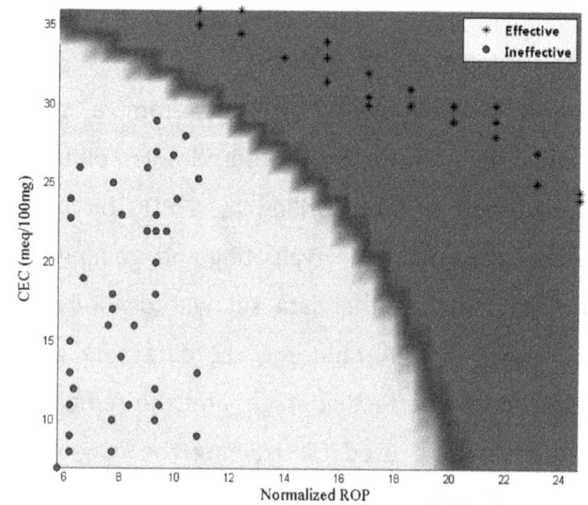

Figure 3: LS-SVM model classifies the drilling data into effective and ineffective for the inputs of CEC-ROP_n.

Figures 4 to 6 represent a general template detecting drilling ineffectivity. In addition, they can be used as a graphical correlation which utilizes $ROP_{normalized}$, D_{cut}, and the reciprocal of E_s as tools for determining the functionality of drilling.

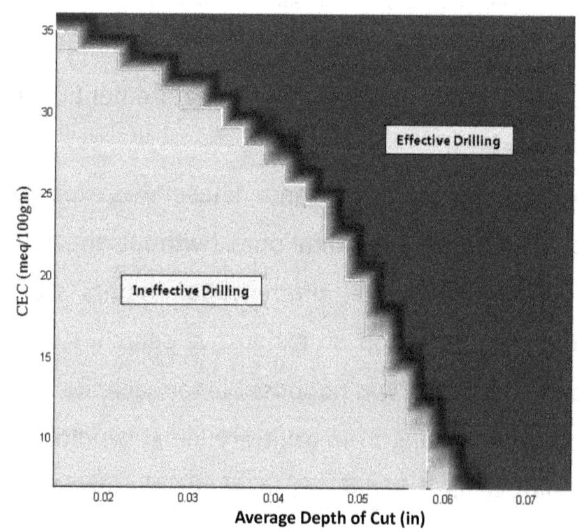

Figure 4: General template or graphical correlation for diagnosing drilling effectivity by inputs of CEC-D_{cut}.

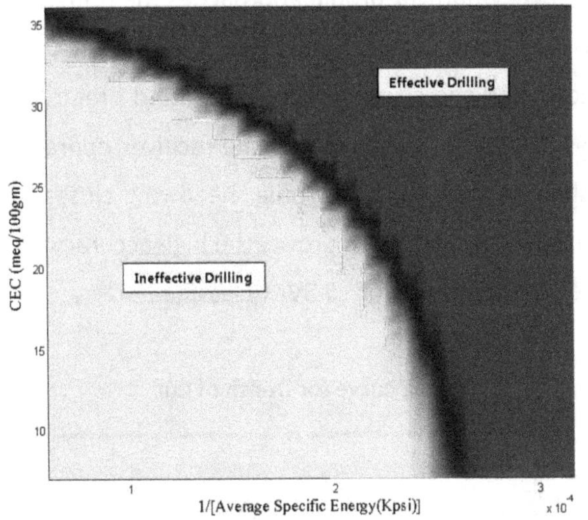

Figure 5: General template or graphical correlation for diagnosing drilling effectivity by inputs of CEC-1/Es.

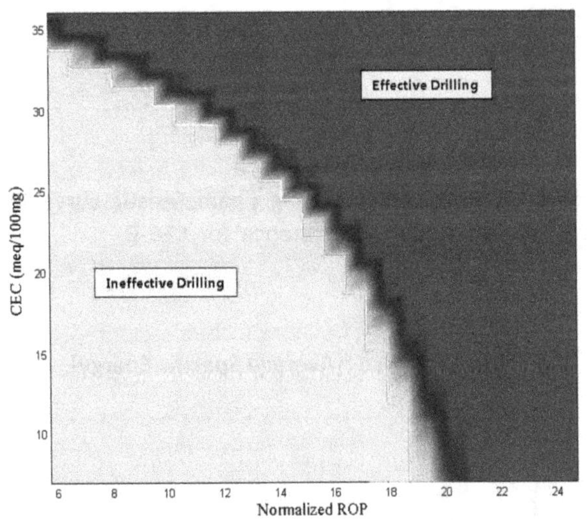

Figure 6: General template or graphical correlation for diagnosing drilling effectivity by inputs of CEC-ROP_n.

Table 2 performs a comparison between the observed and simulated drilling manner, where the inputs are D_{cut}- CEC. The predicted data show good agreement with the observed data with no false prognostication.

Table 2: Making a comparison between the observed and simulated drilling effectively for the inputs of D_{CUT}-CEC.

NO.	D_{Cut}	CEC	Drilling Observed	Drilling Simulated
1	0.037535	36	Effective	Effective
2	0.023459	11	Ineffective	Ineffective
3	0.023459	8	Ineffective	Ineffective
4	0.037535	35	Effective	Effective
5	0.05161	30	Effective	Effective
6	0.028151	10	Ineffective	Ineffective
7	0.029324	22	Ineffective	Ineffective
8	0.032843	35	Effective	Effective
9	0.042226	33	Effective	Effective
10	0.070377	25	Effective	Effective
11	0.030028	26.8	Ineffective	Ineffective
12	0.032843	9	Ineffective	Ineffective
13	0.05161	30.5	Effective	Effective

Also, Table 3 shows a comparison between the observed and simulated drilling effectively for the inputs of 1/[Es]-CEC.

Table 3: A comparison between the observed and simulated drilling effectively for the inputs of 1/[E_s]-CEC.

NO.	1/[E_s]	CEC	Drilling Observed	Drilling Simulated
1	0.000277	28	Effective	Effective
2	0.000238	33	Effective	Effective
3	0.00012	27	Ineffective	Ineffective
4	9.8E-05	16	Ineffective	Ineffective
5	0.000159	36	Effective	Effective
6	8.01E-05	24	Ineffective	Ineffective
7	8.01E-05	9	Ineffective	Ineffective
8	8.01E-05	9	Ineffective	Ineffective
9	0.000238	30	Effective	Effective
10	0.000218	32	Effective	Effective
11	0.000218	30.5	Effective	Effective
12	0.0001	25	Ineffective	Ineffective

Moreover, Table 4 shows a comparison between the observed and simulated drilling effectively for the inputs of 1/[Es]-CEC.

Table 4: A comparison between the observed and simulated drilling effectively for the inputs of $ROP_{Normalized}$-CEC.

NO.	$ROP_{Normalized}$	CEC	Drilling-Observed	Drilling-Simulated
1	21.58898	26	Effective	Effective
2	12.33656	35	Effective	Effective
3	12.33656	36	Effective	Effective
4	13.87863	35	Effective	Effective
5	18.50484	33	Effective	Effective
6	15.4207	34	Effective	Effective
7	9.25242	17	Ineffective	Ineffective
8	6.16828	15	Ineffective	Ineffective
9	4.62621	19	Ineffective	Ineffective
10	9.715041	20	Ineffective	Ineffective
11	10.79449	9	Ineffective	Ineffective
12	7.71035	11	Ineffective	Ineffective

Furthermore, this comparison has been performed for 1/[Es]-CEC and $ROP_{normalized}$. For both, the results elucidate the high powerful ability of LS-SVM classifier.

ROC curves are used to exhibit the performance and accuracy profile of classifications. The *x*-axis was set to (1-specificity), while *y*-axis was assigned to be sensitivity. As stated by Equations 19 and 20, one may obtain the following relations at the upper left-corner:

$$Sensitivity = \frac{TP}{TP+FN} = 1 \xrightarrow{Then} FN = 0 \tag{26}$$

$$1\text{-}Specificity = 1\text{-}\frac{TN}{FP+TN} = 0 \xrightarrow{Then} FP = 0 \tag{27}$$

Hence, whatever profile approaches the upper-left corner, false negative and false positive come near zero, and the classifier acts ideally and preferably. As displayed in Figures 7-9, the receiver operating characteristic (ROC) profile headway closes to upper-left corner and proves the high accuracy and reliable behavior of LS-SVM classifier.

ROC curve for Depth of Cut

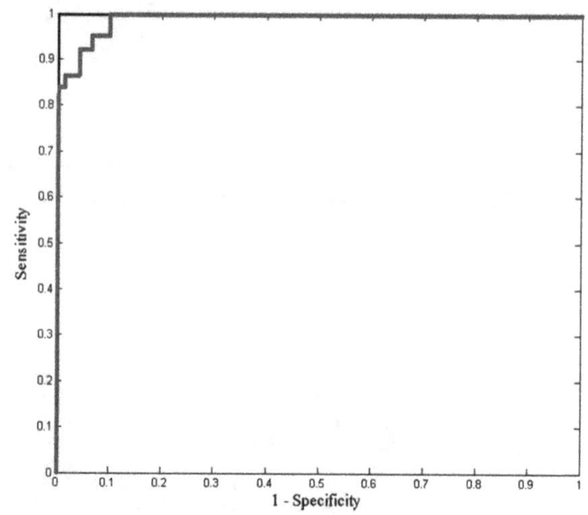

Figure 7: Receiver Operating Characteristic curve for analysis of model performance for CEC-D_{cut}.

ROC curve for 1/[Average Specific Energy]

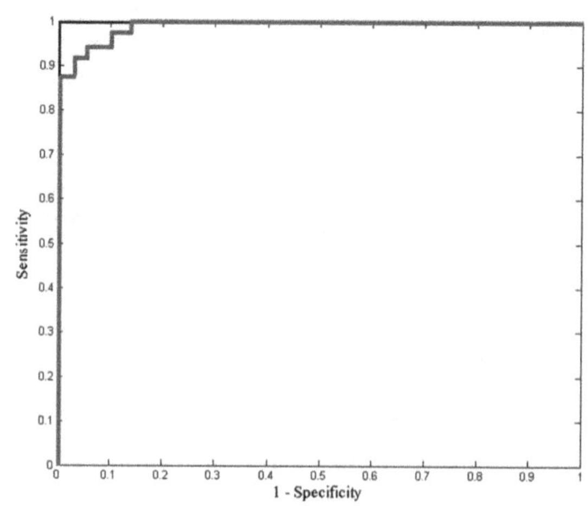

Figure 8: Receiver Operating Characteristic curve for analysis of model performance for CEC-1/Es.

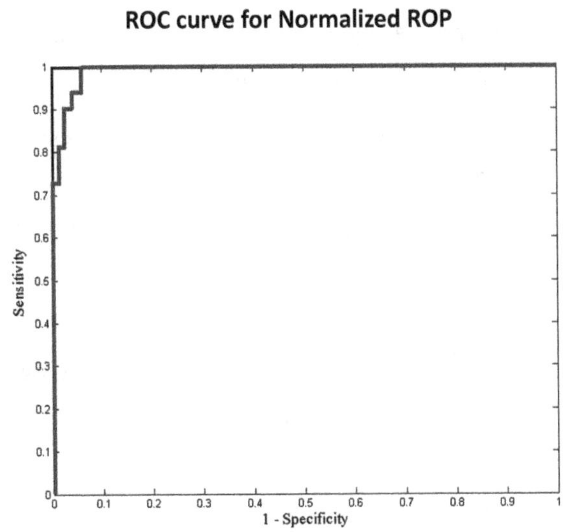

Figure 9: Receiver Operating Characteristic curve for analysis of model performance for CEC-ROP$_n$.

CONCLUSIONS

This study presents a CSA-based approach capable of searching for the optimal parameters of SVM to diagnose drilling ineffectivity. Four parameters (in three pairs) were employed as tools to develop graphical correlations. The data were collected from Iranian oilfields drilled by PDC bit with water-based mud. For evaluating CSA-LS-SVM model, receiver operating characteristic (ROC) was adapted to monitor the model behavior. In addition, a comparison of the obtained results with the observed results demonstrates that the proposed CSA-LS-SVM works functionally and practically. Furthermore, CSA-LS-SVM is an intelligent system with a self-adapting attribute which helps drilling industry to upgrade drilling performance.

APPENDIX

To calculate CEC using log data, R_{sh}, ρ_{ma}, φ_t, F_{sh}, and temperature (T) have to be estimated. First, one should calculate the counterion concentration, Q_v (meq/cc pore volume)as follows:

$$Q_v = \frac{F_{sh}}{R_{sh} B_{\max}} \tag{1}$$

B_{max} can be calculated using the following correlation derived from the experimental data collected by Waxman and Thomas (1975)[58].

$$B_{\max} = -0.0003T^2 + 0.205T - 0.0929 \tag{2}$$

where, T(°C) is the temperature of the zone of interest , and B_{max} (mho/m/(meq/cc)) is the maximum equivalent counterion conductance.

Q_v is related to cation exchange capacity by:

$$CEC = Q_v \frac{100\phi_t}{(1-\phi)\rho_{ma}} \tag{3}$$

where, φ_t ($\varphi_t = \varphi_e + \varphi_B$) is the total porosity, and φ_e is the effective porosity; φ_B is the fractional volume of water bound to shale, and ρ_{ma} (gr./cc) is the density of rock matrix containing shale, and CEC (meq/100 gr.) is the cation exchange capacity. Using the approximation adapted in the dual water model, the total porosity is calculated as an average of density and neutron porosity, i.e.:

$$\phi_t = \frac{\phi_{Neutron} + \phi_{Density}}{2} \tag{4}$$

The shale formation factor is then calculated by:

$$F_{sh} = \phi_t^{-2} \tag{5}$$

ρ_{ma} can be obtained from a core or cutting analysis if it is available; otherwise, the analyst will have to select a representative value. F_{sh} and R_{sh} are the shale formation resistivity factor and resistivity respectively.

NOMENCLATURES

CEC	: Cation Exchange Capacity
CSA	: Coupled Simulated Annealing
d_b	: Bit Diameter
Es	: Specific Energy (kpsi)
FN	: False Negative
FP	: False Positive
KKT	: Karush-Kuhn-Tucker
MBT	: Methylene Blue Test
ROC	: Receiver Operating Characteristic
ROP	: Rate of Penetration [L/T]
ROP_n	: Normalized Rate of Penetration
rpm	: Rotary Speed [L/T]
SVR	: Support Vector Regression
TN	: True Negative
TP	: True Positive
WOB	: Weight on Bit

REFERENCES

1. Aarts E. and Korst J., "Simulated Annealing and Boltzmann Machines," New York: John Wiley and Sons, **1989**, 415-417.
2. Abe S., "Support Vector Machines for Pattern Classification," New York: Springer-Verlag *Springer*, **2005**, 1-467.
3. Aghassi A., "Investigation of Qualitative Methods for Diagnosis of Poor Bit Performance Using Surface Drilling Parameters," MSc Dissertation, *Louisiana State University*, **2003**, 1-135.
4. Al-Anazi A. F. and Gates I. D., "Support Vector Regression for Porosity Prediction in a Heterogeneous Reservoir: a Comparative Study," *Computations and Geosciences*, **2010**, *36*, 1494–1503.
5. Allred R. B. and McCaleb S. B., "Rx for Gumbo Shale Drilling," SPE 4233, The Six[th] Conference on Drilling and Rock Mechanics of Society of Petroleum Engineers of AIME, Dallas, TX, **1973.**
6. Bible M., Lesage M., and Falconer I., "Method for Detecting Drilling Events from Measurement While Drilling Sensors," USA Patent 4876886, Anadrill Inc., **1989**.
7. Bland R., Haliday B., Illerhaus R., Isbell M., and et al., "Drilling Fluid and Bit Enhancement for Drilling Shales," AADE Annual Technical Forum-Improvements in Drilling Fluids Technology, Houston, Texas, **1999**.
8. Bourgoyne Jr., A. T., Chenevert M. E., Milheim K. K., and Young Jr. F.S., "Applied Drilling Engineering," SPE Text Book Series, Richardson, TX, **1991**.
9. Burges C. A. "Tutorial on Support Vector Machines for Pattern Recognition," *Data Mining Knowledge Discovery 2*, **1998**, 121-167.
10. Chesser B. G. and Perricone A. C., "A Physicochemical Approach to the Prevention of Balling of Gumbo Shale," SPE 4515, *48th Annual Fall Meeting of the Society of Petroleum Engineers of AIME*, Las Vegas, NV, **1973**.
11. Cheatham C. A. and Nahm J. J., "Bit Balling in Water-Reactive Shale During Full-Scale Drilling Rate Test," IADC/SPE 19926, IADC/SPE Conference, Houston, **1990**.
12. Cheatham C. A., Nahm J. J., and Heikamp N. D. "Effect of Selected Mud Properties on Rate of Penetration – Full-Scale Shale Drilling Simulations," SPE/IADC 13465, SPE/IADC Drilling Conference, New Orleans, LA. March, **1985**.
13. Chamkalani A., "Application of LS-SVM Classifier to Determine Stability State of Asphaltene in Oilfields by Utilizing SARA Fractions," *Petroleum Science and Technology*, **2011**, 6-11.

14. Copper G. A. and Roy S., "Prevention of Bit Balling by Electro-Osmosis," SPE 27882, SPE Western Regional Meeting, Long Beach, CA., **1994**.

15. Demircan G., Smith J. R., and Bassiouni Z., "Estimation of Cation Exchange Using Log Data: Application to Drilling Optimization, Society of Petrophysicists and Well Log Analysts (SPWLA)," *41st Annual Logging Symposium*, Dallas, Texas, June, **2000**.

16. Du C. J. and Sun D. W., "Multi-Classification of Pizza using Computer Vision and Support Vector Machine," *Journal of Food Engineering*, **2008**, *86*, 234-242.

17. Falconer I. G., Burgess T. M., and Sheppard M. C., "Separating Bit and Lithology Effects from Drilling Mechanics Data," IADC/SPE 17191, Drilling Conference, Dallas, TX, **1988**.

18. Fawcett T., "An Introduction to ROC Analysis," Pattern Recognition Letters 27, **2006**, 861-874.

19. Gao G. H., Zhang Y. Z., Zhu Y., and Duan G. H., "Hybrid Support Vector Machines-Based Multi-Fault Classification," *Journal of China University of Mining Technology*, **2007**, *17*, 246-250.

20. Han J. and Kamber M., "Data Mining: Concepts and Techniques," 3rd ed., Morgan Kaufmann, San Francisco, **2011**, 1-673.

21. Hanley J. A. and McNeil B. J., "The Meaning and Use of the Area Under a Receiver Operating Characteristic (ROC) Curve," *Radiology*, **1982**, *143*, 29-36.

22. Hood J. A., Leidland B.T., Haldorsen H., and Heisig G., "Aggressive Drilling Parameter Management Based on Down-hole Vibration Diagnostic Boosts Drilling Performance in Difficult Formation," SPE 71391, *SPE Annual Technical Conference and Exhibition*, New Orleans, L.A., **2001.**

23. P. A. Watson, "Drilling Fluids," SPE/IADC 21933, Drilling Conference, Amsterdam, Netherlands, **2001**.

24. Horng M. H., "Multi-class Support Vector Machine for Classification of the Ultrasonic Images of Supraspinatus," *Expert System Applications*, *36*, **2009**, 8124-8133.

25. Ipek G., Smith J. R., and Bassiouni Z., "Diagnosis of Ineffective Drilling Using Cation Exchange Capacity of Shaly Formations," *Journal of Canadian Petroleum Technology,* **2006**, *45*, 26-30.

26. Keerthi S. S. and Lin C. J., "Asymptotic Behaviors of Support Vector Machines with Gaussian Kernel," Neural Computation, **2003**, *15*, 1667-1689.

27. King C. H., Pinckard M. D., Krishnamoorthy K., and Benton D. F., "Method of and System for Optimizing Rate of Penetration in Drilling Operations," USA Patent 6155357, Noble Drilling Services Inc., **2000**.

28. King C. H., Pinckard M. D., Sparling D. P., and Weegh A. O. D., "Method of and System for Monitoring Drilling Parameters," USA Patent 6152246, Noble Drilling Services Inc., **2000**.

29. Kirkpatrick S., Gelatt Jr. C., and Vecchi M., "Optimization by Simulated Annealing," *Science,* **1983**, *220*, 671-680.

30. Lee Y. K. and Lee C. K., "Classification of Multiple Cancer Types by Multi-Category Support Vector Machines using Gene Expression Data," *Bioinformatics,* **2003**, *19*, 1132-1139.

31. Ledgerwood L. W. and Salisbury D. P., "Bit Balling and Well-Bore Instability of Down-Hole Shale," SPE 22578, *66th Annual Technical Conference and Exhibition of Society of*

Petroleum Engineers, Dallas, TX, October, **1991.**

32. Li D., Yang W., and Wang S., "Classification of Foreign Fibers in Cotton Lint using Machine Vision and Multi-class Support Vector Machine," *Journal of Computers and Electronics in Agriculture*, **2010**, *74*, 274-279.
33. Mike Stephens, Sandra Gomez-nava and Marc Churan, "Methylene Blue Test for Drill Solids and Commercial Bentonites," Section 12 in: API Recommended Practices 13I: Laboratory Testing of Drilling Fluids, 7th ed. and ISO 10416:2002, *American Petroleum Institute*, **2004**, 34-38.
34. O'Brien D. E. and Chenevert M. E., "Stabilizing Sensitive Shale with Inhibited, Potassium Base Drilling Fluids," *Journal of Petroleum Technology,* **1973**, *25*, 1089-1100.
35. Pessier R. C. and Fear M. J., "Quantifying Common Drilling Problems with Mechanical Specific Energy and a Bit-specific Coefficient of Sliding Friction," SPE 24584, *SPE Annual Technical Conference and Exhibition*, Washington D. C., **1992**.
36. Pinckard M. D., "Method of and System for Optimizing Rate of Penetration in Drilling Operations," USA Patent 6192998, Noble Drilling Services Inc., **2001**.
37. Roy S. and Cooper G. A., "Prevention of Bit Balling in Shale: Some Preliminary Results," IADC/SPE 23870, IADC/SPE Drilling Conference, New Orleans, LA., **1992**.
38. Salzberg S. L., "On Comparing Classifiers: Pitfalls to Avoid and a Recommended Approach," *Data Mining Knowledge Discovery Journal,* **1997**, *1*, 317-327.
39. Smith J. R., "Performance Analysis of Deep PDC Bits Runs in Water-Base Muds," ETCE 2000, ASME - Drilling Technology Symposium," Houston, Texas, **2000.**
40. Smith J. R., "Diagnosis of Poor PDC Bit Performance in Deep Shales," Dissertation, Louisiana State University, **1998**, 1-103.
41. Smith J. R., "Drilling Over-Pressured Shales with PDC Bits: A Study of Rock Characteristics and Field Experience Offshore Texas," PhD Thesis, Louisiana State University, **1995**.
42. Smith J. R. and Lund J. B., "Single Cutter Tests Demonstrate Cause of Poor PDC Bit Performance in Deep Shales," ETCE 2000, ASME -Drilling Technology Symposium, Houston, Texas, **2000.**
43. Smith J. R., "Addressing the Problem of PDC Bit Performance in Deep Shales," IADC/SPE 47814, IADC/SPE Asia Pacific Drilling Conference, Jakarta, Indonesia, **1998**.
44. Smith L., Mody F. K., Hale A., and Romslo N., "Successful Field Application of An Electro Negative 'Coating' to Reduce Bit Balling Tendencies in Water Base Mud," IADC/SPE 35110, IADC/SPE Drilling Conference, New Orleans, LA., March, **1996**.
45. Suykens J. A. K. and Vandewalle J., "Least Squares Support Vector Machine Classifiers," *Neural Processing Letter*, **1999**, *9*, 293-300.
46. Van Oort E., Friedheim J. E., and Toups B., "Drilling Faster with Water-Base Muds," AADE, AADE Annual Technical Forum-Improvements in Drilling Fluids Technology, Houston, Texas, **1999**.
47. Van Oort E., "On the Physical and Chemical Stability of Shales," *Journal of Petroleum Society Engineers,* **2003**, *38*, 213- 235.
48. Vapnik V. N., "The Nature of Statistical Learning

Theory," (2nd ed.), New York, Springer-Verlag, **1995**, 1-279.

49. Vapnik V. and Lerner A., "Pattern Recognition Using Generalized Portrait Method," *Automation Remote Control*, **1963**, *24*, 774-780.
50. Warren T. M. and Sinor L. A., "PDC Bits: What's Needed to Meet Tomorrow's Challenge," University of Tulsa, Centennial Petroleum Engineering Symposium, Tulsa, Oklahoma, SPE 27978, **1994**.
51. Warren T. M. and Armagost W. K., "Laboratory Drilling Performance of PDC Bits," SPE 15617, *61st Annual Technical Conference and Exhibition of Society of Petroleum Engineers*, New Orleans, LA., **1986**.
52. Waxman M. H. and Smits L. J. M., "Electrical Conductivities in Oil Bearing Shaly Sands," *Journal of Society of Petroleum Engineers*, **1968**, *8*, 107-122, .
53. Xavier S. S., Suykens J. A. K., Vandewalle J., and Bolle D., "Coupled Simulated Annealing," *IEEE Transaction on Systems Management Cybernetics B Cybernetics*, **2010**, *40*, 320-335.
54. Yang B. S., Hwang W. W., Kim D. J., and Tan A. C., "Condition Classification of Small Reciprocating Compressor for Refrigerators Using Artificial Neural Networks and Support Vector Machines," *Mechanical Systems and Signal Processing*, **2005**, *19*, 371-390.
55. Yao X. J., Panaye A., Doucet J. P., Chen H. F., and et al., "Comparative Classification Study of Toxicity Mechanisms Using Support Vector Machines and Radial Basis Function Neural Networks," *Anal. Chim. Acta. Computer Tech. Optimization.*, **2005**, *535*, 259-273.
56. Zijsling D. H. and Illerhaus R., "Eggbeater PDC Drill-Bit Design Concept Eliminates Balling in Water-Base," *Society of Petroleum Engineers*, **1991**, 11-14.
57. Zuo R. and Carranza E. J. M., "Support Vector Machine: A Tool for Mapping Mineral Prospectivity," *Computation and Geoscience*, **2011**, *37*, 1967-1975.
58. Waxman M. N. and Thomas C. E., "Electrical Conductivities in Oil-bearing Shaly Sand, The Relation between Hydrocarbon Saturation and Resistivity Index," The tem- perature coefficient of electrical conductivity: *Society of Petroleum Engineers Journal*, **1974**, *14*, 213–225.

Pervaporation (PV) Separation of Methanol/Methyl Tert-butyl Ether Mixtures in Low Permeate Pressure Conditions

Ghasem Khatinzadeh*, Mohammad Mahdyarfar, and Ali Mehdizadeh

Chemical Polymeric and Petrochemical Technology Development Research Division, Research Institute of Petroleum Industry (RIPI), Tehran, Iran

ABSTRACT

Recently, pervaporation separation processes have gained much attention in the separation of azeotropic and close boiling point organic-organic mixtures due to its high separation efficiency, economy, safety, and energy saving potentials. In this work, the effects of experimental factors such as feed composition and operating temperature on the performance of a commercial poly vinyl alcohol membrane in the separation of methanol/methyl tertiary butyl ether (MTBE) mixtures at low permeate side pressures (1-20 mmHg) were evaluated. Separation factor and methanol flux significantly increased by decreasing permeate side pressure, especially to less than 5 mmHg. Therefore, the reduction of pressure from 20 to 1 mmHg at 45 °C at a feed methanol concentration of 5 wt.% increased methanol flux and separation factor from 248 to 412 g/hm^2 and 73 to 211 respectively. In addition, the results indicate that by increasing feed temperature and methanol concentration in the feed, methanol flux is increased, while separation factor dropped.

Keywords: Pervaporation, Separation, Permeate Pressure, MTBE, Methanol

INTRODUCTION

In recent years, methyl tertiary butyl ether has been extensively used as a lead free octane enhancer [1]. It has been shown that the addition of 10-15 wt.% MTBE to gasoline increases the octane number by 2-5 [2]. MTBE is also used as a reagent in fine chemical production. It is manufactured by reacting isobutene with excess methanol over a sulfonic ion exchange resin. The excess methanol then has to be separated from the products by distillation. The operation requires a high capital expenditure and is not energy efficient because of the formation of methanol and MTBE azeotrope [1].

Pervaporation (PV), which is a more energy efficient and cheaper process, is one of the most developing membrane technologies, which has attracted much attention in the separation of azeotropic organic-organic mixtures [3-4]. PV is derived from combining permeation and evaporation, the two mechanisms involved in the process. In this process, the components of a liquid feed permeate through a dense membrane and evaporate into downstream

***Corresponding author**
Ghasem Khatinzadeh
Email: khatinzadehgh@ripi.ir

at different rates. The difference in vapor pressure between the feed liquid and permeate vapor is the driving force for the process [5].

Considering the similar nature of methanol and water, hydrophilic membranes are used for the separation of methanol from MTBE. On this basis, extensive research has been performed to find an optimized membrane material to maximize separation performance in terms of separation factor, flux, and stability [6]. Membrane materials studied included poly vinyl alcohol (PVA) [7], poly vinyl alcohol/cellulose acetate [8], cellulose acetate-poly(N-vinyl-2-pyrrolidone) [9], and poly (ether ether ketone) [10]. The characterization of pervaporation process is usually considered with regards to the changes in different operational variables such as feed composition, temperature, and permeate pressure. Although much research has been conducted in this area [11-12], the studies have mostly been performed at low vacuum, and the effect of pressure on the separation of methanol from MTBE at high vacuum has not considerably been investigated. Permeate pressure is an important operating parameter as high vacuum corresponds to high energy cost. Thus, for the optimization of PV, the investigation of the effects of operating parameters on membrane performance at low permeate pressures seems necessary. In the present work, the separation of methanol-MTBE mixtures by pervaporation using a commercial membrane has been dealt with.

EXPERIMENTAL PROCEDURES

In the pervaporation experiments carried out, the feed mixture was circulated from the feed tank in a temperature range of 20-45 °C and in a permeate pressure range of 1-20 mmHg through the cell using a peristaltic pump at a flow rate of 1.2 L/min. The volume of feed tank was 7 L, which was very big compared with the permeation volume. Therefore, the variation of the feed concentration during the experimental period was negligible. The permeate was condensed and collected in cold traps made of Pyrex glass, immersed in liquid nitrogen. Before starting the PV experiments, the test membrane was equilibrated for about 4 hrs with feed solutions. Figure 1 shows the experimental set-up used.

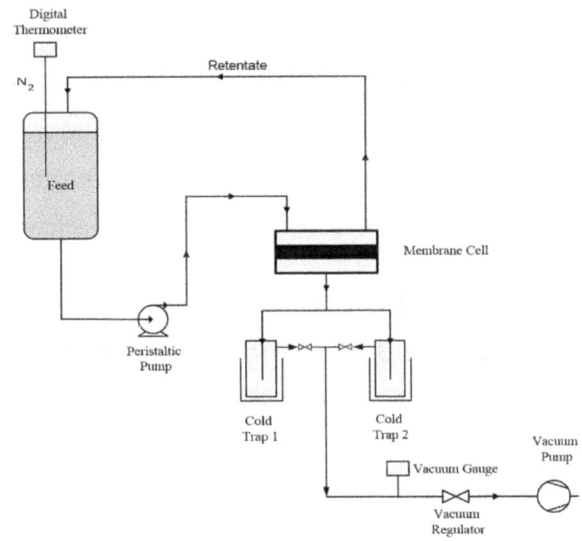

Figure 1: A schematic diagram of experimental apparatus.

In all the experiments, the commercially available PERVAP 2256 membrane from Sulzer Chemtech consisting of PVA active layer (thickness=2 μm) and polyacrylonitrile (PAN) (thickness=80 μm) support has been used.

The membrane module used in this study was of plate and frame type, which was made of stainless steel and supported on a porous disc with an effective membrane area of 177 cm^2. The feed side of the membrane was at atmospheric pressure and permeate side was maintained in a desired pressure range using a vacuum pump. The PV experiments were repeated three times to check for repeatability. Repeatability is expressed as relative standard deviation, ranged between 2 and 4.5%. Permeate fluxes (g/h m^2) were gravimetrically determined at different temperatures and pressures using Equation 1:

$$J = \frac{M}{A.t} \quad (1)$$

where, J is the permeate flux, and M is permeate mass (g); A (m^2) represents the membrane area and t (h) is operation time. Separation factor was calculated using Equation 2:

$$\alpha = \frac{Y_{Me} / Y_M}{X_{Me} / X_M} \quad (2)$$

where, α is separation factor, and *X* and *Y* are the concentrations of components in feed and permeate respectively. Subscripts Me and M represent methanol and MTBE respectively. The analyses of samples were performed using a Varian 3800 gas chromatograph equipped with an FID detector and a capillary column.

RESULTS AND DISCUSSION

Effect of Feed Methanol Concentration

The methanol flux and pervaporation selectivity values are presented as functions of methanol concentration in the feed in Figures 2 and 3. As observed, a rise in methanol concentration from 1 to 10 wt.% in a pressure range of 3-20 mmHg, causes a considerable increase in methanol flux, for example, from 51 to 670 g\hm^2 at 3 mmHg, while separation factor (membrane selectivity to methanol) decreases from 254 to 108 at the same pressure. Feed composition influences the pervaporation performance of a membrane by varying the membrane microstructure, which in turn results in different sorption and diffusion behaviors. Because of a strong interaction between methanol and the membrane, by increasing methanol concentration in the feed, the membrane becomes more swollen. As a result, polymer chains become more flexible and there will be more free volume leading to higher methanol flux [7]. In addition, decreasing separation factor by increasing feed methanol concentration due to the interaction forces of permeating molecules increases the concentration of MTBE molecules compared with methanol in permeate [13].

In any events, as observed in Figures 2-3, within this concentration range, pressure variations in the range of 3-10 mmHg are more effective on flux and selectivity, and thus on membrane performance, than the corresponding variations in the range of 10-20 mmHg.

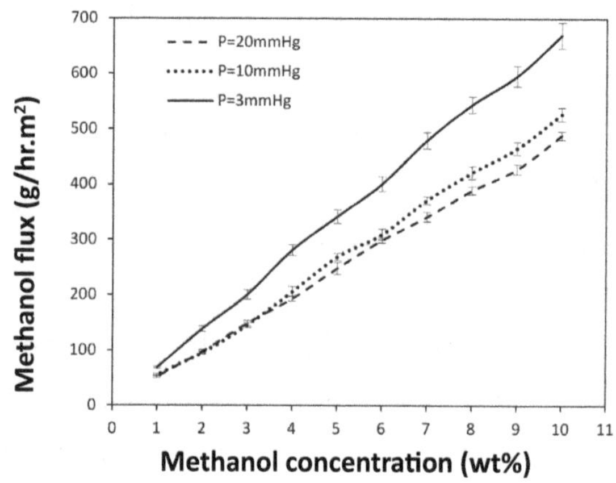

Figure 2: Methanol flux as a function of methanol concentration in the feed (MTBE-methanol mixture) (Temperature: 45°C, Feed flow rate: 1200 mL/min).

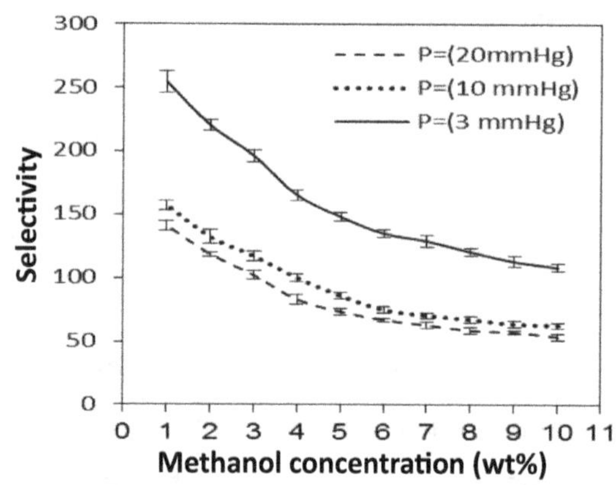

Figure 3: Selectivity as a function of methanol concentration in the feed (MTBE-methanol mixture) (Temperature: 45°C, Feed flow rate: 1200 mL/min).

EFFECT OF PERMEATE PRESSURE

Figure 4 shows the effect of permeate pressure on methanol flux at 45 °C and at feed compositions of 1 and 5 wt.% of methanol. By lowering permeate pressure, flux remarkably increases. The main contribution to the driving force is the activity gradient of the components in the membrane. Since the permeate pressure is directly related to the activity of the components at the permeate side of the membrane, the permeate pressure has a strong influence on the pervaporation performance. Reducing permeate pressure increases the driving force on the diffusing molecules in the membrane and will result in the component being swept out rapidly from downstream side and consequently in higher permeate flux [14].

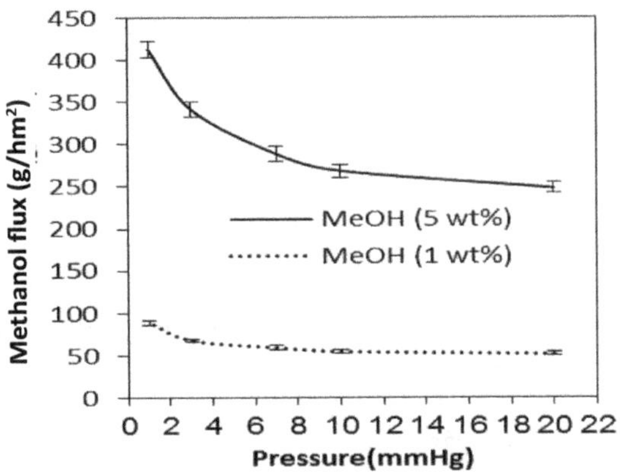

Figure 4: Effect of permeate pressure on methanol flux (Temperature: 45°C, Feed flow rate: 1200 mL/min).

In addition, as shown in Figure 5, the separation factor considerably drops at higher permeate pressures, especially at permeate pressures of over 10 mmHg. In fact, diffusion through the membrane is a rate-determining step at reduced pressures in PV. The selectivity of the membrane to methanol indicates the larger driving force and desorption rate of methanol molecules compared with MTBE. Increasing permeate pressure lowers the driving force and results in slowing down the desorption rates of the sorbed molecules. In this case, vapor pressures of the two components of the feed mixture govern the membrane selectivity [15]. Therefore, as shown in Figure 5, since methyl tertiary butyl ether has a higher vapor pressure compared with methanol, increasing pressure reduces membrane selectivity to methanol.

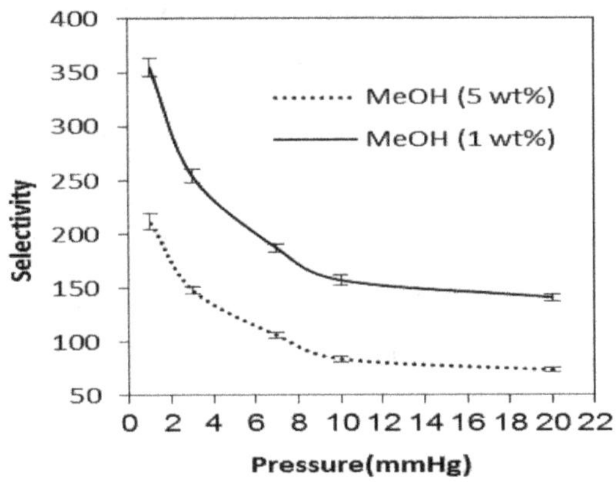

Figure 5: Effect of permeate pressure on membrane selectivity (Temperature: 45°C, Feed flow rate: 1200 mL/min).

The considerable point is the remarkable changes in flux and selectivity in pressures below 10 mmHg, especially those below 5 mmHg. Peivasti et al. [12] have investigated the effect of permeate pressure on the selectivity and flux of PVA membrane in the pressure range of 15-35 mbar in the separation of 10-30 wt.% methanol-MTBE mixtures. The results indicate slight changes in the membrane performance under conditions such as this pressure range, and the greatest variations occur in selectivity and flux from 52 to 60 and 650 to 680 g/h m^2 respectively. However, as Figures 4-5 show, pressure change from 10 to 1 mmHg strongly affects flux and selectivity, thereby improving membrane performance more effectively.

Effect of Feed Temperature

The effects of the operating temperature ranging from 20 to 45 °C on the permeation flux and separation factor are presented in Figures 6-7. It is evident that permeation flux increases, while the separation factor decreases when the temperature rises. Mass fluxes through the membrane are highly dependent on temperature. They increase exponentially according to empirical Arrhenius law [16]. This may result from the rise in diffusion coefficient of the solute as the temperature increases. According to solution-diffusion mechanism, increased temperature causes the solubility on the surface of membrane and diffusion rate in the membrane to increase. Furthermore, the increase of temperature produces a higher vapor pressure of the permeating components on the feed side of membrane, while the pressure on the downstream side remains unchanged. This situation enhances the driving force, leading to a rise in permeation flux.

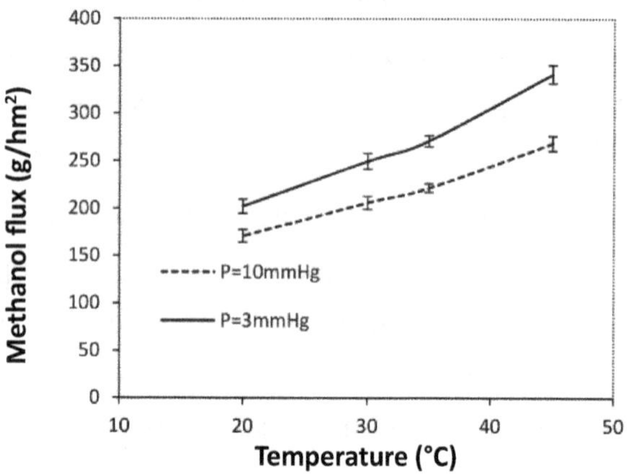

Figure 6: Effect of feed temperature on methanol flux (Feed methanol concentration: 5 wt.%, Feed flow rate: 1200 mL/min).

Additionally, decreasing membrane selectivity to methanol by increasing temperature shows that this parameter affects the diffusion of MTBE molecules into membrane more in comparison with methanol molecules. In other words, the transport of MTBE molecules through the membrane is more sensitive to the operating temperature and becomes easier at higher temperatures. Therefore, the separation factor decreases at an increased temperature.

However, as the results show, the reduction of pressure below 10 mmHg within a wide temperature range considerably improves membrane performance (as previously observed for concentration in the effect of feed methanol concentration), thereby reducing the required membrane area. Thus, given that the membrane is usually the most expensive component in the PV process, PV separation should be run at low permeate pressures in order to minimize the membrane area. As a result, considering the high expense of the process at low pressures, the selection of optimal conditions requires mass, energy, and equipment cost calculation [17].

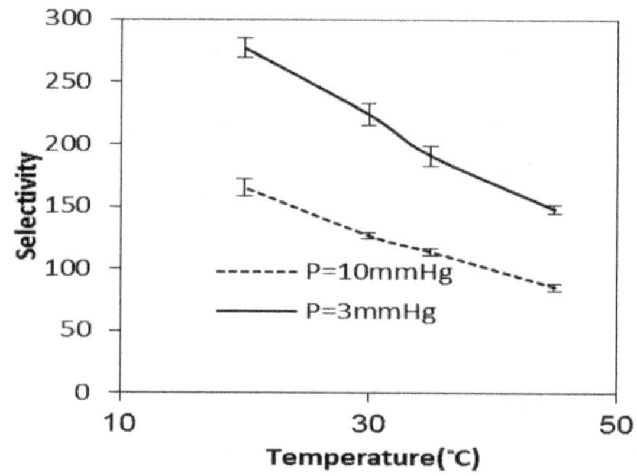

Figure 7: Effect of feed temperature on membrane selectivity (Feed methanol concentration: 5 wt.%, Feed flow rate: 1200 mL/min).

CONCLUSIONS

Increasing methanol concentration in the feed enhances membrane swelling, which results in a rise in flux, but a reduction in selectivity. Both flux and selectivity increase by reducing permeate pressure due to increased driving force and higher volatility of MTBE compared with methanol. In addition, increasing feed temperature causes the enlargement of free volume inside the membrane resulting in a considerably higher flux and a reduction in selectivity. Decreasing pressure effectively increases flux and selectivity in a wide temperature range anyway.

Overall, by decreasing permeate pressure, the membrane performance improves for the separation of methanol-MTBE mixture resulting in a reduced membrane area, especially at permeate pressures of less than 5 mmHg. Low permeate pressures mean high energy consumption; consequently, the key step in running a PV system is the determination of the optimum pressure to establish a balance between membrane area and required energy consumption to achieve the minimum expenses for the scale up of the pervaporation system.

The experimental data obtained in this work can be used to obtain simulation equations, which will be useful in the simulations leading to the design of a scale-up unit. Obtaining supplementary data together with mass, energy, and equipment cost in order to calculate the total separation cost are the future objectives of this group.

REFERENCES

1. Shah V. M., Bartels C. R., Pasternak M., and Reale J., "Opportunities for Membranes in the Production of Octane Enhancers," *AIChE Symp. Ser*, **1989**, *85*, 272.
2. Mohammadi A. T., Villaluenga J. P. G., Kim H. J., Chan T. et al., "Effects of Polymer Solvents on the Performance of Cellulose Acetate Membranes in Methanol/Methyl Tertiary Butyl Ether Separation," *Journal of Applied Polymer Science*, **2001**, *82*, 2882-2895.
3. Garg P., Singh R. P., and Choudhary V., "Pervaporation Separation of Organic Azeotrope using Poly(Dimethyl Siloxane)/Clay Nanocomposite Membranes," *Separation and Purification Technology*, **2011**, *3*, 435-444.
4. Park H. C., Ramaker N. E., Mulder M. H. V., and Smolders C. A., "Separation of MTBE-Methanol Mixtures by Pervaporation," *Separation Science and Technology*, **1995**, *30*, 419-433.
5. Neel J., "Introduction to Pervaporation, In: R. Y. M. Huang (Ed.)," *Pervaporation Membrane Separation Processes*, **1991**, 36-37.
6. Tamaddondar M., Pahlavanzadeh H., Hosseini S. S., Ruan G. et al., "Self-assembled Polyelectrolyte Surfactant Nanocomposite Membranes for Pervaporation Separation of MeOH/MTBE," *Journal of Membrane Science*, **2014**, *472*, 91–101.
7. Hilmioglu N. D. and Tulbentci S., "Pervaporation of MTBE/Methanol Mixtures through PVA Membranes," *Desalination*, **2004**, *3*, 263-270.
8. Zhou K., Zhang Q. G, Han G.L., Zhu A. M. et al., "Pervaporation of Water–Ethanol and Methanol–MTBE Mixtures Using Poly (Vinyl Alcohol)/ Cellulose Acetate Blended Membranes," *Journal of Membrane Science*, **2013**, *448*, 93-101.
9. Wu H., Fang X., Zhang X., Jiang Z. et al., "Cellulose Acetate–Poly (N-Vinyl-2-Pyrrolidone) Blend Membrane for Pervaporation Separation of Methanol/MTBE Mixtures," *Separation and Purification Technology*, **2008**, *64*, 183-191.
10. Zereshki S., Figoli A., Madaeni S. S., Simone S. et al., "Pervaporation Separation of MeOH/MTBE Mixtures with Modified PEEK Membrane: Effect of Operating Conditions," *Journal of Membrane Science*, **2011**, *371*, 1-9.
11. Ray S. and Ray S.K., "Synthesis of Highly Methanol Selective Membranes for Separation of Methyl Tertiary Butyl Ether (MTBE)–Methanol Mixtures by Pervaporation," *Journal of Membrane Science*, **2006**, *278*, 279-289.
12. Peivasti M., Madandar A., and Mohammadi T., "Effect of Operating Conditions on Pervaporation of Methanol/Methyl Tert-butyl Ether Mixtures," *Chemical Engineering and Processing: Process Intensification*, **2008**, *47*, 1069-1074.
13. Matsui S. and Paul D. R., "Pervaporation Separation of Aromatic/Aliphatic Hydrocarbons by a Series of Ionically Crosslinked Poly(N-Alkyl Acrylate) Membranes," *Journal of Membrane Science*, **2003**, *213*, 67-83
14. Duggal A. and Thompson E. V., "Dependence of Diffusive Permeation Rates and Selectivities on Upstream and Downstream Pressures: VI. Experimental Results for the Water/Ethanol System," *Journal of Membrane Science*, **1986**, *27*, 13-30.
15. Neel J., Nguyen Q. T., Clement R., and Lin D. J., "Influence of Downstream Pressure on the Pervaporation of Water-Tetrahydrofuran Mixtures through a Regenerated Cellulose Membrane (Cuprophan)," *Journal of Membrane Science*, **1986**, *27*, 217-232.
16. Cabasso I., Grodzinski J. J., and Vofsi D., "Polymeric Alloys of Polyphosphonates and Acetyl Cellulose: Sorption and Diffusion of Benzene and Cyclohexane," *Journal of Applied Polymer Science*, **1974**, *18*, 2117-2136.
17. Lin L., Kong Y., Yang J., and Shi D., "Scale-Up of Pervaporation for Gasoline Desulphurization Part 1. Simulation and Design," *Journal of Membrane Science*, **2007**, *298*, 1-13.

17

An Experimental Study of the Effect of Amines on Polymer Efficiency and Thermal Stability of Water-based Drilling Fluid

Khalil Shahbazi[1]*, Alireza Arshadi[1], and Majid Valizadeh[2]

[1] Department of Petroleum Engineering, Ahwaz Faculty of Petroleum Engineering, Petroleum University of Technology, Ahwaz, Iran
[2] Petroleum Engineering Research Division, Research Institute of Petroleum Industry (RIPI), Tehran, Iran

ABSTRACT

Amine compounds are believed to have acceptable results regarding their use as clay swelling inhibitor and corrosion inhibitor additives in drilling fluids. It is crucial to know the capability of amine compounds to enhance the thermal stability of drilling fluid; herein, a water-based fluid composed of biopolymer and other additives as the proper representatives of muds used in the pay zone section of a well is used. In order to enhance drilling fluid thermal stability to 250 °F (which is about 200 °F for starch in common drilling fluids), the compatibility of four amine compounds (mono ethanolamine, 1,6-diaminohexane, polyamine, and choline chloride) with three polymers (starch-green, poly-anionic cellulose, and starch-high temperature) is investigated; one of the named polymers is also chosen to examine its thermal stability performance along with the mentioned amine compounds at 250 °F. Two percent by volume concentration (2 Vol.%) of the mentioned amines were used. For these purposes, the rheological and filter loss properties were studied. The results showed that the starch-green is almost compatible with four amines. Moreover, mono ethanolamine and 1,6-diaminohexane acted better in terms of thermal stability at elevated temperatures.

Keywords: Amine, Polymer, Swelling Inhibitor, Thermal Stability

INTRODUCTION

In drilling operations, drilling fluid is one of the main contributors to the overall cost. What makes this component crucial is its direct contact with wellbore formation and almost all the equipment from mud pit to drill bit. It is also considered as a conduit for making changes in the bottom hole condition and for transferring helpful data during drilling operations. However, the mentioned tasks are achievable only if drilling fluid sustains its properties when it undergoes different conditions during drilling such as high temperature and troublesome shales.

Shale makes up 75% of drilled formations and causes 90% of wellbore instabilities [1]. Since the beginning of the drilling industry, people had been mainly focused on how to drill hydratable shales without facing problems such as stuck pipe, hole enlargement and washouts, tight hole, accumulation of fill on the hole bottom after trips,

***Corresponding author**
Khalil Shahbazi
Email: shahbazi@put.ac.ir

solids build up in mud, and poor cementing jobs. These cover 90% of wellbore instability problems and cost annually $500 million [2]. To solve the drilling problems associated with shaly formations, various nonaqueous drilling fluids (NADF) have been used in fields by the operators. In addition to shale inhibition, suitable lubricity and temperature stability were seen by using these systems [3,4]. However, these advantages are realized as the ultimate goal in High Performance Water Base Mud (HPWBM) research. NADF has disadvantages such as high cost, environmental limitations, disposal problems, health, and safety issues [5,6]. High performance (HP) water-based drilling fluids are particularly advantageous in comparison with conventional water-based systems because they provide faster penetration rates, enhanced hole cleaning, greater shale inhibition, and improved wellbore stability [7]. Generally, High Pressure High Temperature (HPHT) wells have been drilled via oil-based muds, formate brines, or high temperature synthetic polymers. These conventional systems can overcome HPHT circumstances through using: 1: Oil-based mud or synthetic muds, which can be formulated to withstand high temperatures over long periods of time because of their high thermal stability; 2: formate brines, which can protect polymers from thermal degradation; 3: synthetic polymers, which are modified chemically in a way to enhance their thermal stability in order to postpone the degradation time [8].

Amine-based drilling fluids, which are considered as one of the examples of high performance water-based drilling fluids (substitution for oil-based muds), are believed to have acceptable applications mainly as clay swelling inhibitor additives. Clay swelling inhibition, a paramount challenge is to overcome low thermal stability when amine compounds in HPWBM are used. Today, with the drilling of the deeper wells, we encounter higher bottom hole pressures and temperatures. Thus, we need mud additives with high thermal stability for the accomplishment of such deep drilling operations.

It is worth mentioning that the component of drilling fluid which is highly vulnerable to high temperature is polymer. Starch is one of the most common natural polymers used as a fluid loss controller additive in drilling fluids. This additive is cost-effective and affluent [9].

Starch is the principal component of the seeds of cereal grains (such as corn, wheat, and rice) and tubers (such as potato and tapioca). Starch is degraded by heat and agitation. With continued circulation in a hole at temperatures of 200 °F (93 °C) and higher, starch breaks down rapidly and loses the sealing action of the filter cake [10].

In addition to all the applications of amine compounds as drilling fluid additives, it can also enhance the thermal stability of mud. The addition of poly-ethoxylated alkyl diamine compound can provide thermal stability at 200 °F [11]. Therefore, the thermal stability of starch should be increased for applications at higher temperatures considered in this study, i.e. 250 °F.

EXPERIMENTAL PROCEDURES

Experimental Equipment and Materials

Four various amine compounds were used; monoethanol amine and choline chloride belong to mono cationic category. Also, 1, 6-diaminohexane belongs to oligomeric group, and the other one belongs to polyamine. All the amine compounds were added to mud at 2% by volume concentration. Three different polymers (starch-green, low viscosity polyanionic cellulose (PAC-LV), and modified starch) were used as the base fluids. The rheological properties of the fluid were measured directly using a rotational viscometer (Fann 35 SA model), and the filtration properties of the fluids were measured by filter press (Fann model). A rolling oven (Fann model) was used for the dynamic heating of the fluids at 250 °F for 16 hours.

METHODOLOGY

In order to find the effects of each amine on the thermal stability of drilling fluid, firstly, it is intended to investigate the compatibility of the amine compounds with the three mentioned polymers. Fifteen samples were prepared for this purpose the detailed compositions of which are tabulated in Tables 1-5. Three of these samples are base fluids (from three polymers) which contain no amine compounds, and the other ones contain only one amine compound at 2 vol.%. Thus, all the samples contain the same amount of an additive like limestone, xanthan gum, and soda ash.

Rheological properties were measured at ambient temperature for each sample after hot rolling. Rheological parameters (*Av, Pv,* and *YP*) were calculated based on Bingham plastic model using the following equations (Equations 1-3):

$$Av = \frac{\theta_{600}}{2} \quad (1)$$

$$Pv = \theta_{600} - \theta_{300} \quad (2)$$

$$YP = \theta_{300} - Pv \quad (3)$$

Filtration tests were run at ambient temperature and at the pressure of 100 psig with a multifilter press, and the filtrate volume collected in 30 minutes was reported as a filter loss of fluid.

As it was mentioned before, in this study, the effect of amine compounds on the compatibility and thermal stability of drilling fluid is presented. The rheological properties and filtrate volume were measured for each sample both before and after hot rolling for 16 hours at 250 °F. It should be noted that all the amine additives were used at two percent by volume concentration (2 vol.%).

Table 1: Base samples containing no amine.

Component \ Sample No.	1	2	3
Saturated salt water (SSW) (cc)	350	350	350
Soda ash (g)	1	1	1
Starch G (g)	8	-	-
PAC-LV (g)	-	8	-
Starch HT (g)	-	-	8
XC polymer (g)	1	1	1
Limestone(LSP)(g)	15	15	15

Table 2: Samples containing MEA.

Component \ Sample No.	4	5	6
Saturated salt water (SSW) (cc)	350	350	350
Soda ash (g)	1	1	1
Starch G (g)	8	-	-
PAC-LV (g)	-	8	-
Starch HT (g)	-	-	8
XC polymer (g)	1	1	1
Limestone(LSP)(g)	15	15	15
MEA (%)	2	2	2

Table 3: Samples containing polyamine.

Component \ Sample No.	7	8	9
Saturated salt water (SSW) (cc)	350	350	350
Soda ash (g)	1	1	1
Starch G (g)	8	-	-
PAC-LV (g)	-	8	-
Starch HT (g)	-	-	8
XC polymer (g)	1	1	1
Limestone(LSP)(g)	15	15	15
Polyamine(%)	2	2	2

Table 4: Samples containing choline chloride.

Component \ Sample No.	10	11	12
Saturated salt water (SSW) (cc)	350	350	350
Soda ash (g)	1	1	1
Starch G (g)	8	-	-
PAC-LV (g)	-	8	-
Starch HT (g)	-	-	8
XC polymer (g)	1	1	1
Limestone(LSP)(g)	15	15	15
Choline chloride(%)	2	2	2

Table 5: Samples containing 1-6, diaminohexane.

Component \ Sample No.	13	14	15
Saturated salt water (SSW) (cc)	350	350	350
Soda ash (g)	1	1	1
Starch G (g)	8	-	-
PAC-LV (g)	-	8	-
Starch HT (g)	-	-	8
XC polymer (g)	1	1	1
Limestone(LSP)(g)	15	15	15
1,6-diaminohexane (%)	2	2	2

RESULTS AND DISCUSSION

Compatibility Investigation

For compatibility purposes, some properties of the aged samples are compared to that sample which contains no amine (base sample) of its own polymer.

The best compatible polymer is the one the properties of which show improvement or have the least deterioration after amine addition. This concept means higher rheological properties like yield point and apparent viscosity, and denotes lesser filtrate volume in the case of filter loss measurement.

Regarding Figure 1 showing the filter loss of different fluid types, the high filtrate volume of sample one, which is the base fluid of starch-G without any amine compounds in its formulation, is noticed. Comparing this column with the other columns of the same polymer (starch-G) shows that except choline chloride, all the other three amines actually decreased filtrate volume and had a better impact on filter loss.

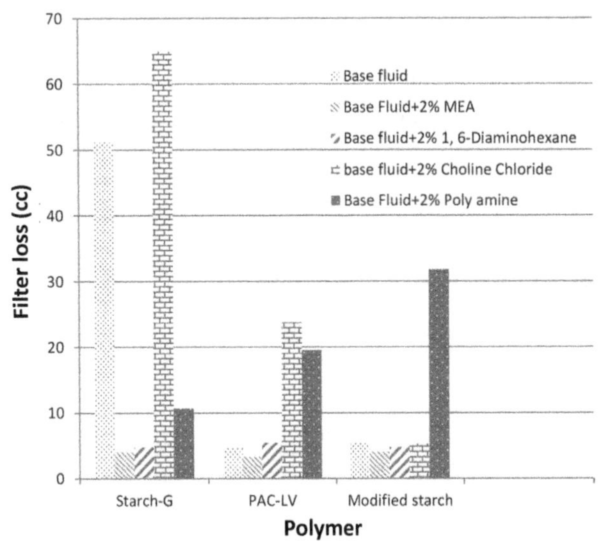

Figure 1: Filter loss of different fluid types after 16 hours hot rolling at 250 °F.

Regarding PAC-LV and modified starch, almost MEA and 1,6-diaminohexane made an improvement in filter loss; however, the base fluid of these two polymers had much lower filtrate volume than starch-G.

Figure 2 shows that both starch-G and modified starch had a better improvement regarding apparent viscosity after amine addition. However, MEA and 1,6-diaminohexane showed a more viscous behavior with PAC-LV.

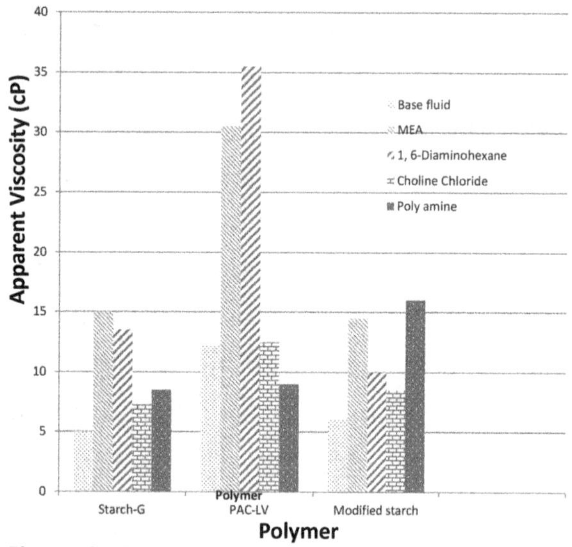

Figure 2: Apparent viscosity of different fluid types after 16 hours hot rolling at 250 °F.

Also, Figure 3 shows that the yield point of the base fluid of starch-G is zero, which is considerably improved with all the amines after amine addition.

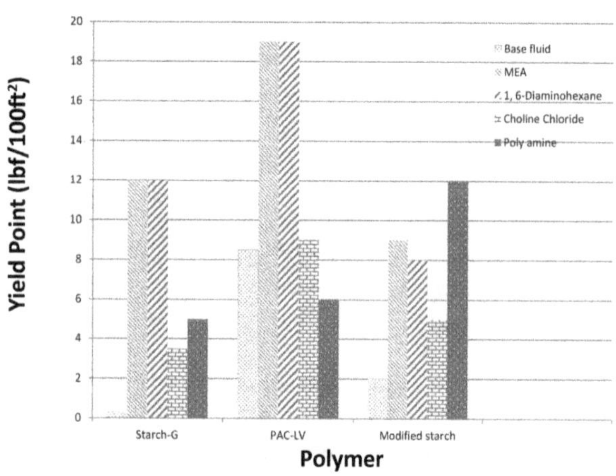

Figure 3: Yield point of different fluid types after 16 hours of hot rolling at 250 °F.

Thermal Stability

The investigation of the impact of adding amine on thermal stability needs a comparison between the two states of (before and after) hot rolling data of the same amine type. Just like the pervious section, concerning filter loss, less filtrate volume is desirable. It should be emphasize that the more stable rheological properties such as apparent viscosity, yield point and gel strength is also desirable for drilling fluids.

According to Figure 4, except choline chloride that almost made no difference to the base fluid (no amine), all the other amine types acted better at maintaining their filter loss property especially MEA and 1, 6-diaminohexane; this means that better thermal stability was observed after the addition of these amines.

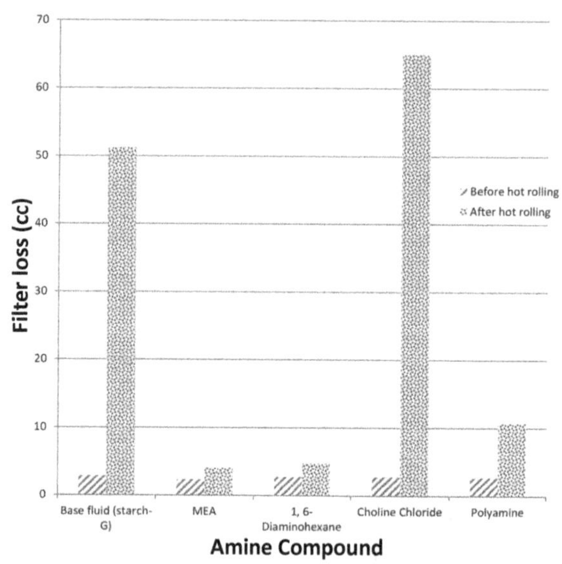

Figure 4: Filter loss comparison before and after 16 hours hot rolling at 250 °F (the base of starch-G).

Figure 5 accordingly shows that MEA and 1,6-diaminohexane were able to maintain their apparent viscosity better than polyamine and choline chloride respectively because hot rolling decrease apparent viscosity less, and apparent viscosity is close to the value obtained before hot rolling.

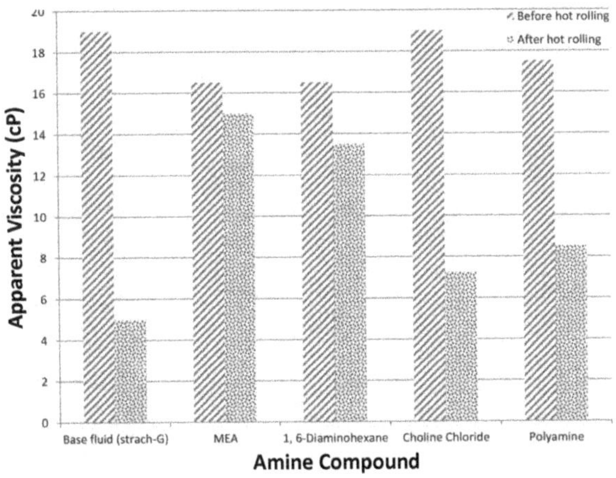

Figure 5: Apparent viscosity comparison before and after 16 hours hot rolling at 250 °F (the base of starch-G).

Regarding the yield point, Figure 6 shows that MEA and 1,6-diaminohexane were able to maintain their yield point better than polyamine and choline chloride respectively because after hot rolling the yield point is higher and closer to the value obtained before hot rolling.

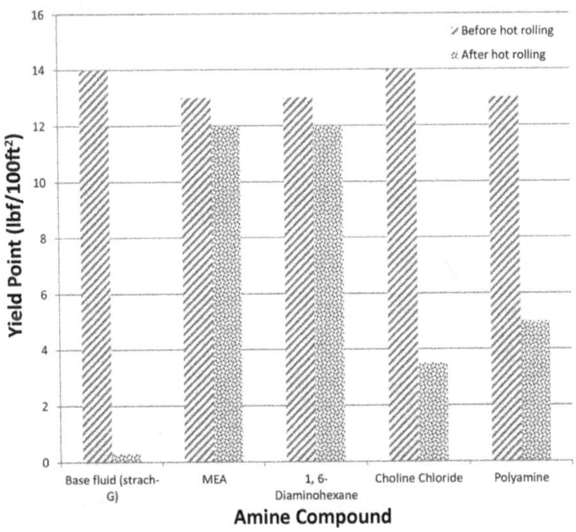

Figure 6: Yield point comparison before and after 16 hours hot rolling at 250 °F (the base of starch-G).

Regardless of the better compatibility of amines with starch-G polymer, thermal stability was also investigated for the other two polymers (PAC-LV and modified starch), and the compared results are shown in Figures 7-12.

Figures 7 to 9 also show that, similar to the starch-G results, MEA, and 1,6-diaminohexane had better thermal stability regarding filtrate volume, apparent viscosity, and yield point.

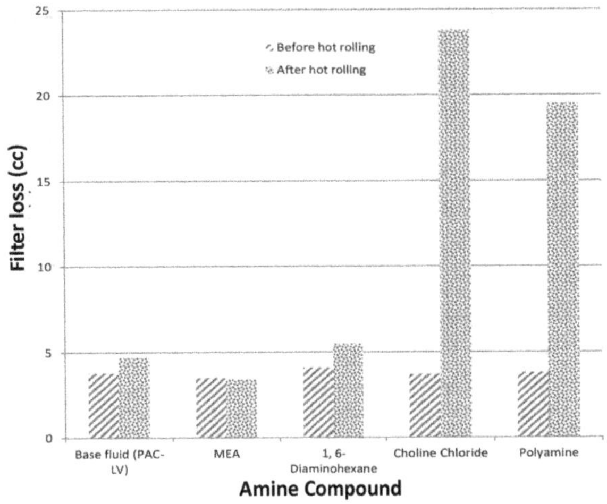

Figure 7: Filter loss comparison before and after 16 hours hot rolling at 250 °F (the base of PAC-LV).

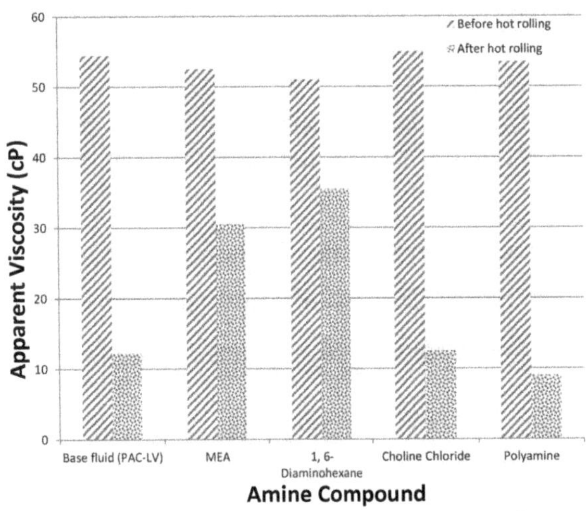

Figure 8: Apparent viscosity comparison before and after 16 hours hot rolling at 250 °F (the base of PAC-LV).

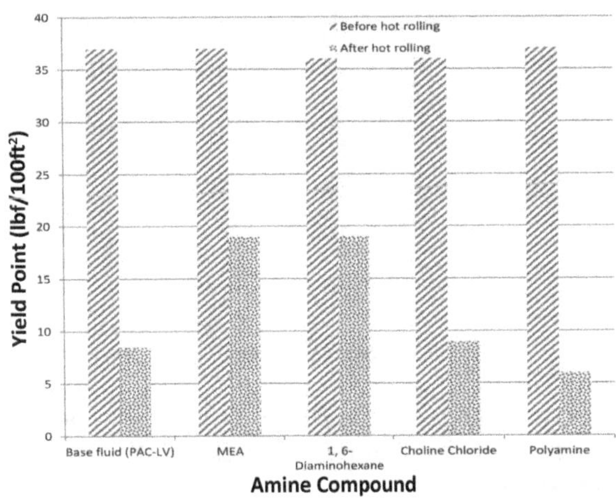

Figure 9: Yield point comparison before and after 16 hours hot rolling at 250 °F (the base of PAC-LV).

Figure 10 shows that filter loss deterioration is so high for polyamine that causes its slight advantage over MEA and 1,6-diaminohaxane regarding rheological properties to pale into insignificance.

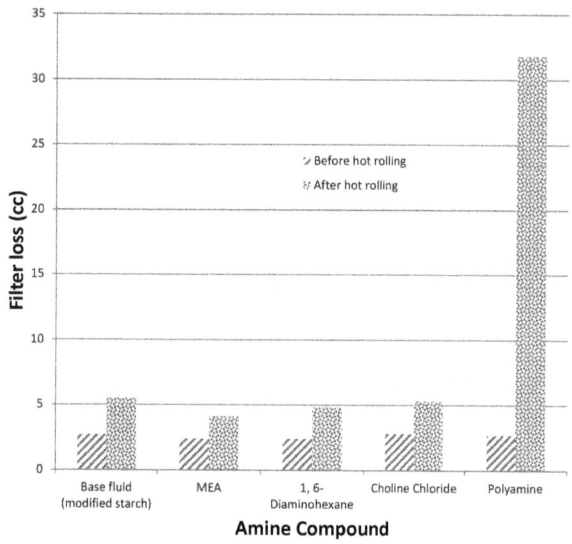

Figure 10: Filter loss comparison before and after 16 hours hot rolling at 250 °F (the base of modified starch).

Moreover, Figures 11 and 12 show that polyamine acted better at maintaining its rheological properties after a rise in temperature.

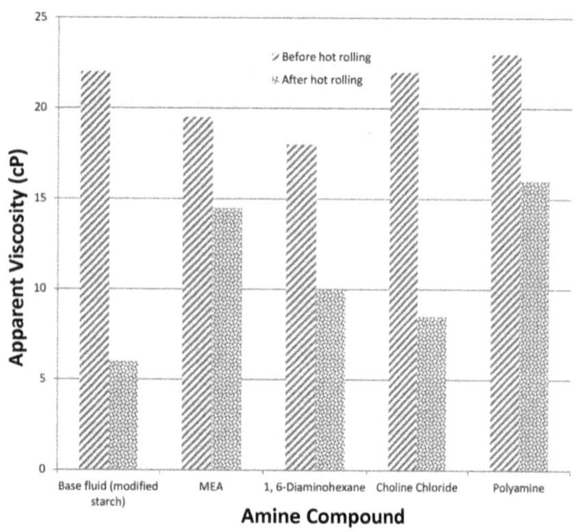

Figure 11: Apparent viscosity comparison before and after 16 hours hot rolling at 250 °F (the base of modified starch).

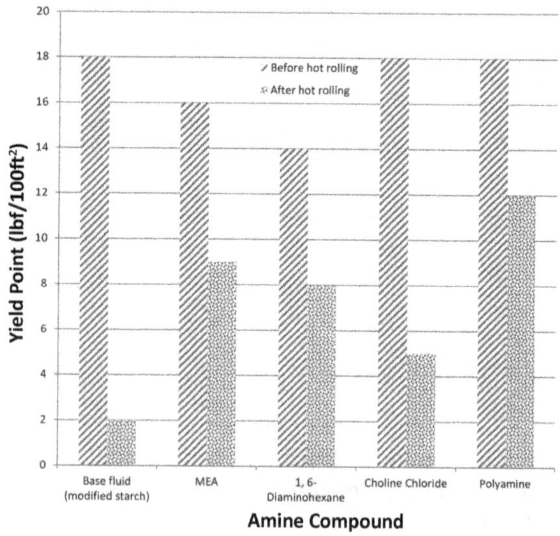

Figure 12: Yield point comparison before and after 16 hours hot rolling at 250 °F (the base of modified starch).

According to the above figures, MEA and 1,6-diaminohexane along with PAC-LV base had better thermal stability than polyamine and choline chloride. However, with modified starch being the base fluid, polyamine had slightly better (higher) gel strength than MEA and 1,6 diaminohexane. The effect of amine compounds on the thermal stability enhancement of water-based drilling fluids was investigated in this study. The results clarified that almost all of the four amines are compatible with starch-green specially MEA and 1,6-diaminohexane. Also, the rheological properties of fluids with the base of PAC-LV are slightly higher because PAC-LV has higher rheological properties compared to starch-green. Moreover, with the presence of amine additives (MEA and 1,6-diaminohexane in this study) in water-based drilling fluids, thermal stability can be reached up to 250 °F at two volume percent (2 Vol.%) of the amine additive.

CONCLUSIONS

One of the main characteristics of amines as clay swelling inhibitors is due to the possession of multiple active ionic sites, usually cationic ones. This feature allows this compound to interact closely with materials; however, amines with higher chain length as MEA (chain with two carbon atoms) and 1,6-diaminohexane (chain with six carbon atoms) act better by maintaining the original properties of drilling fluid as temperature rises. On the other hand, choline chloride simply behaves like potassium in potassium chloride which does not possess those multiple active sites of MEA or 1,6-diaminohexane. Also, polyamines are believed to have a plenty of active sites; however, their bigger size compared to linear amines lowers their efficiency in improving rheological properties at their original state by rising temperature.

Considering the experimental data, the better compatibility of amine compounds with green starch is mainly ascribed to the presence of stronger hydrogen bonding between amine molecules and starch molecules, which causes persistence in the rheological behavior of the whole mud.

ACKNOWLDEGMENTS

The authors are thankful to the Drilling and Well Completion Technologies and Research Group of the Research Institute of Petroleum Industry (RIPI).

NOMENCLATURES

Av	: Apparent Viscosity (cP)
HP	: High Performance
HPHT	: High-Pressure and High-Temperature
θ_{600}	: Dial reading at 600 rpm
θ_{300}	: Dial reading at 300 rpm
MEA	: Mono ethanolamine
NADF	: Non-aqueous Drilling Fluids
Pv	: Plastic Viscosity (cP)
PAC-LV	: Poly anionic cellulose-low viscosity
Starch-G	: Green starch
YP	: Yield Point (lbs/100ft^2)

REFERENCES

1. Hanyi Z., Qiu Z., Huang W., Huang D., and et al., "Successful Application of Unique Polyamine High Performance Water-based Drilling Fluid in Bohai Bay Shale formations," Presented at *IPTC (International Petroleum Technology Conference)*, Beijing, China, **2013**.
2. Tas B. T., Hakki G. I., and Parlaktuna M., "An Experimental Investigation of the Shale Inhibition Properties of a Quaternary Amine Compound," Presented at *19th International Petroleum and Natural Gas Congress and Exhibition of Turkey*, Turkey, **2013**.
3. Friedheim J. E., Hans G. J., Park A., and Ray C. R., "An Environmentally Superior Replacement for Mneral-oil Drilling Fluids," Presented at Offshore Europe, *Journal Society of Petroleum Engineers*, Aberdeen, United Kingdom, **1991**, 299-312.
4. Friedheim J. E. and Conn H. L., "Second Generation Synthetic Fluids in the North Sea: are they Better?" Presented at *SPE/IADC Drilling Conference. Society of Petroleum Engineers*, New Orleans, Louisiana, **1996**, 215-227.
5. Beihoffer T. W., Dorrough D. S., Deem C. K., Schmidt D. D., and et al., "Cationic Polymer Drilling Fluid can Sometimes Replace Oil-based Mud," *Oil and Gas Journal*, United States, **1992**, *90*(11), 1-12.
6. Patel A. D., Stamatakis E., Davis E., and Mlli C., "Shale Hydration Inhibition Agent and Method of Use," United States Patent 6 247 543, **2001**.
7. Galindo K. A., Zha W., Zhou H., and Deville J. P., "High Temperature, High Performance Water-Based Drilling Fluid for Extreme High Temperature Wells," Presented at *SPE*

International Symposium on Oilfield Chemistry, Texas, USA, **2015**, 1-9.

8. Mojammadi A., Kalhor M., Taraghikhah S., and Tahmasbi Nowtaraki K., "A Brief Introduction to High Temperature and Foam Free Water Based Drilling Fluids," IADC/SPE Asia Pacific Drilling Technology Conference, *Society of Petroleum Engineers*, Singapore, **2016**, 1-7.
9. Annis M. R. and Smith V., "Drilling Fluids Technology," Revised Edition Exxon Company, **1996**.
10. Darley H. C. H. and Gray G. R., "Composition and Properties of Drilling and Completion Fluids," *Gulf Professional Publishing*, **1988**.
11. Gholizadeh-Doonechaly N., Tahmasbi K., and Davani E., "Development of High-performance Water-based Mud Formulation Based on Amine Derivatives," Presented at *SPE International Symposium on Oilfield Chemistry, Society of Petroleum Egineers*, Texas, USA, **2009**, 1-8.

18

A Review of Aerogel Applications in Adsorption and Catalysis

Davood Karami
Department of Chemical and Petroleum Engineering, Schulich School of Engineering, University of Calgary, Calgary, AB, Canada.

ABSTRACT

Aerogels are a special class of porous material, which have excellent physicochemical properties such as low density, high porosity, high surface area and adjustable surface chemistry. Aerogels were first prepared several decades ago, but never truly commercialized due to their high cost. Technological advancements in the production and quality of different types of aerogel cut costs down and market factors increase demand. Therefore, the viability of this porous material on the several applications has been widely studied. Among the various high-performance applications, aerogel has attracted significant attention as a chemical sorbent and catalyst for CO_2 capture and conversion as well as methane conversion to value-added products. Although aerogel performance for zero CO_2 emission to the environment is promising, some drawbacks of aerogels such as complicated drying process, mechanically weak structure and processing cost should also be considered in material development. Ultimately this review article will cover an overview regarding the aerogels synthesis, processing and their recent applications in CO_2 capture and conversion.

Keywords: Aerogel, Supercritical Drying, Catalyst, Sorbent, Carbon Capture

INTRODUCTION

Aerogel is generally referred to any material derived from organic, inorganic or hybrid precursors that are prepared by a sol-gel process and a proper drying technology rapidly to extract the liquid in an alcogel and replace the liquid to make a three-dimensional and the highly porous network. Aerogels were firstly introduced by S. Kistler at 1930's when he extracted the pore-filing liquid of wet gels using a supercritical drying method [1-2] to obtain an air-filled solid material with nearly same dimensions as their original wet gel. The difficult multistage processing steps developed by Kistler made aerogels obscure material for about 30 years. However, during the last decades, noticeable advancement in aerogel synthesis and their drying technologies, different types of aerogels are produced including inorganic (such as SiO_2, TiO_2, Al_2O_3, ZrO_2, etc.), organic (i.e. resorcinol-formaldehyde (RF), polyurethane, polyimide, polystyrene, etc.) [3] and carbon (i.e. carbon, carbon nanotubes, graphene) [4-5], and more recently SiC-based aerogels [6-7]. Furthermore, the combination of aerogels (aerogel composites)

***Corresponding author**
Davood Karami
Email: dkarami@ucalgary.ca

has often given an additional functionality such as mechanical strength, hydrophobicity and catalytic features to original materials and has extended some high-performance applications of aerogels [8]. Figure 1 shows the development of the aerogel after Kistler's initial work along with increasing the number of publications about aerogels during the last decades. In fact, following the significant advances in the synthesis process, aerogel has attracted significant interests by the researchers due to the promising technological functions of these materials applied for the advanced performances. The unusual aerogel properties such as low density (0.003–0.05 g/cm^3), noticeably high surface area (500–2000 m^2/g), high porosity (90–99%), with special surface chemistry that can be fabricated into a various morphologies and sizes make them suitable for many applications [9]. The aerogels are mostly used in thermal insulations in aerospace and building sectors [10], however, they are also promising materials for chemical sorption and catalysis applications that require high surface area and high porosity [11]. Some applications of aerogels includes adsorption and environmental cleanup [12], chemical sensors [13], acoustic transducers [14], energy storage devices [15], metal casting molds [16], water repellant coatings and biomedical and pharmaceutical applications [17]. Due to the above-mentioned properties, the catalytic feature of aerogels is rapidly developing for high-temperature performances in the materials and energy-related fields. The carbon capture application of aerogels is relatively a novel field and aims at cleaning flue gases from CO_2 and establishing clean energy technologies with zero CO_2 emissions. The CO_2 release to the atmosphere is of major pollutants that cause the serious environmental problems such as global warming and hazards for human health. CO_2 conversion to fuels and chemicals is another fascinating process by which aerogel can be utilized as a catalytic material. The combination of both CO_2 capture process and CO_2 conversion process can contribute to ideal CO_2 recycling concept. There are only a few review articles and book chapters that focus on the aerogels for the outlook of catalytic, energy and environmental-related applications [18-19]. This review paper will provide a comprehensive overview of the recent the development in aerogel materials that covers the synthesis and processing and the applications of almost all aerogels and their composites from different environmental and catalytic aspects.

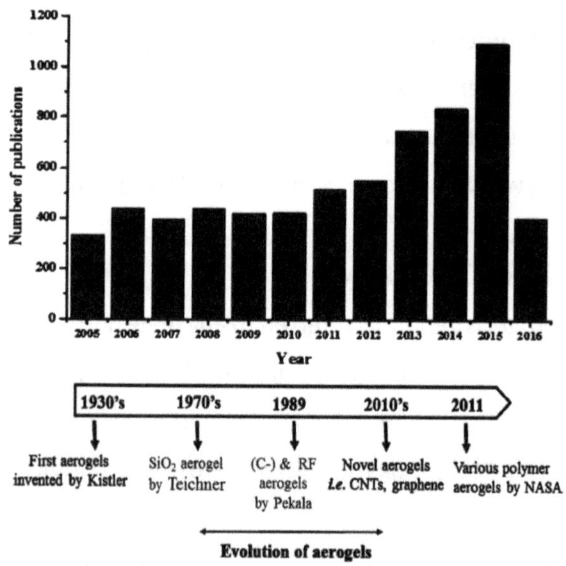

Figure 1: Number of publications (ScienceDirect record) during the last ten years containing aerogel in the content along with the evolution pattern of aerogels after invention. (Date of search: 13 Feb.2016).

EXPERIMENTAL PROCEDURES

Aerogel General Preparation

Preparation of aerogels is general between almost all types of aerogels. Preparation method of aerogels involves four significant steps: 1) sol-gel process, 2) aging, 3) drying and 4) thermal treatment. Regarding aerogel activation for catalytic reactions and the use of aerogel as a support, additional thermal treatment such as calcination and reduction steps are conducted after drying [20]. The possible applications of aerogels extremely depend on the network structure and surface groups of aerogels that may be controlled by the sol-gel process parameters [21]. Therefore, it is essential to determine a precise synthesis and processing conditions for aerogels to obtain the required properties for a specific application.

Sol-gel Process

Nearly all aerogels are prepared by the sol-gel approach but via different starting precursors and conditions. Sol is prepared from a mixture of main compounds, water, solvents, co-solvents and catalysts due to hydrolysis and condensation reactions. Usually by altering the reaction conditions (e. g. pH, temperature), the colloidal particles of sol link to form a three-dimensional and interconnected porous network, which is the most important feature of aerogel fabrication. Depending on its pore-filling solvent, gel mixture can be named alcogel and acetogel using alcohol and acetone as a pore solvent, respectively. While the hydrogel is usually prepared in the aqueous solution. During the sol-gel process, depending on the hydrolysis growth process and the rate of condensation, network particles sizes in either the polymeric or the colloidal state are formed in the range of 1–10 nm [22]. For instance, the sol–gel reaction pathway to prepare a silica aerogel network using tetraalkoxysilanes involves the hydrolysis reaction replacing the OR alkoxides with OH hydroxyl groups and, then, during condensation reactions, the silanol groups produce siloxane bonds along with alcohol and water by-products. In most cases, condensation starts simultaneously with hydrolysis reactions and continues during the whole sol-gel process. The sol-gel process allows modifying the gel nanostructure by adjusting the reaction crucial parameters. There are many parameters that influence the sol-gel reactions and significantly control the nanostructure of the network and final material properties specifically 1) the concentrations of precursors to the solvent, 2) solvent and co-solvent types and ratios, 3) ratio of water to precursors, 4) temperature and 5) pH [23]. In addition, there is a possibility for incorporation of extra phase or a molecular compound in sol-gel reaction to provide a special feature on gel network either chemically by using suitable organofunctional alkoxide components, or physically by introducing additives/dopants in the porous network [24-25].

Aging

The chemical reactions continue to proceed even after the gel formation [26]. It is due to the existence of an additional possibility of gel network to further grow in the solvent. This process is called aging period. In fact, the solvent inside the pores contains a reactive species (such as -OH) or unreacted monomers which can equally condense in the network. Usually, the whole process of aging takes from hours to days by immerging the gel in the initial sol or suitable solvent and under the controlled conditions [27]. The major function of the aging is to improve the mechanical strength of the fragile network of aerogels. In the aging process, the network particles undergo several phenomena such as Ostwald ripening or coarsening. Several parameters namely pH, time and temperature are the most significant factors which can change the kinetics of aging process [28]. In addition, most of the textural properties, such as pore size, porosity, and surface area of synthesized gels are changed during aging. Aging is considered as an easy and strong technique for network strengthening of the as-prepared gels [29].

Drying

The process of drying of gels without any collapse of the original porous structure is an important step during aerogel preparation. Three drying procedures are used for the obtained gels such as 1) supercritical drying (using i.e. alcohol, acetone, or CO_2) [30], 2) ambient pressure drying, and 3) freeze-drying [31]. Normal pressure drying induces tension in the pores liquid surface at solid-liquid-vapor interface, which forms a capillary tension in pores walls. This is the main cause of the structural collapse and dimensional shrinkage in a dried gel. For an efficient drying of gels, the gradient of the capillary stresses developed in pores walls must impartially be avoided. The most efficient way to dry the gels, is extracting the pores liquid under the supercritical condition. In supercritical drying, the gel is placed inside the batch pressure vessel, then the pressure and temperature of vessel surpass the critical point (TC, PC) of the solvent. Due to no distinct separation between liquid and vapor phases, the capillary stresses are not built up; therefore, a crack-free solid gel is finally obtained. The critical drying condition is very different depending on the type of solvent [32]. The most commonly used solvent is CO_2 because of its mild critical temperature (TC=31°C) in supercritical CO_2 ($scCO_2$) drying. The operation of this drying technique is safer compared to the supercritical drying of other solvents (for example, ethanol TC = 240 °C) [33]. Normal drying is a promising, simple and safe technique to dry the gels under the room temperate conditions and may be a suitable way for the mass industrial productions. However, in this method, the surface of the pore walls inside the gel must be chemically treated with some non-polar groups to prevent further condensations in surface after being compressed by capillary stresses. In addition, the pore-filling solvent must be exchanged with a hydrocarbon or a solvent with less surface tension to avoid the capillary stresses tolerable by the pore walls. Freeze drying is a simple drying method used to prepare aerogels with practical porous structures. The liquid inside the gels is first frozen and removed by a sublimation process under very low pressures. The final materials are called cryogel with maximum 80% porosity and only half of the surface area of its aerogel counterpart. A rapid freezing rate often results in the small ice crystals and therefore, products are formed with small pore sizes and high pore surface area. The cryogels have more high volume of macroporosity, a large extent of shrinkage and low specific surface area compared to their $scCO_2$ counterparts [34].

Thermal Treatment

In addition to the processing steps, mixed metal oxide aerogels, which are prepared for sorption and catalysis purposes, undergo an extra processing step of calcination and activation. In the calcination process, aerogel is exposed to the high temperature (usually above 600 °C) and atmospheric pressure with/without an inert atmosphere of N_2 or Ar to decompose some volatile materials and enhances the network with the pure metal oxides.

Adsorption Applications

A promising way to reduce greenhouse gases release in the atmosphere is to abate the emission of CO_2 [35]. The CO_2 capture technology must be able to be implemented both economically and on a very large scale [36]. The current technologies depend on the selective capture of CO_2 from flue gas and air by liquid amines and by membrane separation [37]. However, these developed technologies suffer from some problems such as energy-intensive capture and regeneration process, high corrosion of process equipment, oxidative degradation of the aqueous amines and highly possible production of toxic compounds during the CO_2 capture [38]. Porous solid materials such as zeolites, metal-organic frameworks (MOFs), activated carbon, and calcium oxides, etc. are considered as the emerging technologies for post-combustion and atmospheric CO_2 capture replacing the above-mentioned technologies [39]. Although the selectivity was low, high extent of porosities in these sorbents give a high adsorption capacity around 10 mmol/g at temperatures of about 500 °C and a moderate adsorption capacity of 6 mmol/g at room temperatures. Figure 2 shows CO_2 capacity and adsorption/desorption temperatures of well-known solid sorbents. To increase the CO_2 capacity of these porous sorbents at low temperatures, the active materials such as potassium carbonate, amines [40] and ionic liquid-amino acid moieties [41-42] impregnate on the surface of porous adsorbents such as conventional mesoporous silica (such as SBA-15, MCM-41). Sometimes the amine dispersion on the surface via the impregnation method is not homogenous; therefore, it significantly affects the sorption capacity results [43]. Besides above-mentioned porous sorbents, aerogels application for capturing CO_2 is quite promising due to their numerous advantages such as low density, high porosity, and high surface area. In addition, the flexibility of the sol-gel process, the microstructural form, as well as the composition and surface chemistry of aerogel can be modified intentionally for CO_2 capture [44]. For this purpose, recent investigations have focused on using aerogels as new CO_2 capture materials by changing the physicochemical properties of different aerogels.

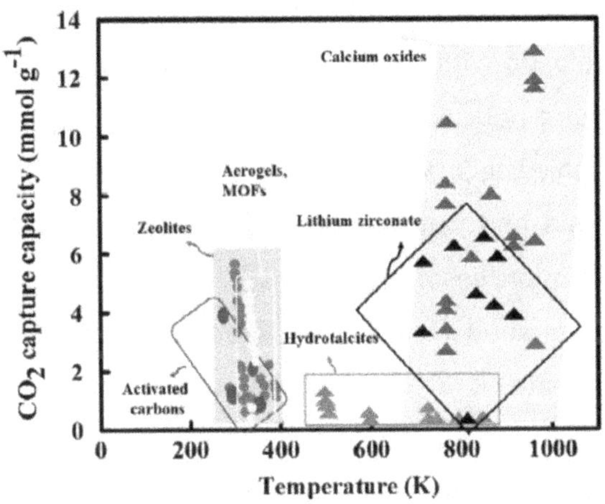

Figure 2: CO_2 capturing properties versus sorption temperature for various solid porous sorbents.

In this regard, various amines modified SiO_2 aerogels [45] have been studied with the CO_2 capacity of 6.97 mmol/g and 7.9 mmol/g under simulated flue gas conditions (in pre-humidified 10% CO_2), respectively [46]. Furthermore, the other parameters that vastly impact on the rate of CO_2 uptake are the temperature and partial pressure of CO_2, sorbent particle size, the density of the sorbent, gas/solids contacting, equipment responsiveness, etc. The partial pressure of CO_2 and temperature of sorption are important parameters regarding the kinetics of sorption, as at the optimal pressure, the kinetics of the sorption is usually slow at low temperature [47]. Generally, the newly developed sorbents must exhibit excellent capture properties such as CO_2 capacity, H_2O capacity, kinetics (rate of CO_2 uptake) and cyclic stability (over thousands of cycles) [48]. The co-adsorption of the low concentration water present in the flue gas may lead to an increase in the sorbent stability. However, adsorption of water at a large quantity could severely affect sorbent capacity and require the excess energy to vaporize the adsorbed water during the regeneration process. Amine-grafted sorbent using commercially available 10-micron size aerogel (Cabot Corp. Nanogel@) was developed with a sorption capacity of around 6.1 mmol/g [49]. Also, this sorbent had a fast kinetic by achieving 90% of its equilibrium capacity within 10 min and can maintain its capacity up to 10 adsorption/desorption cycles.

Besides supported amines, supported potassium carbonate, K_2CO_3, was proposed as a promising sorbent for low-temperature CO_2 capture. This sorbent adsorbs CO_2 in the presence of H_2O at 60-100 °C by the following reaction:

$$K_2CO_3 + CO_2 + H_2O \leftrightarrow 2KHCO_3$$

Regeneration of K_2CO_3 occurs at 120-300° C with the release of CO_2 and H_2O. Based on the above reaction, the theoretical amount of CO_2 adsorbed per gram of potassium carbonate is calculated at 7.2 mmol/g. Our patented θ-alumina aerogel with a surface area around 420 m^2/g used as a support for 50% K_2CO_3 shows the highest CO_2 capture efficiency at the above-mentioned condition (50). Using conventional alumina as support, sorbent efficiency is lower than our aerogel sorbent due to excess water adsorption by hydrophilic γ-alumina and low potassium carbonate dispersion on the low surface area of θ-alumina (51).

At higher temperatures (over 600 °C), the capture capacity of calcium oxide has been applied for CO_2 removal through carbonate formation. The calcium oxide weak structure is likely to sinter. Therefore, capacity decreases significantly during an extended cyclic performance. Many approaches have been proposed to maintain this capacity for a long period of time but the problem remains unsolved due to the higher efficiency and excellent stability cannot be attained. While capturing techniques appear promising and have favorable thermodynamics, the cost has been substantial, due to the capture sorbents must be very finely divided for the maximum surface area. Moreover, these reactions are non-catalytic and depend entirely on molecular reactions at the surface of the sorbents. This is a real and unsatisfied need for improved capture capability of sorbents which have enhanced capture efficiencies. The novel calcium-oxide-based refractory sorbents synthesized by flame spray pyrolysis (FSP) for CO_2 capture was developed [52]. FSP was used for converting precursor droplets into solid nanoparticles in flames. FSP allows for the controlled synthesis of nanoparticles with high specific surface areas. The most stable sorbent (40 wt% ZrO_2-60 wt% CaO) gives a CO_2 capacity of 10.76 mmol/g in an extended cyclic operation of 50 without any activity loss. Carbonation was conducted at 700 °C in 30% CO_2 for 30 min. Despite many preceding efforts to produce a high CO_2 capacity sorbent, a practical amount of 10.78 mmol/g has never been exceeded during extended cyclic operation. In accordance with our invention, a CaO aerogel sorbent comprised of a uniform dispersion of individual nanoparticles of calcium oxide was synthesized as substantially fluffy clusters of particles. Our patented sorbents had a

BET surface area in the range of about 70 to about 150 m^2 /g, a pore volume in the range of about 1 to about 6 cm^3/g and a bulk density in the range of about 0.02 to about 0.05 g/cm^3. The CaO aerogels were mixed with metal oxide stabilizer (specially zirconia) by three methods of 1) Mixing stabilizer during alcogel preparation, 2) Impregnation of aerogel with stabilizer precursor solution, and 3) Shelling aerogel or calcined aerogel surface by core-shell method. Our capturing results show the CaO-ZrO_2 aerogel sorbent had a CO_2 capacity of 15.6 mmol/g with activity loss of 10% during 30 extended cyclic operations [53]. The other proposed process for combustion with a direct CO_2 collection is chemical-looping combustion (CLC) which produce streams of relatively pure carbon dioxide and condensable steam [54]. Preparation of suitable oxygen carriers is crucial for the long-term operation of a CLC process. Nickel and iron oxides (NiO and Fe_2O_3) are promising materials for CLC. However, they exhibit a slow solid-state reaction between active metal oxides and support material at high temperature. Our highly effective alumina aerogel support calcined at over 800°C was used to fabricate NiO- and Fe_2O_3-supported compositions via the incipient wetness impregnation (IWI) method. These oxygen carriers exhibit the promising activities towards CLC process [50].

Catalysis Applications

A comprehensive review on aerogels in catalysis up to the year 1990, covered preparation, physicochemical properties, and catalytic applications of mainly high-temperature aerogels such as partial oxidation, hydrogenation and oxidative coupling [55]. Binary mixed oxide aerogels that are mostly NiO, CuO, Li_2O, and Fe_2O_3 oxides associated mainly with Al_2O_3, SiO_2, ZrO_2, and MgO were developed for the above reactions. Commercially available metal alkoxides, acetates or acetylacetonates are used as primary sol-gel precursors. The method of the preparation is the co-gelling of both metal derivatives in suitable organic solvents (or dispersing agents) by reaction with water in stoichiometric amounts. The supercritical drying is performed with respect to the solvent (dispersant) which have the highest critical temperature when two different solvents or dispersants have been selected. The first Fe_2O_3-Al_2O_3 aerogels were prepared from the mixing of two butanolic solutions of iron (III) acetylacetonate and of aluminum s-butylate in s-butanol [56]. Fe_2O_3-SiO_2 aerogel catalyst showed much higher productivities (300 times more) than the unsupported Fe_2O_3 or the conventional xerogels [57]. Moreover, the mixed aerogel did not deactivate by carburization or through the formation of inactive carbon species, due to the oxidized form which was the best way to obtain such high activities in hydrocarbons, and that only the aerogel was able to maintain an adequate state of oxidation with time on stream. The same trend was also reported when alumina used as support [58]. Binary and ternary aerogels containing copper catalysts were tested in the synthesis of methanol with carbon dioxide and hydrogen feeds [59]. As with Fischer-Tropsch reactions, pre-oxidized aerogels were considered more efficient in the synthesis than the pre-reduced aerogels. Pure ZrO_2 aerogel is also a good catalyst for the methanol synthesis. The presence of zirconia is positive for the formation of methanol due to an electronic interaction between copper and zirconia. Recently a series of mixed silica aerogels containing lanthanum and niobium oxides was prepared with BET surface areas of 700 to 800 m^2/g before reaction [60]. These catalysts were used for the oxidative coupling of methane with a methane-to-oxygen ratio of 3:1 at atmospheric pressure at 800 °C. For La-SiO_2 aerogels, a methane conversion of 20% was achieved and the selectivity to C_2 was 30%. Nb-SiO_2 were as active as the La-SiO_2 catalysts but less selective in C_2 in the range of 8%. Ternary aerogels La-Nb-SiO_2 yielded to the same result (selectivity in C_2 40% and methane conversion 13%). Since the aerogel method seems to be unique for producing solid with large textural properties stable at the temperatures that are generally selected for catalytic reactions. Aerogels are suggested to be evaluated in hydrotreatment processes such as hydrodenitrogenation (HDN) and hydrodesulphurization (HDS) because of their very large pore volumes and good stabilities [61]. Aerogels can find applications as components in solid superacid catalysts [62] using large surface area aerogel supports. Aerogels provide a way to make very advanced catalysts for problematic reactions. Metal-on aerogel support catalysts are very attractive since the sol-gel step allows to disperse the metal as a thin film or in any desired shape [63]. Aerogels can show a monodisperse distribution of particle which is easily fluidized. Therefore, they can be used in real conditions and are not restricted to the laboratory scale. The mixed aerogels or their parent gels yield to very homogeneous mixtures originating from chemical interactions at the molecular level in the liquid phase (or dispersed in a liquid phase). Aerogels are a sort of a three-dimensional dry structure of the sol-gel product with all their catalytic capabilities intact and available for discovery. Because aerogels exhibit high solid-state chemical reactivities, they can also be applied to synthesize new catalysts with large surface areas and pores at moderately lower temperatures than the sintering temperatures which deteriorated the mixtures of powders. Aerogels show great promise in catalytic applications. Compared to conventional catalysts, their activity and selectivity in several catalytic reactions appear to be much higher. Besides, in some catalytic reactions, common catalysts become deactivated with time on stream. Currently, aerogel applications in CO_2 conversion to the value-added products have been attracted many interests, because CO_2 is a cheap, non-toxic and abundant C_1 feedstock. CO_2 chemical utilization is a challenge and an important topic [64]. Thus, any efficient reaction using CO_2 as a raw material has positive consequences on efforts towards carbon management. However, CO_2 is a thermodynamically stable compound, and then its reduction requires high energy. It is probably the reason why currently the toxic carbon monoxide is used mostly in industry. CH_4 and CO_2 are relatively inexpensive due to their natural abundance. The reaction between CO_2 and CH_4 to produce synthesis gas ($CO + H_2$) can be used in chemical energy transmission systems or utilized in the Fischer-Tropsch reaction to produce liquids. Many researches have been conducted on the catalytic performance of noble metals such as Rh, Ru, Pd, Pt and Ir for CO_2 reforming of CH_4 [65]. Rh and Ru are generally accepted to show both high activity and stability in CH_4 dry reforming. Typically, the catalyst activity was in the order of Rh ≈ Ru > Ir, Pt, and Pd. The nature of support may have some effects on the activity of noble metals. Ni-based catalysts have been broadly investigated [66-67] due to their similar activity and relatively low price when compared with noble metals. However, Ni-based catalysts were readily deactivated by carbon deposition and sintering. Thus, more efforts were made for the development of a stable Ni-based catalyst with high activity. Al_2O_3, MgO, CaO were used as a support to prepare Ni-based catalysts for CH_4 reforming reaction. The pre-oxidized support with Lewis basic site is observed to enhance the carbon-resistant ability of the catalyst for the strong adsorption capacity of CO_2. Moreover, strong interaction might promote the dispersion of Ni on the support, while much stronger interaction may decrease the reducibility of Ni on the catalyst surface. Besides, the formation of fibrous carbon is significantly influenced by the metal particle size which indicates that the support can affect the catalytic stability of the catalyst. The other potential use of CO_2, as an alternative feedstock replacing CO in the methanol synthesis has received attention as a second effective way of CO_2 utilization in the past decades. Under proper conditions, methanol made from atmospheric CO_2 by its reaction with hydrogen is considered the most economical way to reduce the greenhouse effect [68]. Most of the catalysts for CO_2 hydrogenation contained Cu and Zn as the main components together with different modifiers (Zr, Ga, Si, Al, B, Cr, Ce, V, Ti, etc.) [69]. the Cu–ZnO/ZrO_2 prepared by reverse co-precipitation under ultrasound irradiation was found to show a significant improvement in the total surface area of the metal phase, and then the enhanced CO_2 conversion and methanol yield [70]. In the noble metal-based catalysts, the supported Pd catalysts are the most commonly used catalyst, which exhibits considerable activity and selectivity for hydrogenation of CO_2 to methanol, and the supports has a significant effect on the performance of the catalyst. Transitional metal carbides are a type of metal-derived compounds with the incorporation of carbon in the metal lattice. They have an excellent catalytic performance for hydrogenation, which is like those of Pt, Rh, and other noble metals [71]. Thus, the metal carbides can be used as potential alternatives for noble metal catalysts. Due to superior hydrogen adsorption, activation and transfer capabilities of metal carbides, they have potential application in CO_2 hydrogenation reactions. Different carbides were tested for CO_2 hydrogenation, the results indicated that Mo_2C and Fe_3C showed high CO_2 conversion and good methanol selectivity [72]. Practically, a highly active, selective and poisoning-resistant catalyst for methanol synthesis is desirable.

CONCLUSIONS

Technological advancements in the production and quality of basic materials are fundamental foundations for more innovation and growth. New advanced materials arising from developments in the material sciences offer new growth prospects through the support of new existing industrial and commercial products and processes. Two cutting-edge processes of 1) CO_2 (dry) reforming of methane and 2) oxidative coupling of methane using CO_2 and O_2 as an oxidant have been developed for CO_2 utilization and value-added products production. In these promising reactions, high surface area θ-alumina (over 400 m^2/g), α-alumina (over 200 m^2/g) aerogels, mixed alumina-magnesia and alumina-zirconia and silica aerogels support will be used to prepare very active catalysts. Ni-Rh (or Ru) supported α-alumina or magnesia aerogel will be the potential candidates for CO_2 reforming reaction.

In the search for excellent catalysts, hundreds of materials have been tested for oxidative coupling of methane. Among them, Li-magnesia aerogel, Re_2O_3 (Re: La, Sm, Gd)-silica aerogel and Mn-Na_2WO_4-silica aerogel which showed higher performance than most other materials will be applied for oxidative coupling reactions in future.

NOMENCLATURES

CLC:	Chemical-Looping Combustion
FSP:	Flame Spray Pyrolysis
HDN:	Hydrodenitrogenation
HDS:	Hydrodesulphurization

REFERENCES

1. Kistler S. S., "Coherent Expanded Aerogels and Jellies," *Nature*, **1931**.
2. Kistler S. S., "Coherent Expanded-Aerogels," *J. Phys. Chem.*, **1932**, *36* , 52-64.
3. Moner M. A., Molins E., and Liber D., "Sol-Gel Route to Direct Formation of Silica Aerogel Microparticles Using Supercritical Solvents," *J. Sol-Gel Sci. Technol.*, **2003**, *26*, 645-649.
4. Aegerter M. A., Leventis N., and Koebel M. M. (3rd ed.), Aerogels Handbook, Springer, USA, **2011**.
5. Horikawa T., Hayashi J., and Muroyama K., "Size Control and Characterization of Spherical Carbon Aerogel Particles from Resorcinol-formaldehyde Resin," *Carbon*, **2004**, *42*, 169-175.
6. Kong Y., Zhong Y., Yang M., Teng K., and et al., "Facile Synthesis of Resorcinol–formaldehyde/ Silica Composite Aerogels and their Transformation to Monolithic Carbon/silica and Carbon/silicon Carbide Composite Aerogels," *J. Non-Cryst. Solids.*, **2012**, *358*, 3150-3155.
7. Worsley M. A., Satcher Jr. J. H., and Baumann T. F., "Synthesis and Characterization of Monolithic, High Surface Area SiO_2/C and SiC/C Composites," *J. Mater. Chem.*, **2010**, *20*, 4840-4844.
8. Wang J., Kuhn J., and Lu X., "Monolithic Silica Aerogel Insulation Doped with TiO_2 Powder and Ceramic Fibers," *J. Non-Cryst. Solids.*, **1995**, *186*, 296-300.
9. Maleki H., Duraes L., and Portugal A., "An Overview on Silica Aerogels Synthesis and Different Mechanical Reinforcing Strategies," *J. Non-Cryst. Solids.*, **2014**, *385*, 55-74.
10. Baetens B. P. J. R. and Gustavsen A., "Aerogel Insulation for Building Applications: A State of the Art Review," *Energy Build*, **2011**, *43*, 761-769.
11. Kistler S. S., Swann S. J., and Appel E. G., "Aerogel Catalysts Thoria: Preparation of Catalyst and Conversions of Organic Acids to Ketones," *Ind. Eng. Chem.*, **1934**, *26*, 388-391.
12. Matias T., Marques J., Gando-Ferreira L., and Valente A. J. M., "Silica-based Aerogels as Adsorbents for Phenol-derivative Compounds," *Colloids Surf. A.*, **2015**, *480*, 260-269.
13. Plata D. L., Wolfe R. L., Carroll M. K., and Bakrania S. D., "Aerogel-platform Optical Sensors for Oxygen Gas," *J. Non-Cryst. Solids.*, **2004**, *350*, 326-335.
14. Nagahara H., Suginouchi T., and Hashimoto M., "P1M-8 Acoustic Properties of Nanofoam and its Applied Air-Borne Ultrasonic Transducers," *Proc. IEEE Ultrason. Symp.*, **2006**, *3*, 1541.
15. Long J. W., Fischer A. E., Bourg M. E., and Lytle J. C., "Self-limiting Electropolymerization en Route to Ultrathin, Conformal Polymer Coatings for Energy Storage Applications," *PMSE Preprints*, **2008**, *99*, 772.

16. Steinbach S.and Ratke L., "The Microstructure Response to Fluid Flow Fields in Al-cast Alloys," *Trans. Indian Inst. Metals.*, **2007**, *60*, 167-171.

17. Yin W. and Rubenstein D. A., in: Koebel M.M. (5th ed.), Aerogels Handbook, *Springer*, USA, **2011**, 683-694.

18. Pajonk G. M., "Aerogel Catalysts," *Appl. Catal.*, **1991**, *72*, 217-266.

19. Ulker Z., "Supercritical Fluid Technology for Energy and Environmental Application," *Journal of Supercritical Fluid Technology for Energy and Environmental Application*, **2014**, *157*, 269-275.

20. Zuo L., Zhang Y., Zhang L., Miao Y.E., and et.al., "Polymer/Carbon-Based Hybrid Aerogels: Preparation, Properties and Applications," *Materials*, **2015**, *8*, 6806-6848.

21. Hüsing N. and Schubert U., "Aerogels-Airy Materials: Chemistry, Structure, and Properties," *Angew. Chem. Int. Ed.*, **1998**, *37*, 22-45.

22. Brinker C. J., Keefer K. D., Schaefer D. W., and Ashley C. S., "Sol-gel Transition in Simple Silicates," *J. Non-Cryst. Solids.*, **1982**, *48*, 47-64.

23. Brinker C. J. and Scherer G. W., GW Scherer "Sol-Gel Science, (1st ed.)," *Academic Press*, New York, **1990.**

24. Rao A., Pajonk G. M., and Koebel M. M., Aerogels handbook, *Springer*, USA, **2011.**

25. Maleki H., Duraes L., and Portugal A., "Synthesis of Lightweight Polymer-reinforced Silica Aerogels with Improved Mechanical and Thermal Insulation Properties for Space Applications," *Micropor. Mesopor. Mater.*, **2014**, *197*, 116-129.

26. Hench L. L. and West J. K., "The Sol-gel Process," *Chem. Rev.*, **1990**, *90*, 33-72.

27. Iler R. K., Hench L. L., and Ulrich D. R., "Science of Ceramic Chemical Processing," *Wiley*, **1986**, 140-147.

28. Davis P. J., Brinker C. J., and Smith D. M., "Pore Structure Evolution in Silica Gel during Aging/drying I. Temporal and Thermal Aging," *J. Non-Cryst. Solids.*, **1992**, *142*, 189-196.

29. Strom R. A., Yasmine M., and Petermann G., "Strengthening and Aging of Wet Silica Gels for Up-scaling of Aerogel Preparation," *J. Sol-Gel Sci. Technol.*, **2007**, *41*, 291-298.

30. Maleki H., Duraes L., and Portugal A., "Development of Mechanically Strong Ambient Pressure Dried Silica Aerogels with Optimized Properties," *J. Phys. Chem. C.*, **2015**, *119*, 7689-7703.

31. Chen H. B., Huang W., and Schiraldi D. A., "Fabrication and Properties of Irradiation-Cross-linked Poly (Vinyl Alcohol)/Clay Aerogel Composites," *ACS Appl. Mater. Interfaces.*, **2014**, *6*, 16227-16236.

32. Kawagishi K., Saito H., and Furukawa H., "Superior Nanoporous Polyimides via Supercritical CO_2 Drying of Jungle-Gym-Type Polyimide Gels," *Macromol. Rapid Commun*, **2007**, *28*, 96-100.

33. Matson D. W. and Smith R. D., "Supercritical Fluid Technologies for Ceramic-Processing Applications," *J. Am. Ceram. Soc.*, **1989**, *72*, 871-881.

34. Kalinin S. V., Mamchik A. I., and Vertegel A. A., "Influence of the Drying Technique on the Structure of Silica Gels," *J. Sol-Gel Sci. Technol.*, **1999**, *15*, 31-35.

35. Zeng S., Zhang X., Dong H., and Zhang X., "Efficient and Reversible Capture of SO_2 by Pyridinium-based Ionic Liquids," *Chem. Eng. J.*,

2014, *251*, 248-256.

36. Begag R., "Super-hydrophobic Aerogel as Sorbent Material for CO_2 Capture," *Aspen Aerogels*, **2011**, 17-29.
37. Inoue S., Itakura T., Furukawa Y., and Yamanaka Y., "Experimental Study on CO_2 Solubility in Aqueous Piperazine/alkanolamines Solutions at Stripper Conditions," *Energy Procedia*, **2013**, *37*, 1751-1759.
38. Kong Y., Jiang G., Fan M., Shen X., and et al., "Use of One-pot Wet Gel or Precursor Preparation and Supercritical Drying Procedure for Development of a High-performance CO_2 Sorbent," *RSC Adv.*, **2014**, *4*, 43448-43453.
39. Choi S., Drese J. H., and Jones C. W., "Adsorbent Materials for Carbon Dioxide Capture from Large Anthropogenic Point Sources," *ChemSusChem*, **2009**, *2*, 796-854.
40. Yue M. B., Sun L. B., Cao Y., and Wang Z. J., "Efficient CO_2 Capturer Derived from As-Synthesized MCM-41 Modified with Amine," *Chem. Eur. J.*, **2008**, *14*, 3442-3451.
41. Uehara Y., Karami D., and Mahinpey N., "Roles of Cation and Anion of Amino Acid Anion-Functionalized Ionic Liquids Immobilized into a Porous Support for CO_2 Capture," *Energy Fuels*, **2018**, *32*(4), 5345-5354.
42. Uehara Y., Karami D., and Mahinpey N., "Effect of Water Vapor on CO_2 Sorption–Desorption Behaviors of Supported Amino Acid Ionic Liquid Sorbents on Porous Microspheres," *Ind. Eng. Chem. Res.*, **2017**, *56*(48), 14316-14323.
43. Xu X., Andresen J. M., Miller B. G., and Scaroni A. W., "Novel Polyethylenimine-Modified Mesoporous Molecular Sieve of MCM-41 Type as High-Capacity Adsorbent for CO_2 Capture," *Energy Fuels*, **2002**, *16*, 1463-1469.
44. Rechberger F., Ilari G., and Niederberger M., "Assembly of Antimony Doped tin Oxide Nanocrystals into Conducting Macroscopic Aerogel Monoliths," *Chem. Commun.*, **2014**, *50*, 13138-13141.
45. Qi G., Estevez L., Jones C. W., and Giannelis E., "High Efficiency Nanocomposite Sorbents for CO_2 Capture Based on Amine-functionalized Mesoporous Capsules," *Energy Environ. Sci.*, **2011**, *4*, 444-452.
46. Linneen N., Pfeffer R., and Lin Y. S., "CO_2 Capture Using Particulate Silica Aerogel Immobilized with Tetraethylenepentamine," Micropor. Mesopor. Mater., **2013**, *176*, 123-131.
47. Nguyen S. T., Le T., and Duong H. M., "Cellulose Aerogel from Paper Waste for Crude Oil Spill Cleaning," *Ind. Eng. Chem. Res.*, **2013**, *52*, 18386-18391.
48. Begag R., Rhine W., Gould G., and Nahass P., "Superhydrophobic Amine Functionalized Aerogels as Sorbents for CO_2 Capture," *Greenhouse Gas Sci. Technol.*, **2013**, *3*, 30-39.
49. Lin J. Y., Pfeffer R., "in: AIChE-Global Congr. on Process Safety-Annual Meeting," US, **2012**.
50. Karami D., Bararpour S. T., and Mahinpey N., "Highly Active Sorbents and Oxygen Carriers Supported by Calcined Alumina Aerogel for Low-temperature Carbon Capture and Chemical-looping Combustion of Methane", A patent was filed in US patent office, V812601USP, **2016**.
51. Bararpour S. T., Karami D., and Mahinpey N., "Effect of Various Types of Alumina Supports on the Carbonation Behavior of K_2CO_3: Comparing Physical Mixing and Incipient Wetness Impregnation Methods" *Chemical Engineering Journal*, **2018**, *137*, 138-157.

52. Howard C. J., "Sulfur Tolerant Highly Durable CO_2 Sorbents," US patent, 2010/0139486A, **2017**.
53. Karami D. and Mahinpey N., "The Preparation of very Efficient and Stable Zirconia Stabilized CaO Aerogels for CO_2 Capture at High Temperatures," US Patent, 0346547A, **2010.**
54. Karami D. and Mahinpey N., "Utilization of Alumina Aerogel as High Surface Area Support in the CLC Process," *Materials Chemistry and Physics*, 99-111.
55. Bianchi D., Landoulsi H., Pajonk G. M., and Teichner S. J., "Aerogel catalysts," *Applied Catalysis Journal*, **1991**, *72*, 217-266.
56. Blanchard F., Pommier B., Reymond J. P., and Teichner S. J., "On the Mechanism of the Fischer-Tropsch Synthesis Involving Unreduced Iron Catalyst," *J. Mol. Catal.*, **1982**, *17*, 171-181.
57. Pommier B., Reymond J. P., and Teichner S. J., "Fischer-Tropsch Synthesis on Oxidized Supported Iron Catalysts," *Zeitschrift für Physikalische Chemie Journal*, **1985**, *144*, 203-222.
58. Blanchard F., Pommier B., and Teichner S. J., "New Fischer-Tropsch Catalysts of the Aerogel Type, in Studies in Surface Science and Catalysis," Preparation of Catalysts III, Elsevier, Amsterdam, **1983**, *16*, 395-407.
59. Pommier B. and Teichner S. J., "On the Mechanism of the Fischer-Tropsch Synthesis Involving Unreduced Iron Catalyst," *in Proc. 9th Int. Congress on Catalysis,* The Chemical Institute of Canada, Ottawa, **1988**.
60. Droege M. W. and Westbrook C., Lawrence Livermore National Laboratory UCRL100768, **1989**.
61. Armor J. N., Carlson E. J., Zambri P. M., "Aerogels as Hydrogenation Catalysts," *Appl. Catal.*, **1985**, *19*, 339-348.
62. Yamaguchi T., "Recent Progress in Solid Superacid," *Appl. Cat.*, **1990**, *61*, 1-25.
63. Ma J., Sun N., and Sun Y., "A Short Review of Catalysis for CO_2 Conversion," *Catalysis Today*, **2009**, *148*, 221-231.
64. Erdohelyi A., Cserenyi J., and Solymosi F., "Activation of CH_4 and its Reaction with CO_2 over Supported Rh Catalysts," *J. Catal.*, **1993**, *141*, 287-299.
65. Bitter H., Seshan K., Lercher J.A., "The state of zirconia supported platinum catalysts for CO_2/CH_4 reforming," *J. Catal.*, **1997**, 171, 279-286.
66. McGee W. D., Pan Y., and Riley D. P., "Highly Selective Generation of Urethanes from Amines, Carbon Dioxide and Alkyl Chlorides," *J. Chem. Soc. Chem. Commun.*, **1994**, *6*, 699-700.
67. Corthals S., Nederkassel J., Noyen J. V., Moens B., and et al., "Influence of Composition of $MgAl_2O_4$ Supported $NiCeO_2ZrO_2$ Catalysts on Coke Formation and Catalyst Stability for Dry Reforming of Methane," *Catal. Today.*, **2008**, *138*, 28-32.
68. Nitta Y., Suwata O., and Okamoto Y., "Copper-zirconia Catalysts for Methanol Synthesis from Carbon Dioxide: Effect of ZnO Addition to Cu-ZrO_2 catalysts," *Catal. Lett.*, **1994**, 26, 345-354.
69. Arena F., Barbera K., Italiano G., Bonura G., and et al., "Synthesis, Characterization and Activity Pattern of Cu–ZnO/ZrO_2 Catalysts in the Hydrogenation of Carbon Dioxide to Methanol," *J. Catal.*, **2007**, *249*, 185-194.
70. Dubois J. L., Sayama K., and Arakawa H., "CO_2 Hydrogenation over Carbide Catalysts," *Chem. Lett.*, **1992**, *21*, 5-8.

Permissions

All chapters in this book were first published in JPST, by Research Institute of Petroleum Industry; hereby published with permission under the Creative Commons Attribution License or equivalent. Every chapter published in this book has been scrutinized by our experts. Their significance has been extensively debated. The topics covered herein carry significant findings which will fuel the growth of the discipline. They may even be implemented as practical applications or may be referred to as a beginning point for another development.

The contributors of this book come from diverse backgrounds, making this book a truly international effort. This book will bring forth new frontiers with its revolutionizing research information and detailed analysis of the nascent developments around the world.

We would like to thank all the contributing authors for lending their expertise to make the book truly unique. They have played a crucial role in the development of this book. Without their invaluable contributions this book wouldn't have been possible. They have made vital efforts to compile up to date information on the varied aspects of this subject to make this book a valuable addition to the collection of many professionals and students.

This book was conceptualized with the vision of imparting up-to-date information and advanced data in this field. To ensure the same, a matchless editorial board was set up. Every individual on the board went through rigorous rounds of assessment to prove their worth. After which they invested a large part of their time researching and compiling the most relevant data for our readers.

The editorial board has been involved in producing this book since its inception. They have spent rigorous hours researching and exploring the diverse topics which have resulted in the successful publishing of this book. They have passed on their knowledge of decades through this book. To expedite this challenging task, the publisher supported the team at every step. A small team of assistant editors was also appointed to further simplify the editing procedure and attain best results for the readers.

Apart from the editorial board, the designing team has also invested a significant amount of their time in understanding the subject and creating the most relevant covers. They scrutinized every image to scout for the most suitable representation of the subject and create an appropriate cover for the book.

The publishing team has been an ardent support to the editorial, designing and production team. Their endless efforts to recruit the best for this project, has resulted in the accomplishment of this book. They are a veteran in the field of academics and their pool of knowledge is as vast as their experience in printing. Their expertise and guidance has proved useful at every step. Their uncompromising quality standards have made this book an exceptional effort. Their encouragement from time to time has been an inspiration for everyone.

The publisher and the editorial board hope that this book will prove to be a valuable piece of knowledge for researchers, students, practitioners and scholars across the globe.

List of Contributors

Gao Dapeng
Institute of Mechanics, Chinese Academy of Sciences
School of Earth and Space Science, Peking University, Beijing, China

Liu Yuewu, Hou Shaoji, Li Qi and Han Guofeng
Institute of Mechanics, Chinese Academy of Sciences

Ye Jigen and Hu Yunpeng
PetroChina Research Institute of Petroleum Exploration & Development, Beijing, China

Mohammad Emami Niri
Institute of Petroleum Engineering, University of Tehran, College of Engineering, Tehran, Iran

Karim Akbari Aghdam
Petroleum and Gas Engineering Department, Pars Oil and Gas Company (POGC), Asaluyeh, Bushehr, Iran

Seyyed Saeed Ghorashi
Research Institute of Petroleum Industry (RIPI), Tehran, Iran

Vahid Azamipour and Mehdi Assareh
Thermodynamics Research Laboratory, School of Chemical Engineering, Iran University of Science and Technology, Tehran, Iran

Ehsan Khamehchi, Seyed Arman Hosseini Kaldozakh and Ali Alizadeh
Department of Petroleum Engineering, Amirkabir University of Technology, Tehran, Iran

Mahdi Assareh
Thermodynamics Research Laboratory, School of Chemical Engineering, Iran University of Science and Technology, Tehran, Iran

Ahmed Zoeir
Faculty of Petroleum and Natural Gas Engineering, Sahand University of Technology, Tabriz, Iran

Jafar Qajar
School of Chemical and Petroleum Engineering, Shiraz University, Shiraz, Iran

Zhang Jun and Zeng Jia
School of Petroleum Engineering, Northeast Petroleum University, Heilongjiang, China

Mohammad Hossein Zare and Gholam Abbas Safiyan
National Iranian South Oilfield Company, Ahwaz, Iran

Abbas Helalizade
Department of Petroleum Engineering, Ahwaz Faculty of Petroleum Engineering Petroleum University of Technology (PUT), Ahwaz, Iran

Ehsan Taheri
Faculty of Engineering, Tarbiat Modares University, Tehran, Iran

Amoirodin Sadrnejad and Hasan Ghasemzadeh
Faculty of Civil Engineering, K. N. Toosi University of Technology, Tehran, Iran

Bader H. Albusairi and Adam Al-Mulla
Department of Chemical Engineering, Kuwait University, Kuwait

Yaser Jahanpeyma and Saeid Jamshidi
Chemical and Petroleum Engineering Department, Sharif University of Technology, Tehran, Iran

Yanfeng Liu
Sinopec Petroleum Exploration and Production Research Institute, Beijing, China

Yuetian Liu and Lu Sun
China University of Petroleum (Beijing), Beijing, China

Jian Liu
PetroChina Research Institute of Petroleum Exploration and Development, Beijing, China

Mohammad Parvazdavani
Reservoir Studies Research Division, Research Institute of Petroleum Industry (RIPI), Tehran, Iran

Behjat Kari
Petroleum Engineering Department, Amirkabir University of Technology, Tehran, Iran

Mahmood Dinmohammad and S. A. Mousavi Dehghani
Upstream Faculty, Research Institute of Petroleum Industry (RIPI), Tehran, Iran

Abdolhamid Sameni
Chemical Engineering Faculty, Tarbiat Modares University, Tehran, Iran

Ali Chamkalani
Petroleum University of Technology, Ahwaz, Iran

Ghasem Khatinzadeh, Mohammad Mahdyarfar and Ali Mehdizadeh
Chemical Polymeric and Petrochemical Technology Development Research Division, Research Institute of Petroleum Industry (RIPI), Tehran, Iran

Khalil Shahbazi and Alireza Arshadi
Department of Petroleum Engineering, Ahwaz Faculty of Petroleum Engineering, Petroleum University of Technology, Ahwaz, Iran

Majid Valizadeh
Petroleum Engineering Research Division, Research Institute of Petroleum Industry (RIPI), Tehran, Iran

Davood Karami
Department of Chemical and Petroleum Engineering, Schulich School of Engineering, University of Calgary, Calgary, AB, Canada

Index

www.ingramcontent.com/pod-product-compliance
Lightning Source LLC
LaVergne TN
LVHW060543210926
848485LV00005B/1247
9781639893775